Residential Steel Design
and Construction

Other McGraw-Hill Titles of Interest

Dodge Cost Books from McGraw-Hill

Residential Steel Design and Construction

**Energy Efficiency,
Cost Savings,
Code Compliance**

John H. Hacker

Julie A. Gorges

McGraw-Hill

New York San Francisco Washington, D.C. Auckland Bogotá
Caracas Lisbon London Madrid Mexico City Milan
Montreal New Delhi San Juan Singapore
Sydney Tokyo Toronto

Library of Congress Cataloging-in-Publication Data

Hacker, John H.
    Residential steel design and construction / John H. Hacker, Julie
A. Gorges
       p.   cm.
    Includes index.
    ISBN 0-07-025475-3 (hardcover). — ISBN 0-07-025476-1 (pbk.)
    1. Building, Iron and steel. 2. Dwellings. I. Gorges, Julie A.
II. Title.
TH1611.H23   1997
693'.71—dc21   97-41818
                                                 CIP

McGraw-Hill

A Division of The McGraw·Hill Companies

    4 5 6 7 8 9 0   KGP/KGP   9 0 2 1 0 9

ISBN 0-07-025475-3 (HC)

ISBN 0-07-025476-1 (PBK)

*The sponsoring editor for this book was Zoe G. Foundotos, the editing
supervisor was Bernard Onken, and the production supervisor was
Pamela A. Pelton. It was set in Times by John Hacker and Julie Gorges.*

Printed and bound by Quebecor / Kingsport.

McGraw-Hill books are available at special quantity discounts to use
as premiums and sales promotions, or for use in corporate training programs. For more information, please write to the Director of Special
Sales, McGraw-Hill, 11 West 19th Street, New York, NY 10011. Or contact your local bookstore.

This book is printed on acid-free paper.

Contents

Chapter 10. Materials and Tools 93

Chapter 11. Step-by-Step Framing Details 107

Acknowledgements

We are very grateful to the following people and the firms they represent for their help, insight, input, and review of this book: Zoe Foundotos, our editor at McGraw-Hill, Bill Farkas at National Home Builder's Association Research Center, Carmen Gravley at American Iron and Steel Institute, Linda Connell at Pacific Northwest Laboratories.

While developing the building methods outlined in this book, Chuck and Lolly Booth, owners of Omega Transworld, LTD, Linda and Millard Fuller, cofounders of Habitat for Humanity, and Kitti Thanakitamnuay of Noble Steel Tech in Thailand have been tremendously supportive and instrumental in putting to practice our innovative ideas. We would also like to thank the following magazines for their support: *Automated Builders, Construction Marketing Today, Metal Home Digest, Permanent Building and Foundations Magazine,* and *Western HVACR News*. American Institute of Architecture, the Building Industry Association, the State of California Energy Commission, Southern California Edison and the Gas Company, along with HUD, have been forerunners in encouraging energy efficient building practices and through their awards have supported our efforts. Our gratitude also to Ray House for providing photographs of the Habitat for Humanity houses.

I would also like to thank my family for their various contributions and for their support: My loving wife, Carmen, who has put up with me all these years, my son, Dan who provided technical support, extensive computer knowledge, and countless hours drafting plans and designing our web page, my daughter, Joanie, for providing some of the photographs used in this book and for her support at various conferences and home shows, and my youngest daughter, Nicole, along with her husband, Bill. Finally, I must thank Julie Gorges for using her talent as a writer to make this manuscript readable and easy to understand. We pulled through this together, and she's still talking to me!

John H. Hacker

* * *

In addition to the people listed above, I would like thank my husband, Scott, and my children Jonathan and Christopher, for their patience, understanding, and support, during this long and difficult period. As any freelance writer knows, the road to success is tremendously hard and sometimes painful. My mother, Carmen, has been a pillar of support over the years, and I want to express my appreciation for her continuing faith and encouragement in my writing. I would also like to thank my agent, Jennifer Flannery, for her help and advice as I try to break into fiction. Last, but not least, this book would not have been possible without my coauthor, John Hacker. His engineering background, innovative ideas, and technical knowledge, along with many long, hard hours in his "retirement years" contributed to the resourceful and innovative techniques found in this book.

Julie A. Gorges

Abbreviations

ACCA The Air Conditioning Contractors of America.

ACEEE American Council for an Energy-Efficient Economy

AFUE Annual Fuel Utilization Efficiency

AHAM The Association of Home Appliance Manufacturers

AISI American Iron and Steel Institute

ARI The Air-Conditioning and Refrigeration Institute

ASTM formerly American Society for Testing and Materials, now ASTM

btu the British thermal unit

CABO Council of American Building Officials

CFC Chlorofluorocarbons

CFL Compact fluorescent lamps.

cfm Cubit feet of air per minute

DOE Department of Energy

EER Energy efficiency ratio

EIFS Exterior insulated finish system

EPA Environmental Protection Agency

EPS Expanded polystrene rigid insulation

Et/Ec Thermal efficiency/combustion efficiency factor

FPSF Frost-protected shallow foundation

FWA Forced warm air

HCFC Hydrochloro-fluorocarbons

HID High intensity discharge bulbs

HRU Waste heat recovery systems

HSPF Heating system performance factor, a method of rating the efficiency of heating equipment

HUD Housing and Urban Development

ICBO International Conference of Building Officials

MEC Model Energy Code

MILS 1/1000 of an inch

MEPS Molded expanded polystyrene

NAHB National Association of Home Builders

O.C. On center

ORNL Oak Ridge National Lab

OTW Other Than Wood

RF Radio frequency lamps

SEER Seasonal energy efficiency rating

SERP Super Efficient Refrigerator Program

UBC Uniform Building Code

XEPS Extruded polystyrene

PART 1

Basic Principles of Energy-Efficient Steel Framing

1

Introduction

Wood frame construction has been the unchallenged norm for residential building in North America for decades. Over the last few years, however, the high and fluctuating price of wood, declining quality of framing lumber, and environmental concerns have sent homebuilders scrambling for building alternatives. Because of the many advantages of building with steel, more and more builders are turning to steel framing.

However, builders have found one deterrent to building with steel: energy efficiency. A 20-gauge (0.033 inch thick) standard steel stud conducts roughly 10 times more heat than a 1½-inch thick wood stud. Therefore, a poorly designed steel framing system increases heating and cooling costs considerably. How can you make up for this efficiency loss without paying a fortune?

This book addresses that issue, proving that steel can be both energy efficient and cost-effective. In the following chapters, you'll find tested building methods that will result in a house or commercial structure with unparalleled energy savings at a reasonable cost.

First though, let's discuss why you should consider building with steel.

Why Build Homes Made of Steel?

To answer that question, let's examine why homes were primarily made of wood or why our ancestors made their homes of clay. Put simply, they were using the materials and technology available at that time. Homes were made of clay because it was available and they didn't have the technology to build any other way. Nomadic people built tents from animal skins because they had access to wild game and needed structures that could be constructed fast and moved. When pilgrims settled in the new colonies, log homes were naturally selected because of the availability of vast forests. As technology progressed, the use of early forms of wood frame and "stud wall" construction became dominant over leaky and wasteful log homes.

We must now ask if wood construction is still the best answer, taking into consideration current availability and technology. Each year, in the United States, over 1.3 million residential structures are built. A typical wood-framed 2000-ft² house requires the wood from between 40 and 50 trees. That means each year over 50 million trees are cut from forests in the United States and Canada for new residential construction alone.

The *Los Angeles Times* on April 15, 1996 reported that in the Umpaqua National Forest, near Roseburg, Oregon, "some (trees) 800 to 1,000 years old are yellow-tagged for harvest under the salvage-logging rider authorized by Congress last year." You've probably read about recent protests against "yellow-tagging" redwood trees in the magnificent Headwaters Forest located in northern California.

Although many argue that wood is a renewable resource, traditional harvesting practices have raised environmental questions. According to *Environmental Building News (EBN)* (volume 3, number 4), "forests that are managed primarily to maximize timber supply tend to perform poorly in terms of wildlife habitat, recreational uses, and even global climate control." The forests are a vital link to the water cycle of our environment. Many experts feel that the devastating 1995–1996 floods in the Pacific Northwest, which cost a combined $385.5 million damage in Washington State alone, were caused by the clear cuts of the mountainsides (Figure 1.1) that destroyed the watershed. Additionally, significant cutting in the 1970s and 1980s led to shortages in timber supply in some areas, leading some to question whether U.S. timber demands can be met in a sustainable manner.

FIGURE 1.1 Clearcutting in Washington State has raised environmental questions.

In contrast, steel can be reused indefinitely. While it takes 40 to 50 trees to frame an average house in wood, it takes only six junked cars to frame in steel; an average of 46 percent of all steel in the United States is recycled, and virtually all the light-gauge steel used in residential construction consists of some recycled steel. Heavy steel items made in mini-mills often have as much as 91 percent recycled content. As more mini-mills begin producing sheet steel, the average recycled content of steel framing materials will increase significantly.

Another factor we should consider is economics. As the cost of wood rises, the percentage of housing costs in relation to the average person's income has disproportionately increased. The high cost of construction along with the high costs of heating and cooling have all but eliminated the hope and dream of home ownership for millions of people. Because the prices of steel are stable, the future for steel housing looks bright. If you add the benefits of reducing energy bills, the marketability of these homes is outstanding.

Steel housing also has some very important international implications. In the past, there has been great availability of lumber for wood framing in the United States. Many other countries, however, did not enjoy this readily available supply of lumber. They have been importing much of their lumber at significantly higher costs. In addition, a lack of water to fight fires and a notable problem with termites are other handicaps. Since steel is manufactured all over the world, steel is even more important as an answer to housing needs in other countries.

Advantages of Steel Construction

Not only do steel frame houses and commercial structures eliminate the need to cut down trees, but there are a number of other advantages to steel frame houses as compared to wood frames.

- Steel framing can be manufactured to custom lengths, resulting in fewer wasted materials than wood.
- Steel won't warp, bow, rot, or fall victim to termites. Homes are also less likely to be invaded by rodents.
- Steel studs, joists, and trusses weigh as much as 30 percent less than wood, making assembled walls easier to handle.
- Cost of steel is stable in comparison with volatile wood prices.
- Steel has superior fire ratings compared to wood.
- Screws and anchors offer stronger seismic resistance than nails.
- Because steel members can withstand extremely heavy wind loads, homes are more hurricane- and typhoon-resistant.
- Some people with chemical sensitivities prefer steel because the terpenes emitted by softwoods can be irritating and because chemical treatment with preservatives aren't necessary. In addition, avoiding insecticides and soil treatments with termiticides contributes to indoor air quality.

Fire storms and earthquakes in California during recent years have exposed deficiencies in wood framing while clearly demonstrating the advantages of building with steel, as listed above. For example, after the 1992 Landers/Big Bear earthquake in California, I was hired by various insurance companies to provide forensic reports for earthquake-damaged homes. Wood framing contributed to a substantial amount of the damage. Several older homes failed because termites weakened the lower walls. In multistory houses, buildings had collapsed because plywood shear panels were nailed into two-by-four studs too close to the edge and the entire stud split. Dryrot was evident in connections between the roof and wall where leaking had occurred.

To give you another example, in 1993 a fire virtually destroyed a subdivision I provided engineering for in 1965, located in Laguna Beach, California. The fires left 200 families homeless. The wood-framed houses fueled the fire storm and prevented the fire department from fighting the brush fires. The one house that survived the fire was not of wood frame construction.

No doubt about it, considering the high and unpredictable price of lumber, the need to preserve our forests, and all the advantages of steel frame construction previously discussed, steel is a superior way to frame a house or commercial structure. That's why more and more people are turning to steel frame construction.

According to a recent survey by *Builder,* 54 percent of builders currently use some form of lumber substitute and plan to use more steel framing and other lumber alternatives during the next five years. Although builders have a reputation of resisting change, the survey showed those involved in construction are using new materials and techniques in order to meet new market demands, code requirements, and cost challenges. Builders are definitely on the lookout for more product innovation. Top interests include steel framing, structural insulated panels, new concrete products, and energy-efficient products.

The American Iron and Steel Institute (AISI) estimates that steel was used in the construction of 50,000 to 75,000 residences in 1994 and that, by the year 2000, 25 percent of all new houses will be constructed with some steel. To the average person, these figures may seem high. Most people don't recognize a steel frame home. If you mention a steel dwelling, people often conjure up a steel box made in a factory or a metal barn. Steel frame homes don't scream out that they are made of metal. They are beautiful homes that look like any other conventional construction (Figure 1.2). Both the interior and exterior of these homes appear standard because the frame is not visible. Gypsum board, paneling, roofing, and siding materials are easily attached or joined to steel framing members.

However, as we mentioned at the start, while the advantages of building with steel are many, steel has one glaring handicap: the energy loss that occurs. Can a builder overcome this disadvantage?

FIGURE 1.2 Steel homes look like any other conventional home.

Steel Framing Can Be Energy Efficient

With over 40 years experience in designing energy efficient steel buildings, I have seen it proved time and again: You *can* compensate for the energy lost through steel construction.

In 1970, I designed an assembly hall that seated over 2000 near Los Angeles using a large-span steel moment frame that exceeded energy standards at that time. This design was used on four more assembly halls: two in California, one in Oregon, and another near Vancouver, British Columbia. In 1973, a similar design was used for the Parducci Winery Warehouse in Ukiah, California, where it was necessary to maintain the temperature of the stored wine within a limited temperature range. The design concept used on these buildings has stood the test of time, and they are still energy efficient by today's standards.

In the early 1990s, as construction began to slow down and the price of lumber went up, interest in steel framing began to develop. While designing steel-framed churches in third-world countries where lumber is hard to come by, I saw the need to apply steel frame construction to housing in the United States. Since energy loss was making many builders apprehensive about using steel, I wanted to produce a building that would be *more* energy efficient than ordinary wood construction.

In 1993, I built my first steel frame duplex in Southern California, using a patented foam-insulated steel foundation system I invented, Insu-Form and a roof framing system using light-gauge steel (Figure 1.3).

FIGURE 1.3 Steel-framed duplex in Cathedral City, California.

Although the house easily surpassed energy requirements, I continued to look for ways to improve the building system. In 1994, I joined forces with Omega Transworld Inc. (OTW), located in New Kensington, Pennsylvania, a manufacturer of block composed of Portland cement, fly ash, and polyester fibers. By using these materials and applying the other energy saving methods you'll find in this book, I was able to meet my goal of building an energy-efficient steel frame home superior to a typical wood frame house.

To verify the energy saving, a study was made during the summer months of 1995. The electric bills of the prototype were compared with a standard wood frame house of similar size and window area built to meet minimum energy requirements. The electric bills for the steel frame prototype for June through September totaled $352.60 or 28¢/ft² (Figure 1.4a). The electric bills for the standard wood frame house for the same four months totaled $988.17 or 76¢/ft². (Figure 1.4b). When comparing the electric bills of the two homes for a 21-month period, the electric bill for the prototype averaged $41.54 per month, while the standard wood frame house averaged $171.00 per month .

The Certificate of Compliance, using the State of California-approved compliance method "Micropas," showed that a prototype built in Cathedral City, California, had a calculated annual energy use of 35.42 kbtu/sf-yr. In comparison, the maximum permitted standard is 50.65 kbtu/sf-yr. (Fig 1.5). This is a calculated savings of 15.23 kbtu/sf-yr or 19,307 kbtu for a home in the desert area, where the electric company charges up to 14¢/kWh.

In addition, based on the Comply4 Calculations, the house was rated a plus 35 points. If you're not familiar with the point system, each component of a structure is evaluated and is assigned plus or minus "points" depending on the heat loss or gain in the winter time and the cool air retention or loss in the summer. For construction to meet the mandated requirements of the law, a structure must rate zero points. Structures with positive or plus points are considered energy efficient. As you can see, 35 points indicates an extremely energy efficient house.

Southern California Edison
P.O. Box 600   ROSEMEAD CA 91771-0001

Name and service address	Account number	24-hour billing or service questions
HACKER, JOHN 68255 CORTA CTHDRLCY CA 92234	54-79-640-1477-01 000-8	☎ CALL (619) 324-8500
	Meter number 8-866649	Next scheduled meter reading SEPT 01, 1995
Rate schedule D-APS	Billing Period 07/05/95 - 08/03/95 Summer Season	Date bill prepared AUG 07, 1995

Previous Charges and Credits	Amount of previous bill 07/05/95	$	73.78
	Payment received 08/01/95 - Thank you	-	73.78
	Balance	$	.00

New Charges and Credits	Energy charge: baseline (1062 kWh x 12.419¢)	$	131.89
	Air conditioner cycling credit	-	16.70
	State energy tax (1062 kWh x 0.02¢)	+	.21
	Current charges now due ▶ ▶ ▶ ▶	$	115.40

Southern California Edison
P.O. Box 600   ROSEMEAD CA 91771-0001

Name and service address	Account number	24-hour billing or service questions
HACKER, JOHN 68255 CORTA CTHDRLCY CA 92234	54-79-640-1477-01 000-8	☎ CALL (619) 324-8500
	Meter number 8-866649	Next scheduled meter reading OCT 03, 1995
Rate schedule D-APS	Billing Period 08/03/95 - 09/02/95 Summer Season	Date bill prepared SEPT 07, 1995

Previous Charges and Credits	Amount of previous bill 08/03/95	$	115.40
	Payment received 08/17/95 - Thank you	-	115.40
	Balance	$	.00

New Charges and Credits	Energy charge: baseline (763 kWh x 12.419¢)	$	94.76
	Air conditioner cycling credit	-	17.28
	State energy tax (763 kWh x 0.02¢)	+	.15
	Current charges now due ▶ ▶ ▶ ▶	$	77.63

Southern California Edison
P.O. Box 600   ROSEMEAD CA 91771-0001

Name and service address	Account number	24-hour billing or service questions
HACKER, JOHN 68255 CORTA CTHDRLCY CA 92234	54-79-640-1477-01 000-8	☎ CALL (800) 655-4555
	Meter number 8-866649	Next scheduled meter reading NOV 02, 1995
Rate schedule D-APS	Billing Period 09/02/95 - 10/03/95 Summer to Winter Season	Date bill prepared OCT 05, 1995

Previous Charges and Credits	Amount of previous bill 09/02/95	$	77.63
	Payment received 09/14/95 - Thank you	-	77.63
	Balance	$	.00

New Charges and Credits	Energy charge: baseline (824 kWh x 12.419¢)	$	102.33
	Air conditioner cycling credit	-	16.70
	State energy tax (824 kWh x 0.02¢)	+	.16
	Current charges now due ▶ ▶ ▶ ▶	$	85.79
	Past due if not paid by 10/24/95		

FIGURE 1.4a Electric bills for the steel-framed prototype.

Southern California Edison
P.O. Box 600 ROSEMEAD CA 91771-0001

Name and service address	Account number	24-hour billing or service questions
POPE, WILLLIAM 68260 BELLA VISTA CTHDRLCY CA 92234	56-79-612-2050-06 000-5	☎ CALL (619) 324-8500
	Meter number 8-625124	Next scheduled meter reading SEPT 06, 1995
Rate schedule DOMESTIC (D)	Billing Period 07/07/95 - 08/08/95 Summer Season	Date bill prepared AUG 09, 1995

Previous Charges and Credits

Amount of previous bill 07/07/95	$	201.72
Payment received 07/20/95 - Thank you	-	201.72
Balance	$	.00

New Charges and Credits

Energy charge: baseline	(1366 kWh x 12.419¢)			
+ over baseline	(850 kWh x 14.277¢)	=	$	291.00
State energy tax	(2216 kWh x 0.02¢)	+		.44
Current charges now due	▶ ▶ ▶ ▶		$	291.44

Southern California Edison
P.O. Box 600 ROSEMEAD CA 91771-0001

Name and service address	Account number	24-hour billing or service questions
POPE, WILLLIAM 68260 BELLA VISTA CTHDRLCY CA 92234	56-79-612-2050-06 000-5	☎ CALL (619) 324-8500
	Meter number 8-625124	Next scheduled meter reading OCT 05, 1995
Rate schedule DOMESTIC (D)	Billing Period 08/08/95 - 09/07/95 Summer Season	Date bill prepared SEPT 11, 1995

Previous Charges and Credits

Amount of previous bill 08/08/95	$	291.44
Payment received 08/29/95 - Thank you	-	291.44
Balance	$	.00

New Charges and Credits

Energy charge: baseline	(1281 kWh x 12.419¢)			
+ over baseline	(878 kWh x 14.277¢)	=	$	284.44
State energy tax	(2159 kWh x 0.02¢)	+		.43
Current charges now due	▶ ▶ ▶ ▶		$	284.87

Southern California Edison
P.O. Box 600 ROSEMEAD CA 91771-0001

Name and service address	Account number	24-hour billing or service questions
POPE, WILLLIAM 68260 BELLA VISTA CTHDRLCY CA 92234	56-79-612-2050-06 000-5	☎ CALL (800) 655-4555
	Meter number 8-625124	Next scheduled meter reading NOV 06, 1995
Rate schedule DOMESTIC (D)	Billing Period 09/07/95 - 10/10/95 Summer to Winter Season	Date bill prepared OCT 12, 1995

Previous Charges and Credits

Amount of previous bill 09/07/95	$	284.87
Payment received 09/29/95 - Thank you	-	284.87
Balance	$	.00

New Charges and Credits

Energy charge: baseline	(1271 kWh x 12.419¢)			
+ over baseline	(364 kWh x 14.277¢)	=	$	209.81
State energy tax	(1635 kWh x 0.02¢)	+		.33
Current charges now due	▶ ▶ ▶ ▶		$	210.14
Past due if not paid by 10/31/95				

FIGURE 1.4b Electric bills for a standard wood-framed house of similar size and window area.

```
COMPUTER METHOD SUMMARY                              Page 1          C-2R
  ===========================================================================
  Project Title.......... SCE REBATE              Date........ 04/28/94
  Project Address........ 68255 CORTA             ------------------------
                          CATHEDRAL CITY
  Documentation Author... JOHN H HACKER           | Building Permit #    |
  Company................ JOHN H HACKER           |----------------------|
  Telephone.............. (619) 324-0216          | Plan Check / Date    |
                                                  |----------------------|
  Compliance Method...... MICROPAS4 by Enercomp, Inc.| Field Check/ Date |
  Climate Zone........... 15                      ------------------------
  ===========================================================================
      |    MICROPAS4 v4.01  File-STEEL  Wth-CTZ15S92  Program-FORM C-2R   |
      |        User#-MP0643  User-JOHN H HACKER  Run-STEEL FRAME HOUSE     |
      -------------------------------------------------------------------


      ======================================================================
      =                    MICROPAS4 ENERGY USE SUMMARY                    =
      =                    ----------------------------                    =
      =                                                                    =
      =  Energy Use                Standard    Proposed    Compliance      =
      =  (kBtu/sf-yr)              Design      Design      Margin          =
      =  ------------------------  ----------  ----------  ----------      =
      =  Space Heating..........     2.35        0.96        1.39          =
      =  Space Cooling..........    32.23       21.44       10.79          =
      =  Water Heating..........    16.07       13.02        3.05          =
      =                            --------    --------    --------        =
      =                Total        50.65       35.42       15.23          =
      =                                                                    =
      =     *** Building complies with Computer Performance ***            =
      ======================================================================


                             GENERAL INFORMATION
                             -------------------


             Conditioned Floor Area.....  1248 sf
             Building Type..............  Single Family Detached
             Construction Type .........  New
             Building Front Orientation.  Front Facing 0 deg (N)
             Number of Dwelling Units...  1
             Number of Building Stories.  1
             Weather Data Type..........  ReducedYear

             Floor Construction Type....  Slab On Grade (Package D)
             Number of Building Zones...  1
             Conditioned Volume.........  9984 cf
             Footprint Area.............  1248 sf
             Ground Floor Area..........  1248 sf
             Slab-On-Grade Area.........  1248 sf
             Glazing Percentage.........  11.2 % of FA
             Average Ceiling Height.....  8 ft
```

FIGURE 1.5 The California Energy Commission's Title 24 energy calculations for the prototype.

The house was so efficient, in fact, that the first steel frame duplex won a first place award for construction of one of the most energy efficient houses in the Leading Edge Contest given by the Los Angeles and Orange County Chapter of the American Institute of Architecture, the Building Industry Association, Southern California Edison, the State of California Energy Commission and The Gas Company (Figure 1.6). The award included a $2000 cash prize in addition to a $2200 Welcome Home Rebate.

FIGURE 1.6 Hacker's 1993 award from the Leading Edge Energy Responsive Design Competition.

"Metal is not as energy efficient as wood by itself," Walter Jones, one of the judges from Southern California Edison said in *The Desert Sun*, "but it's what John Hacker has done with it with the insulation that makes it a very, very good job."

Since the building of the first prototype, sixteen homes have been completed around the United States, as well as in Mexico and Thailand. Based on over 3 years of monitoring, the homes have proved to be energy-efficient and maintenance-free. In fact, Habitat for Humanity, the nonprofit housing concern, was so impressed with the energy savings, they started construction on their fourth home utilizing my building methods in November 1996. The house was constructed in only six days, built entirely by women volunteers (Figure 1.7).

FIGURE 1.7 Steel-framed Habitat for Humanity "blitz build," utilizing Hacker's building methods.

Together with Habitat, we won an award from Housing and Urban Development (HUD), for their "Building Innovation for Homeownership Program," recognizing our contribution to "expanding affordable homeownership opportunities through the use of innovative technology and developmental techniques (Figure 1.8)."

U.S. DEPARTMENT OF HOUSING AND URBAN DEVELOPMENT
THE SECRETARY
WASHINGTON, D.C. 20410-0001

OCT 16 1996

Mr. John H. Hacker
Insu-Form, Inc.
2721 Thornhill Road
Puyallup, WA 98374

Dear Mr. Hacker:

It gives me great pleasure to inform you that your submission for the **Habitat For Humanity** development has been selected for award and recognition by the National Partners in Homeownership as a **Building Innovation For Homeownership Project**. The Partnership recognizes your project's important contribution to expanding affordable homeownership opportunities through the use of innovative technologies and developmental techniques.

Your project, along with other award-winning projects selected under the BIH program, will be recognized and showcased around the country. On behalf of the Partnership, HUD will be issuing press notices announcing your selection. Information regarding additional Partnership activities in support of BIH awardees will be sent to you at a later date.

The National Partners are committed to working with all developers receiving awards to assure that these projects are completed and sold as soon as possible. If you need assistance or otherwise are interested in learning more about possible services the National Partners may be able to offer, please contact Jacqueline Kruszek of my staff at: (202) 708-4370, Extension 141.

I applaud you for your efforts. Your project is a fine example of how the best of American creativity and innovation can help expand homeownership opportunities for all Americans.

Sincerely,

Henry G. Cisneros

FIGURE 1.8 The notification of the HUD award for "Building Innovation for Homeownership Program."

Noble Tech Steel in Thailand has built two model homes and plans to start construction on 120 more homes using my energy-saving methods (Figure 1.9).

FIGURE 1.9 Steel-framed homes in Thailand.

The Bottom Dollar

Perhaps you think a fortune was spent on energy-efficient products and systems for the award-winning prototype. You'd be wrong. The total cost for the first prototype, a 2200-ft² duplex with two double car garages, cost less than $100,000, including fencing and landscaping, excluding lot and building fees. It is important to recognize that while steel framing materials are generally less expensive than wood, other costs might be incurred, such as on-the-job training and engineering fees for truss systems and load-bearing walls. However, as workers become more familiar with steel, labor time and costs will be significantly reduced, and standardized load and span tables, details, and fastening requirements now available (see Appendix B) may eliminate the need for engineering.

For those of you who are environmentally motivated, we've included information on the new state-of-the-art "super windows," and the latest air conditioners with SEERs (the seasonal energy efficiency ratio) of 15 or higher. However, as often is the case, budgetary limits were a consideration when building the prototype, so standard, modestly priced double-pane vinyl windows and an air conditioner with a SEER of 11.0 were used. Contrary to popular belief, you don't have to invest an enormous amount of money to achieve efficiency. In fact, many of our recommendations in this book won't cost you a penny. For example, it usually doesn't cost extra to reduce window areas on east and west sides, install ducts in a conditioned area, and orient the house in a manner that saves energy. Moreover, minimizing steel frame members and properly sizing your air conditioner will actually save you money. We'll consider which areas cause the greatest energy losses, always keeping cost versus benefit in mind.

You Too Can Build an Energy-Efficient Steel-Framed Home

The first section of this book provides an overview of design and construction techniques that increase energy efficiency in steel frame homes. We'll cover the entire process, from structural design to landscaping. The following chapters discuss thermal breaks, heat transfer considerations, insulation, duct work, air conditioning, even appliances in the home. You'll see, it's the details that make the difference.

The second section starts by giving you names of simple programs you can use to conduct an energy analysis and takes you step-by-step through one of the programs. The following chapters describe the tools and materials needed for light-gauge steel framing and provides a sample design of a house with step-by-step instructions. Standardized load and span tables, details and fastening requirements are also included, as well as examples of steel-framed homes that may eliminate the need for an engineer.

Let us say here, just because a wood frame home claims a R-19 wall rating and an R-38 roof rating doesn't mean the house is energy efficient. It's ironic when supporters of wood framing make statements about losses through steel wall framing when the major losses are in other places, such as the infiltration, the slab edge, and windows. True, if the wood houses used the components we did, they would be more energy efficient, but the truth of the matter is they generally don't. As you'll discover, a well-designed home can more than compensate for energy lost through steel construction.

Why don't we get started?

2

Structural Design

Energy efficiency building standards require steel frame systems to meet a minimum thermal efficiency requirement. This requires construction technologies different from traditional wood frame construction.

Because steel studs are thin compared to wood studs, some designers and builders assume the difference in thickness is more than enough to offset the conductive characteristic of steel. Despite their thin web, however, steel studs are much more conductive than wood studs.

Calculations show that the web of an 18-gauge steel stud is about 31 times thinner than a "two-by" wood stud; however, steel conducts heat 310 times more efficiently than a typical wood stud. If the wood-framing members of an assembly are replaced with steel stick for stick, the overall R-value of the assembly can be reduced by 50 percent or more. To help you understand the importance of finding ways to make up for this efficiency loss, let's take a few minutes to discuss heat transfer and thermal bridging.

Heat Transfer and Thermal Bridging

Heat seeks a balance with surrounding areas. Therefore, heat moves from the inside to the outside during the winter and from the outside to the inside during the hot summer. Heat can also occur between two mediums, moving from the warmer medium to the cooler medium. Therefore, heat not only flows through the assembly from inside to outside of the building envelope but also moves from the center of the insulated cavity to the framing members.

Resistance (R) is the measure of a material's ability to resist the flow of heat through it, so the higher the R-value of a material, the greater its resistance to heat flow. Another measure of heat flow through materials is known as U, which refers to the ability of an entire built-up section, such as the wall or window section, to permit the flow of heat. The U-value is important in calculating heat loss (see Chapter 7) and when comparing window units (see Chapter 5). U is the inverse of the total R, so, the lower the U-value, the higher the insulating value.

The transfer of heat takes place by one or more of three methods – conduction, convection, and radiation. These natural forces work separately and together in transferring heat between objects.

Conduction is the transfer of heat through a solid material or from one material to another where their surfaces are touching, from the warmest part to the coolest part (Figure 2.1). Dense materials such as concrete, metal, or glass conduct heat more rapidly than porous materials such as wood or fiber products. Any material will conduct some heat when a temperature difference exists. Conductive losses are most prevalent where the same material is exposed to both inside and outside temperatures. Areas of particular concern include built-up headers, studs, corners, window sash and glazing, plates and band joists, and concrete floors. The conduction of heat through building materials is a major source of heat loss.

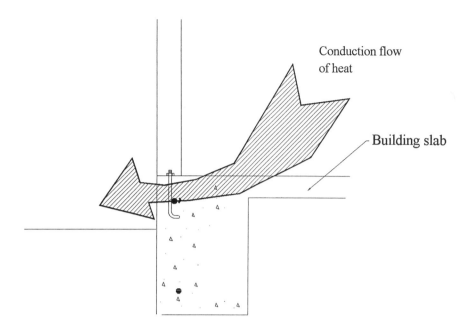

FIGURE 2.1 Heat transfer through conduction.

Convection is the transfer of heat due to the movement of a fluid, typically air or water. Molecules of gas (air) or a liquid (water) become less dense and lighter when heated and rise. As the warmer molecules rise, the cooler ones sink to create convection currents (Figure 2.2). Convective heat transfer usually works in conjunction with conductive heat transfer.

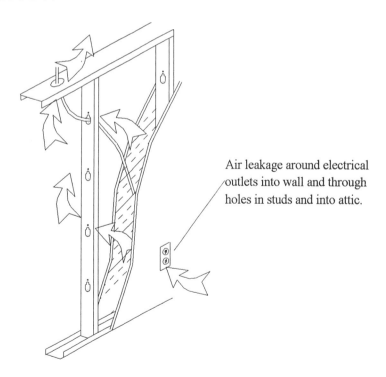

FIGURE 2.2 Heat transfer through convection.

One of the most significant areas of convective loss is infiltration, or air leakage that occurs through cracks in the building. (See Chapter 4 for more information.) Wind is another form of convective air movement that affects heat loss in a home. A building fully exposed to cold winter winds requires more energy for heating then a house sheltered from the wind. (Chapter 6 discusses techniques for siting a building to provide protection from cold winter winds.)

Radiation takes place without a medium. Radiant energy directly transfers heat through space by electromagnetic waves, traveling at the speed of light until absorbed by a solid or reflected. After absorbing heat, all objects radiate heat, assuming the object is hotter than the air surrounding it. The higher the temperature of the object, the greater the quantity of heat radiated. The amount of radiant heat given off depends not only on the temperature of the radiating object but also on the extent and nature of its surface. A rough, dark surface, for example, radiates more heat than a smooth, bright surface at the same temperature (Figure 2.3). Building materials emit radiant heat in all directions, to any surfaces that are at a lower temperature. Windows and glazing contribute significantly to radiant heat loss.

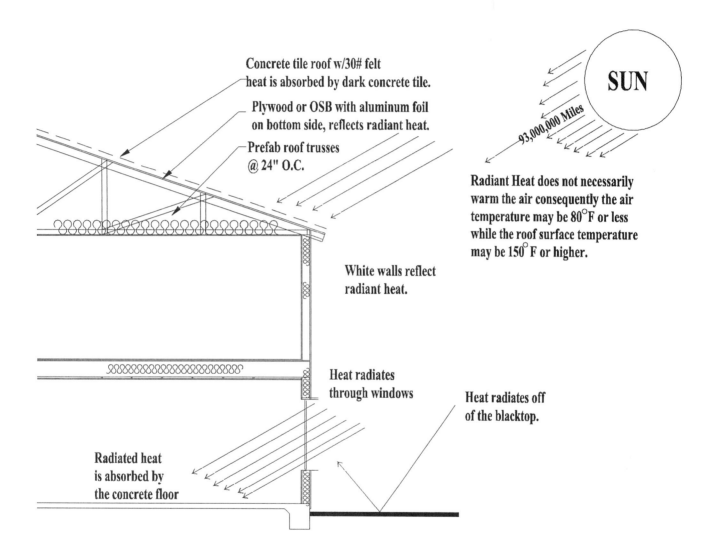

Concrete tile roof w/30# felt heat is absorbed by dark concrete tile.

Plywood or OSB with aluminum foil on bottom side, reflects radiant heat.

Prefab roof trusses @ 24" O.C.

SUN

93,000,000 Miles

Radiant Heat does not necessarily warm the air consequently the air temperature may be 80°F or less while the roof surface temperature may be 150° F or higher.

White walls reflect radiant heat.

Heat radiates through windows

Heat radiates off of the blacktop.

Radiated heat is absorbed by the concrete floor

FIGURE 2.3 Heat transfer through radiation.

When you examine the typical commercial structural system, you'll see why these three methods of heat transfer can cause problems. Buildings generally consist of steel frames with steel purlins used for roof members and steel girts used for wall members. Insulation is often compressed over secondary structural members. Fasteners or connectors provide a metal heat loss path through the compressed insulation. Metal in the girts, purlins, and frames creates a highly conductive path. Heat losses can occur due to conduction through the insulation, conduction through the region adjacent to the Z-Girt, and convection within the wall, enhanced by the air columns created by the wall geometry.

What happens when a highly conductive steel stud breaks the continuity of the cavity insulation? Heat has a tendency to choose the most conductive path, the path of least resistance, to bypass the insulation. This creates an effect called *thermal bridging* (Figure 2.4).

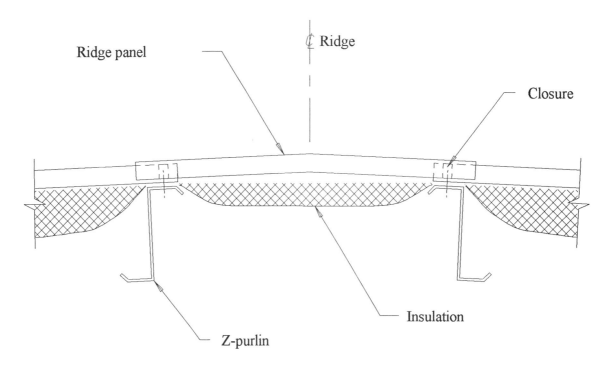

FIGURE 2.4 Thermal bridging.

If appropriate measures aren't taken, these phenomena can cause a number of problems: excessive energy use for heating and cooling, the need for larger space conditioning equipment to handle the larger loads, and condensation of moisture on the warm side of a wall, leading to dust or mildew stains.

What can you do about heat transfer and thermal bridging? The solution is to interrupt the path along which heat flows. This can be accomplished in several different ways.

Keeping Steel to a Minimum

"The way we're building with steel today is not correct," says Andre Desjarlais of Oak Ridge National Lab (ORNL). "We're building with steel as a direct substitute for wood."

When you make the mistake of replacing wood with steel stick for stick, you unnecessarily add energy penalties to the house. Not only the steel studs in the walls but the interface details must also be taken into consideration. Wall framing for windows and doors, building edges and corners where walls meet, and the connections of roofs and floor to the walls all have high concentrations of framing. Substantial heat transfer takes place through these areas, especially when highly conductive steel framing is used.

For these reasons, you'll want to space the studs as far apart as possible without affecting the integrity of the structure. Experts recommend framing exterior walls at 24 in on center as a minimum. Some pre-engineered packages allow even wider spacing. A wall with 24 in on center steel framing performs about 10 percent better than a wall framed 16 in on center. However, verify specifications with manufacturers, since some types of siding/sheathing material may require 16-in centers.

Keep in mind, framing systems are often overdesigned for higher structural strength than required. For example, some builders use as many as three trimmer studs underneath on both sides of the windows. Don't use steel headers as if they were wood headers. Take advantage of the strength and strong connections of steel. Calculate the load bearings for connections and don't make the stud any larger than necessary.

Constructing Your Own Trusses

Because steel is much stronger than wood, builders can span lengths of up to 30 ft – lengths impractical with wood – while avoiding the use of truss supports. Often builders will build heavy-duty trusses, running heavy members from the ridge to the corners resulting in a very heavy structure that needs heavier studs for support. Design trusses coming down the slope in hip roofs (Figure 2.5). If you're constructing your own trusses in the field, this is a fairly simple procedure. In fact, untrained women volunteers from a local Home Depot built all the trusses for a Habitat for Humanity house in less than one day.

Use minimum-sized 20-gauge studs for the web members wherever possible. Avoid the 26-in gauge, which tends to strip easily. (For step-by-step instructions on designing your own roof trusses, see Chapter 11.)

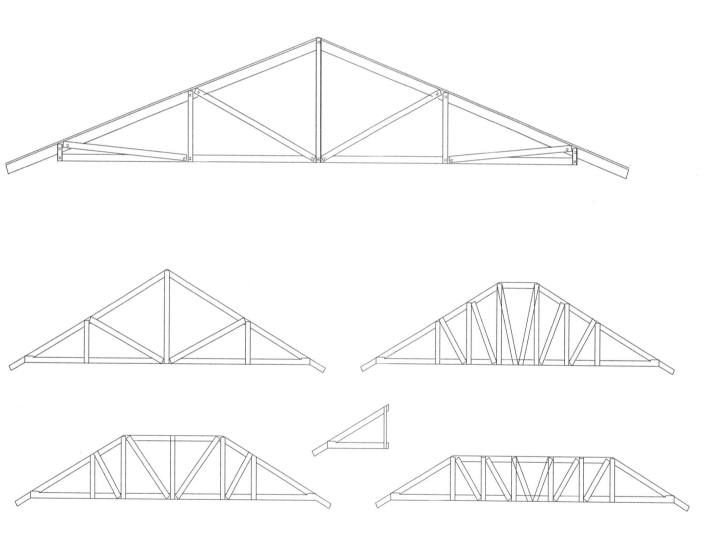

FIGURE 2.5 Trusses coming down the slope in hip roofs.

Using Steel-Framed Fastening Systems

The use of "slammer studs" and Prest-on drywall fasteners can also reduce the number of studs. A typical wall intersection uses three studs, increasing the amount of steel studs by a full 30 percent (Figure 2.6). Slammer studs and Prest-on corner-back fasteners eliminate the use of back-up studs at gable ends, interior walls, and building corners, which saves steel and helps negate the thermal bridging of studs. Using Prest-on drywall fasteners to put on the entire lid of the house creates a thermal break while saving on material. By eliminating studs on corners and wall intersections, you can eliminate up to 50 studs in a typical house. In addition, running continuous drywall on the ceilings and walls creates an airtight structure.

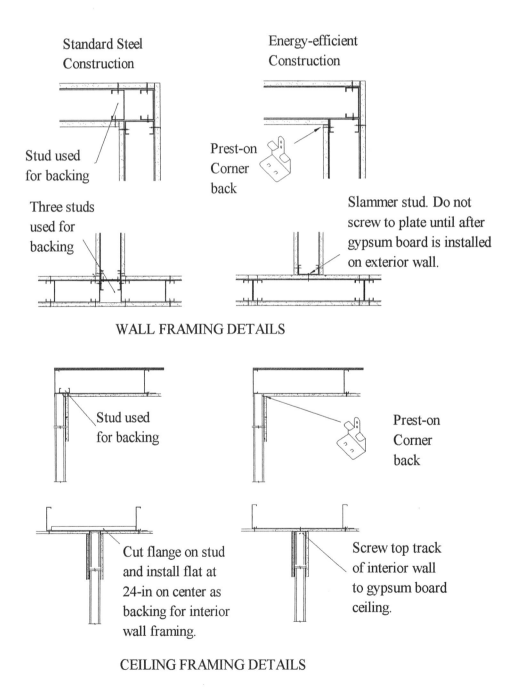

FIGURE 2.6 Slammer studs and Prest-on drywall fasteners.

If this is impractical, consider using the patented Eliminator Track, which eliminates the need for screws at the top of the interior stud and provides a lip to connect the drywall, making an airtight connection.

The interior walls can be constructed with 26-gauge 2½-in studs, which saves steel while providing more living area. The top track is screwed directly to the gypsum board, or the Eliminator Track can be used on the ceiling, eliminating stud backing again. Remember, fewer studs mean less steel, which saves resources and, more important, reduces thermal bridging.

Minimizing the Thickness of Steel

It is also necessary to minimize the thickness of the steel in the framing design. Steel framing members are available in many thicknesses (gauges), and residential designs may use several gauges to accommodate different span and load requirements. Tables 10 to 15 in Appendix B can be used to determine the minimum stud thickness for various loading conditions. *Do not use thicker gauges than necessary.*

An analysis of the framing members with the EZFRAME program (an excellent energy analysis program discussed in Chapter 9) indicates that by using 16 gauge instead of 20 gauge for wall studs, the R-value of the wall assembly decreases by 10 percent.

Punching Large Holes in the Web

Another solution to thermal bridging problems is manufactured modified steel studs for residential use. These products rely on perforations or other gaps in the web of the stud to reduce heat transfer across the cavity (Figure 2.7). Some of these studs also have small nubs on the flanges that reduce the contact area between wall sheathing and the studs.

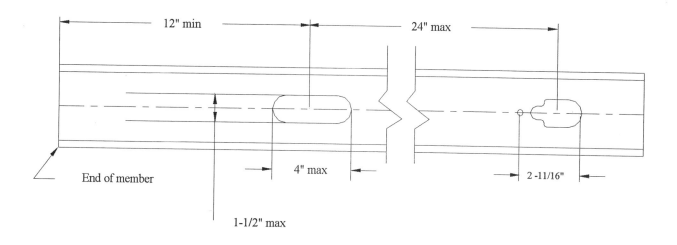

FIGURE 2.7 Punched out webs.

According to the California Energy Commission's Automated Procedure for Calculating U-values of Framed Envelope Assemblies, the percentage of the length of the web which does not conduct heat because of knock-outs in the web is typically 15 percent. Research performed by the National Research Council Canada has shown that minimizing the actual metal in the stud by utilizing a type of lattice design can reduce the thermal bridging up by to 50 percent. However, these studs are generally not as strong, so use with caution.

Installing Horizontal "Hat" Sections

Installing hat sections (Figure 2.8) on one or both sides of studs reduces the contact area between the framing and interior or exterior finishes. However, National Association of Home Builders (NAHB) test data listed in the American Iron and Steel Institute's (AISI) *Thermal Design Guide* shows that the addition of 7/8-in hat channels improves the R-value by less than 4 percent and can also open the wall to infiltration.

FIGURE 2.8 Horizontal hat sections.

Installing Thermal Breaks Between Metal Elements

According to *Metal Elements in the Building Envelope: A Practitioner's Guide,* published by the Department of Energy (DOE), it is sometimes possible to design a nonmetallic thermal break into the wall in order to reduce thermal bridging (Figure 2.9). This approach will be most beneficial when the thermal break is installed in a severely conductive situation, such as a metal truss attached to a metal stud wall.

The danger of relying on this method is that a too-thin thermal break can increase the heat flow as a result of better contact with the metal, which offsets the additional thermal resistance of the insert. To be effective, the thickness of the thermal break material should be no less than 0.40 to 0.50 in thick and should have a conductivity that is at least a factor of 10 lower than the material creating the thermal bridges.

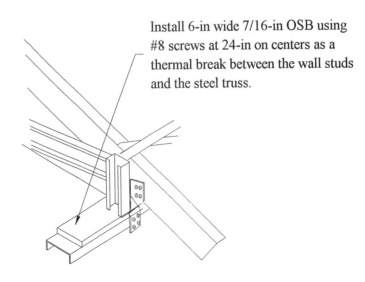

Install 6-in wide 7/16-in OSB using #8 screws at 24-in on centers as a thermal break between the wall studs and the steel truss.

FIGURE 2.9 Installing thermal breaks.

Reducing Convective Air Flow

Since convective airflow can account for a third or more of the wall heat loss for a metal building, including a tight, well-sealed air barrier makes sense. The vapor barrier installed on the warm side of the wall, if sealed adequately, will serve this purpose. (For more information on vapor barriers, see Chapter 4.) Faced metal building insulation, correctly attached at the edges, and the replacement of V-Rib inner metal skins with a flat skin will mitigate this effect as well.

Constructing Double-Stud Exterior Walls

Another way you can reduce thermal bridging is by incorporating double-framed walls. In the 1970s I used 20-gauge, 3-5/8-in exterior walls and 26-gauge 3-5/8-in interior walls with an air gap in the construction of steel-framed commercial buildings. Both the exterior and the interior walls were insulated with R-11 fiberglass insulation (Figure 2.10). This method not only reduces thermal bridging but also increases the level of insulation by insulating both shells.

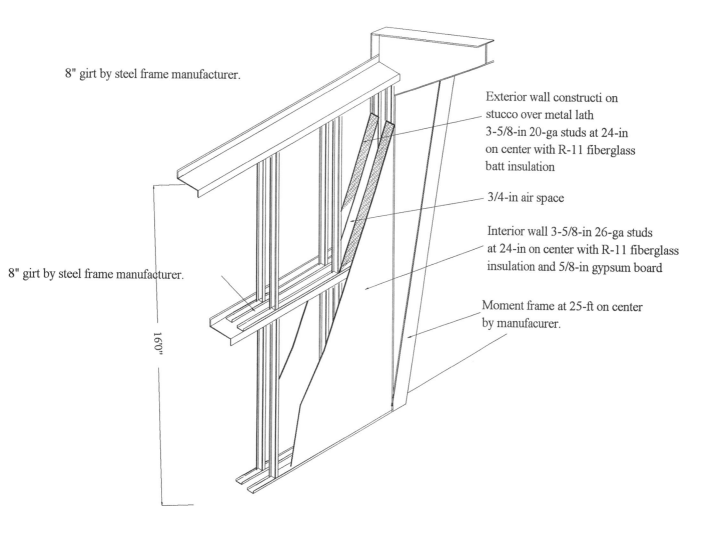

8" girt by steel frame manufacturer.

Exterior wall constructi on
stucco over metal lath
3-5/8-in 20-ga studs at 24-in
on center with R-11 fiberglass
batt insulation

3/4-in air space

Interior wall 3-5/8-in 26-ga studs
at 24-in on center with R-11 fiberglass
insulation and 5/8-in gypsum board

Moment frame at 25-ft on center
by manufacurer.

8" girt by steel frame manufacturer.

16'0"

FIGURE 2.10 Constructing double stud exterior walls.

Installing the Proper Insulation

Installing rigid insulation over the studs, rather than between them, can provide a thermal break between framing members and the exterior substrate that mitigates the impact of thermal bridging. However, exterior foam insulation is not your only choice. A number of innovative wall systems offer advantages discussed in the following chapter.

As shown in Figure 2.3, you can now purchase oriented strand board (OSB) with a layer of aluminum foil on one side, which reflects the radiant heat and is effective in hot climates. This type of sheeting was used on the Habitat house built near Palm Springs, California. The results are now being monitored to compare the energy savings to the additional cost. The energy calculation for the sample house at the end of Chapter 9 indicates R-30 insulation (10-in of insulation) in the attic reduces the energy loss through the ceiling by approximately 15 percent of the total heating losses and 11 percent of the cooling losses. If the additional cost of installing OSB with aluminum foil is over 8¢ a square foot, the sheeting probably won't be cost-effective over well-insulated areas. However, the product works well in uninsulated garage space and may be cost-effective in extremely hot regions.

3

Insulation

As those in the building industry know, insulating materials include materials such as fiberglass (batts and loose), rigid foam boards, cellulose (loose), liquid foams that solidify after installation, reflective aluminum, and vermiculite. When building a home, insulating materials should be compared on the basis of insulating value, ease of installation, cost, availability, and fire and toxicity dangers. Although fiberglass insulation is typically used in wood-framed houses, when building with steel you'll want to look at different options. Because of the high conductivity of steel discussed in the last chapter, your choice of insulation is one of the most important decisions you'll make.

We'll consider various insulation materials for steel framed buildings and compare both existing and new materials.

Special Batt Insulation

To make up for energy losses, a designer could use larger studs and add more insulation. Special batt insulation, sometimes known as steel stud insulation, is wider than the material used with wood studs and available in values from R-8 to R-20.

However, as we've discussed, the advantage of thicker walls or higher-density insulation drops off quickly when the walls are short-circuited by steel studs. In fact, there's a diminishing return on each additional R-value of cavity insulation added over 6 in thick. The resulting thicker walls also cut down the amount of usable living space inside the building.

A new product, now on the market is Icynene, a low-density, open-cell modified polyurethane that is typically foamed into open cavities. The product has a very high expansion rate. Within seconds, the foam expands to its full thickness, filling (at times overfilling) the open cavity, sealing joints and crevices against air and moisture infiltration and filling hard-to-reach gaps between building material. The Icynene Insulation System has an R-value of about 3.6 per inch. From an environmental standpoint, Icynene contains no ozone-depleting chemicals. Water serves as the foaming agent, reacting with other components to generate CO_2, which expands the foam.

However, as the *Environmental Building News*, (volume 4, number 5), points out, using Icynene for insulating metal-framed buildings has a few drawbacks: "Although there is some benefit to tightly filling steel-stud cavities (which can be difficult to do with fiberglass) using an expensive insulation such as Icynene is difficult to justify economically. With steel framing, it makes more sense to use as inexpensive an insulation material as possible to fill the stud cavities, then invest any additional available budget in rigid insulative sheathing." Although Icynene is effective in sealing air leaks, any spray foam cavity insulation will help prevent air infiltration.

Exterior Insulation Sheathing

Studies show that installing exterior insulation is a practical and cost-effective way to enhance thermal performance. By insulating the exterior, the entire mass of wall materials stay at a more constant temperature. This slows indoor temperature changes throughout the day and night. The improved comfort is particularly noticeable in the summer and as a reduction in drafts in the winter.

In fact, recent tests indicate that exterior insulation may actually increase the wall's R-value as much as 20 percent more than the R-value of the added insulation. While this increased R-value by itself does not solve the thermal bridging problem, it does suggest that current estimates should be revised upward by about R-1 when exterior insulation is used over steel studs.

Rigid board pressed into preformed sheets are typically used and can be fabricated with three types of foam cores: molded expanded polystyrene (MEPS), extruded polystyrene (XEPS), and urethane (such as polyurethane and polyisocyanurate). The various types of foam boards available differ as to R-value, availability, thickness, resistance to moisture, fire resistance and toxity, strength, durability, and cost.

Since R-factors vary substantially, *Thermal Resistance Design Guide*, Publication RG-9405 by the AISI, is a useful publication. The manual provides designers and contractors with suggested insulation levels for steel-framed construction and comparison charts based on research at the NAHB Research Center. You can purchase the manual at a minimal cost by calling 1-800-79-STEEL. The results of the AISI-sponsored testing provides R-values on 21 different types of wall construction, including both 3-5/8-in and 6-in framing members sheathed singularly or in combination with plywood, gypsum wallboard, extruded polystyrene board, and polyisocyanurate sheathing. Tests showed that energy-efficient homes may be framed with light-gauge steel at little or no extra cost over wood framing. Depending on the region of the country, just ½ in or 1 in of exterior insulating sheathing used with full width fiberglass insulation batts may be all that is needed to meet requirements of the Model Energy Code (MEC). In fact, for most southern states, exterior insulation sheathing may not be needed at all.

A cause for concern is the use of CFCs (chlorofluorocarbons) and HCFCs (hydrochloro-fluorocarbons) as blowing agents, which causes release of chlorine molecules into the atmosphere, contributing to ozone depletion. In addition, CFCs and HCFCs are prone to "R-value drift" as the blowing agents leak out of the cell structure and air leaks in. Expanded polystyrene rigid insulation (EPS) performs similarly to extruded polystyrene and is the only common rigid foam board stock insulation made without CFCs or HCFCs. EPS typically uses pentane as a blowing agent and has some recycled content. Although HCFCs are not as damaging to the ozone as CFCs, foam insulation manufacturers are looking for zero-ozone-depletion alternatives, some of which are already available.

Additional cost of foam insulation, fasteners, and labor and attachment methods are other possible drawbacks. According to field evaluations prepared by NAHB Research Center for HUD, some builders have problems with current methods of attaching foam board insulation and working with different types of siding. For instance, in places where stucco is popular, attaching the metal lath to foam board can be difficult. Other builders expressed concern with attaching vinyl siding over foam board, stating that many siding installers charge more to apply vinyl. These problems may be resolved as alternative fastener methods are developed and subcontractors become familiar with steel.

Installing Insulative Exterior Sheathing

Most framers install X-bracing using 16- or 18-gauge, 6-in flat strapping material. The sheet of insulative sheathing can be cut with a circular or table saw. Generally, it is applied to the frame in a similar manner as other exterior sheathing materials. One note of caution: By fastening the insulating sheathing directly to the studs with metal connectors, the screws or other fasteners will continue the short-circuit path of the stud, reducing the effectiveness of the insulation. Instead, install plywood or OSB sheathing to the studs and fasten the insulation to the panels with staples or glue.

Sheathing should be fastened so that it's tight to the frame. Insulative sheathing can be placed over window and door openings and later cut flush to the rough opening with a utility knife and straight edge, or a hand saw.

Exterior finish siding materials are installed over insulative sheathing. Depending on the details and materials selected, siding may be installed before or after the windows, doors, and trim boards. Protect these intersections of siding with doors and window trim against air leakage by accurate cuts and caulking. Look for other areas where foam can increase energy efficiency. Be sure to cover the bottom of cantilevers and below steel floors joists over unheated garages and vented crawl spaces.

EIFS, also known as *exterior insulated finish system*, is a term sometimes used to describe a modern stuccolike construction technique where the exterior wall surface is covered with expanded polystyrene foam over which a base coat is troweled into a fiberglass mesh to create a monolithic wall surface. A finish coat is then applied to this exterior wall surface. This coat is typically colored and may be textured to appear as though it is a stuccoed surface. Decorative bands, quoins, reveals, and shapes may be attached to EIFS surfaces to create a variety of unique design appearances.

If you plan to use this method, be aware of a change made at a recent hearing on EIFS by the International Conference of Building Officials (ICBO) ES Evaluation Committee. As of July 1, 1997, weather protection for Type V construction with EIFS on wood and steel framing must comply with Uniform Building Code (UBC) section 1402, using paper or felt. The weather protection additionally must comply with section 2506.4 and the intent of the third paragraph of section 2506.5 in draining trapped water to the exterior of the building.

Section 1402 states: "All weather-exposed surfaces shall have a weather-resistive barrier to protect the interior wall covering. Such barrier shall be equal to that provided for in UBC Standard 14-1 for kraft waterproof building paper of asphalt-saturated rag felt. Building paper and felt shall be free from holes and breaks other than those created by fasteners and construction system due to attaching of the building paper, and shall be applied over studs or sheathing of all exterior walls. Such felt or paper shall be applied horizontally, with the upper layer lapped over the lower layer not less than 2 inches (51 mm). Where vertical joints occur, felt or paper shall be lapped not less than 6 inches (152 mm)."

Concrete Blocks and Panels

Exterior sheathing is not your only choice. A number of innovative wall systems, including concrete blocks and panels (Figure 3.1) are on the rise. In the past, the use of concrete in residential construction has been limited to moderate climates because of concrete's relatively low insulative properties. Adding insulation was difficult and expensive. However, with the recent development of insulated systems, concrete now offers new opportunities for residential construction. Generally, blocks are insulated by: (1) placing a fill material in the blocks as the wall is built, (2) putting plastic inserts into the cores of conventional blocks or using blocks with specially designed cores, or (3) changing the properties of the concrete itself by mixing Portland cement with components such as expanded polystrene beads, additives, sand, and/or water.

By using rigid foams and designs that form thick layers of nearly uninterrupted insulation, today's concrete homes can achieve high R-values. The homes increase energy efficiency further by taking advantage of the high mass of concrete to even out temperature swings, reducing heating and cooling loads. Thus, concrete homes consume less heating and cooling than frame houses that have the same R-value. More and more, you'll hear the term "total energy efficiency," since R-values are only one of the major factors that determines the cost of heating and cooling.

The advantages of concrete homes are many. The blocks produce extremely strong, quiet houses with low utility bills year-round. These houses are so strong, in fact, that three different studies concluded that concrete houses in Andrew's path suffered less damage than standard wood-framed homes. Insurance statistics confirm what tests and common sense tell us: Concrete walls have a higher fire survival rate than wood frame. The homes experience less structural damage from pests and water. There are no voids and little air leakage, making homes more energy efficient and decreasing the likelihood of having problems with allergies due to outside sources. James Dulley, syndicated columnist of *Cutting Your Utility Bills*, was so impressed with concrete/foam block construction, he stated, "When I build my own new house, this construction method is at the top of my list."

FIGURE 3.1 Home built with foamed concrete blocks.

Concrete systems come in the form of three types of products: insulated concrete wall forms, insulated concrete wall systems, and welded wire sandwich panels.

Insulated Concrete Wall Forms: Stackable, interlocking blocks of plastic foam insulation generally snap together to form the foundation and walls, much like Legos. Openings for windows and doors are cut, then the entire assembly is reinforced with steel rods in the hollow cavities. Concrete is poured into the cavities, forming a solid strong monolithic insulated concrete wall. With the foam on the interior and exterior surfaces, the walls can be finished by any common method.

Insulated Concrete Wall Systems: These systems typically center a panel of plastic foam insulation between two conventional metal forms. The panel is held in place by proprietary "connectors." The concrete is then cast into the forms.

Welded-Wire Sandwich Panels: These panels combine concrete and insulation into a single panel. The panels are composed of a polystyrene or polyurethane insulation core surrounded by a welded-wire space frame. The panels are sprayed or trowel applied over the wire mesh with a layer of shotcrete. This technology has been most readily adopted in the Far East, Middle East, and the Caribbean, where concrete building products are commonly used in home building. In the United States, use has generally been limited to institutional and commercial construction. Two United States companies and 15 foreign companies manufacture weld-wire sandwich panels.

In the early 1990's, I utilized Rastra™ blocks, my Insu-form™ foundation system and steel trusses on three homes in Desert Hot Springs, California and discovered several advantages: low-energy usage, strength, fire and termite resistance, and noise reduction. However, windows and doors were difficult to install. In addition, the interior finished drywall cannot be fastened with screws and must be glued, a time consuming procedure that provides no shear value. To solve these problems, I began working with Omega Transworld, Ltd. (OTW) to develop a new building system.

In 1984, Burrel Mining personnel combined their underground mining and construction expertise to market a new product for ventilation control in coal mining. The concept was to increase mine air flow, provide easier installation and reduce back injuries due to handling heavy block. In January 1985, preliminary testing began with the concept of using a lightweight cementious material in a single block form. The product was tested and approved by Mine Safety and Health Administration. All aspects of the product proved to be successful and the blocks have a proven record as non-toxic and safe. Several homes have been built with the blocks (Figure 3.2).

FIGURE 3.2 Habitat for Humanity home built with foamed concrete panels.

Making further improvements, we developed poured wall panels insulated with cellular or foamed concrete poured between two steel studs. A 6-in thick x 24-inch wide, 8- to 10-ft high panel weighs an average of 175 pounds and can be easily tilted in place by two women with a third woman guiding the panel and securing it with screws (Figure 3.3).

FIGURE 3.3 Home built with foamed concrete panels.

Based on the experience of building OTW houses in the United States, Mexico, and Thailand, the walls in a normal single-story house can be framed by three men in under 3 days. The block is solid and serves as an excellent backing to reduce damage to the drywall that occurs in many warehouse-type uses. In addition, the product is environmentally friendly, since the block is partially made of waste material, flyash, the waste product of coal burning power plants. In the United States alone, power plants produce more than 70 million tons of fly ash normally disposed of in landfills.

Field tests conducted on an OTW wall panel demonstrated the advantages of concrete systems. Exterior noise measured at 104 decibels was reduced to less than 50 decibels inside the structure. Leakage testing indicated negligible air leakage through the face of the block. Because the system makes a house virtually airtight, saving up to 7,000 BTUs (British thermal units) per hour, it makes the homes extremely efficient. The foamed cement panel walls acts as thermal mass to moderate interior temperature swings in comparison to those on the exterior of the building, reducing heating and cooling demands. The walls work in the same manner as an adobe or solid masonry dwelling but with the obvious added advantage of the closed cell insulative structure of the panel. (See Chapter 13 for construction details.)

In short, concrete systems are ideally suited to insulate, sound-proof, and fire-proof the frame of steel buildings.

Slab Edge Insulation

An important energy saving element, which is often ignored, is slab edge insulation. Concrete slab floors lose heat to the earth and foundation walls by conduction and through exposed edges by convection and radiation. Because the distance the heat travels to the colder ground is further from the center, the greatest heat loss is at the edge of the slab.

The heat loss through uninsulated slab on grade foundations and floors is significant. One meter of exposed uninsulated slab edge loses as much heat as several square meters of insulated wall area. Insulating the slab reduces heat loss and maintains warmer floor surface temperatures inside. In Chapter 12, you'll find a complete set of energy calculations based on a sample house. The figures below are based on those energy calculations. As you can see, the energy savings from an insulated slab are substantial.

HEATING (sample house with slab edge insulation):

Component	Btuh/Sq.Ft	Btuh	% of Btuh	
Floors	17.3	2005	12.3	Slab edge insulation R-7

HEATING (sample house without slab edge insulation):

Component	Btuh/Sq.Ft	Btuh	% of Btuh	
Floors	51.8	6015	20.7	No slab edge insulation

In very cold climates, slab edge insulation also prevents condensation and frost formation along the inside edge of the floor. Perhaps you've heard of frost-protected shallow foundations (FPSF). These systems are not new. FPSFs have been used in northern Europe for over 25 years, in hundred of thousands of homes. Insulation around the foundation perimeter conserves and redirects heat loss through the building slab toward the soils beneath the building foundation. At the same time, geothermal heat resources are directed toward the foundation, resulting in an elevated frost depth around the building. In cold areas of the country, this method eliminates the need for basements and deep foundations with crawl spaces that permit infiltration into the home. In areas requiring deep footing placement to provide frost protection, FPSFs may allow a foundation depth as shallow as 12 in or 305 mm, and can provide immediate savings due to reduced material costs. Since the insulated foundation system serves as the concrete formwork, only one concrete pour is required. This method complies with the requirements of the Model Energy Code (MEC). (See Chapter 11 for further details.)

Working with slab edge insulation for over 15 years, I can personally attest to its effectiveness. In the early 1980s, California's newly formed California Energy Commission required insulated building slabs. Providing slab edge insulation was not easy, so I developed a bracket that connects the foundation to an insulated steel stud, which surrounds the standard concrete slab. The foundation includes styrene foam panels that recess into the ground, helping the slab retain heat (Figure 3.4).

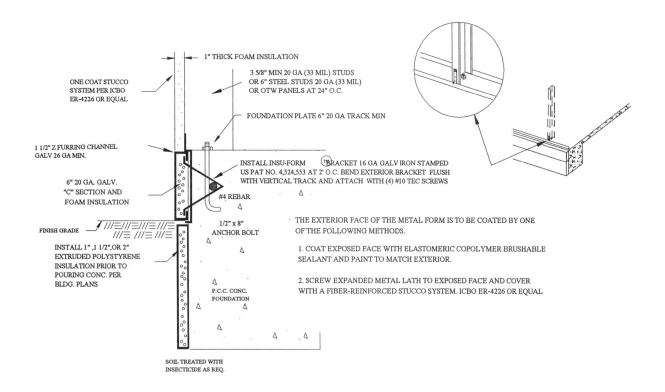

FIGURE 3.4 Patented insulation system, Insu-Form

In 1985, I patented Insu-Form, received building code approval from ICBO and began manufacturing and marketing the system. Approximately 400 homes in the Southern California area used the Insu-Form system. Tests made during the summer months in the desert area of California demonstrated the effectiveness of slab-edge insulation in warmer climates. While outside air temperatures exceeded 116°F and the outside ground temperature adjacent to the foundation exceeded 158°F, the inside slab temperature remained constant at a maximum of 73°F.

When California changed its law in 1987 and slab edge insulation was no longer required in most areas, builders were unwilling to spend the extra money despite a two-year payback in energy savings. However, with the popularity of steel framing and a renewed interest in energy savings, builders are recognizing the value of slab edge insulation. In my homes, Insu-Form brackets are attached to the steel form, which provides (1) a hold-down bolt for the track or bottom plate every 24 in, (2) a strap to anchor the steel studs directly to the foundation, and (3) a holder for the first row of rebar to keep it snug with the anchor bolt (Figure 3.5).

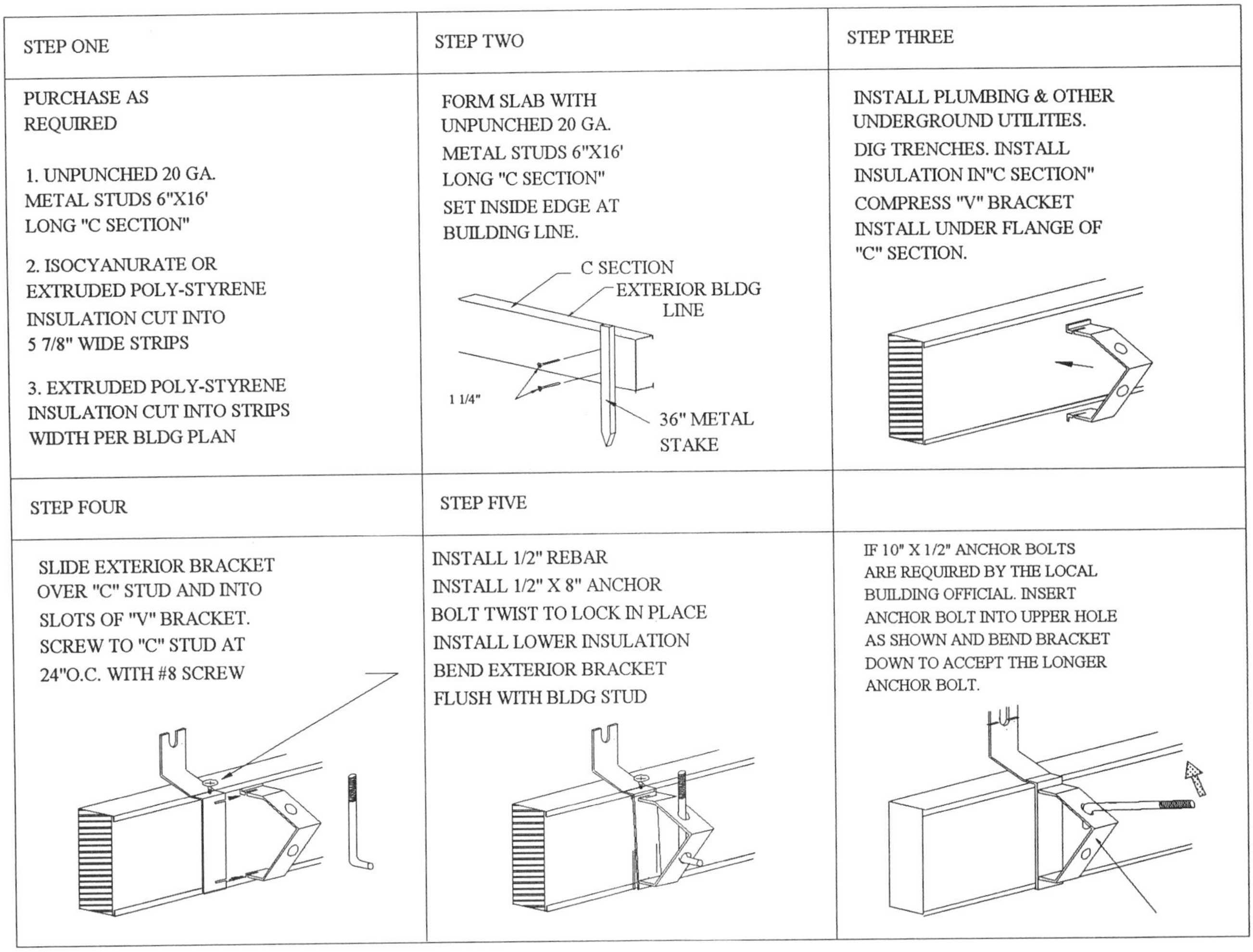

STEP ONE	STEP TWO	STEP THREE
PURCHASE AS REQUIRED 1. UNPUNCHED 20 GA. METAL STUDS 6"X16' LONG "C SECTION" 2. ISOCYANURATE OR EXTRUDED POLY-STYRENE INSULATION CUT INTO 5 7/8" WIDE STRIPS 3. EXTRUDED POLY-STYRENE INSULATION CUT INTO STRIPS WIDTH PER BLDG PLAN	FORM SLAB WITH UNPUNCHED 20 GA. METAL STUDS 6"X16' LONG "C SECTION" SET INSIDE EDGE AT BUILDING LINE. C SECTION EXTERIOR BLDG LINE 1 1/4" 36" METAL STAKE	INSTALL PLUMBING & OTHER UNDERGROUND UTILITIES. DIG TRENCHES. INSTALL INSULATION IN"C SECTION" COMPRESS "V" BRACKET INSTALL UNDER FLANGE OF "C" SECTION.

STEP FOUR	STEP FIVE	
SLIDE EXTERIOR BRACKET OVER "C" STUD AND INTO SLOTS OF "V" BRACKET. SCREW TO "C" STUD AT 24"O.C. WITH #8 SCREW	INSTALL 1/2" REBAR INSTALL 1/2" X 8" ANCHOR BOLT TWIST TO LOCK IN PLACE INSTALL LOWER INSULATION BEND EXTERIOR BRACKET FLUSH WITH BLDG STUD	IF 10" X 1/2" ANCHOR BOLTS ARE REQUIRED BY THE LOCAL BUILDING OFFICIAL. INSERT ANCHOR BOLT INTO UPPER HOLE AS SHOWN AND BEND BRACKET DOWN TO ACCEPT THE LONGER ANCHOR BOLT.

FIGURE 3.5 Attaching Insu-Form bracket to the steel form.

To illustrate how well slab insulation works, the Insu-Form foundation system was used in 1983 for a church in Cathedral City, California. In 1991, when members decided to replace the heat pumps with a higher-efficiency model, they realized that in 8 years the heat had never been turned on. While this building is located in a moderate climate, some days have frost. Slab edge insulation works.

In many areas of the country, a builder can meet the requirements of the Model Energy Code without using exterior insulation board on the studs if the slab is insulated. My patent is a conceptual one, and to my knowledge is the only method that has received ICBO approval. However, in this book we'll show you how standard materials can be used to make the bracket. (Note: These brackets cannot be manufactured and sold to others without a license from John.Hacker.) Chapter 11 provides further details on how to install the foundation system, using either the Insu-Form foundation brackets or the generic system utilizing standard materials.

4

Infiltration and Ventilation

Infiltration is the term used to describe natural air exchange that occurs between a building and its environment when doors and windows are closed. In other words, it is leakage that occurs through various cracks and holes that exist in the building envelope. Next to insulation value, reduced air leaks are the most important factors in preventing rapid heat losses and gains. In fact, air leaks account for 15 to 30 percent of space-conditioning energy loss.

In addition to saving energy, elimination of infiltration of unwanted air reduces indoor air pollution and promotes ventilation. This fact may surprise people who assume a leaky house provides adequate ventilation. "Natural" air leakage is not ventilation; in fact, leaky houses can even be dangerous. Why? A leaky house allows the wind to blow dust, pollen and other particles in through unfiltered openings. Negative air pressure pulls odors, microorganisms, moisture, and radon from the soil through cracks, joints, and pores in the floors and walls that contact the soil. Humid air can cause condensation, mold, and mildew. Indoor air quality is of vital concern to the increasing number of people who are sensitive or allergic to air pollutants

United States Environmental Protection Agency (EPA) scientists reported that dangerous levels of air pollution frequently exist in many homes, even when occupants are unaware of the gases, particles, and microbes they are breathing. Although windows can be opened when more ventilation is needed, the truth is, you can't see or smell many of the dangerous air pollutants commonly found in homes and are probably unaware when ventilation is inadequate. In addition, leaky homes are drafty and uncomfortable, allowing cold air in during the winter and hot air in during the summer.

Because of the health and safety issues involved, as well as the need to eliminate energy waste in steel frame homes, you'll want to carefully consider ways to eliminate air leakage. As discussed in the last chapter, insulation and solid wall systems can reduce air leakage. Other methods of reducing infiltration include proper installation and insulation of ducts, choosing airtight framing materials for windows, careful caulking and proper weather stripping details, and installation of vapor barriers, which will be considered in this chapter.

We'll begin by discussing framing details you can use to reduce air leakage in steel frame homes.

Eliminating Infiltration Through Framing Details

In the August 1995 issue of *Energy Design Update*, the Energy Services Group for Wilmington, Delaware, noted that the measured infiltration rate of steel frame walls in houses was approximately 50 percent higher than the infiltration rate of wood frame walls. Why?

Holes cut in the channels used for top and bottom plates in steel wall construction for electrical and plumbing pipes increased infiltration through the attic. If "hat channels" were used, a slot was created, open to the attic down the full length of the wall, allowing for increased airflow. What can you do to solve these problems?

Minimize mechanical and electrical runs in insulated wall sections. Eliminate holes in the top track by installing a 1 ½-inch conduit to a sub-panel located in an interior wall (Figure 4.1). Install wiring from the sub-panel within the interior walls, running TV and phone lines in interior walls and electrical wiring in raceway conduits from exterior wall locations. If possible, install plumbing within the interior walls. Use sconces to minimize ceiling fixtures (Figure 4.2).

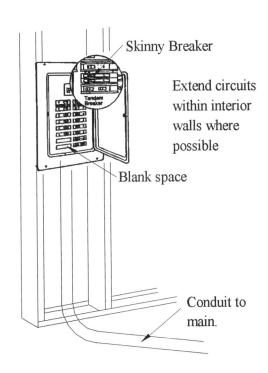

FIGURE 4.1 Installing a conduit to a sub-panel in an interior wall.

FIGURE 4.2 Sconces.

Another way to improve a home's energy efficiency is by reducing wall heights so that less wall area is exposed to the wind. Lowering ceiling heights reduces the volume of air that requires heating or cooling and results in greater comfort due to more even temperatures.

One advantage of steel frame homes helps reduce infiltration: These homes don't settle, which means in 10 or 20 years, the house will still be airtight and the doors and windows will be square and true.

The Importance of Ducts

Until recently, most people who bothered to think about ducts at all considered them a mere medium for transporting air. Problems with ducts were as invisible as the air they carried. However, leaks in ductwork can contribute up to 20 to 60 percent of the air leakage in a house.

The effect of poorly installed ductwork was made evident by the following experience. In 1993, a condominium complex had filed a lawsuit against an air-conditioning contractor. His insurance company hired me as an expert witness. Several of the homeowners complained their electric bill was excessive and the cooling system didn't work properly. An air-conditioning technician tested one of the units. The air conditioning unit was designed to provide approximately 3500 cubit feet of air per minute (cfm), yet tests indicated the unit was providing only 2200 cfm. I reviewed the plans prepared by the mechanical engineer and found the system was properly designed and sized.

However, while visiting the homes, the problem soon became evident. Someone had stepped on the supply duct to the living room. Since the supply duct in the attic was wrapped with duct tape instead of mechanically joined with screws, the duct connection had separated. Consequently, the duct system was sending cool air into the hot attic area in some of the condominium units. The main supply duct in the garage was also leaking supply air where summer heat had caused the duct tape to fail. These problems were easily fixed. One owner told us there was insufficient cooling in the upstairs bedroom. He had installed plush carpet, which sealed the crack under the door. When the door was closed, it sealed the room, pressurizing the duct system. When the door was cut, the system worked properly.

The original contractor repaired the duct systems, cut doors, balanced the air flow, and replaced filters. The homeowners dropped their complaints. As you can imagine, the attorney was upset. He said I was hired to evaluate the problems, not fix them. I just laughed and told the attorney, "So, sue me."

The point is, good system design and improved duct and sealing materials, combined with proper installation saves energy and money. Accordingly, electric utilities can avoid building costly new fossil-fueled power plants, reducing growth in emissions of carbon dioxide to the atmosphere. A growing number of regulatory and energy conservation agencies throughout the nation are actively seeking to eliminate poor ductwork systems as a serious health, safety, and energy issue.

As the above example demonstrates, some design strategies are simple, such as allowing adequate airspace under interior doors or properly sealing duct connections. There are several other ways you can avoid losing energy through duct leaks, but before we list them, let's briefly discuss how a duct system works.

A typical system delivers air from the furnace, air conditioner, or heat pump to a home's living space, then back to the equipment to be heated or cooled again. The system is designed to be a closed, pressure-balanced loop, with the same amount of air entering and leaving the conditioned zones through a network of ducts (Figure 4.3).

If there are leaks in the supply or return ducts, the central system constantly loses air to the outside, creating a negative pressure in the house. The balance is disrupted, and the zones become pressurized or depressurized. When depressurization occurs, the return air plenum is consequently starved for air and pulls air from wherever it can. As a result, air is exchanged with outside air and hotter attic air or even the flue from a gas water heater, which is why some water heaters show burn marks on the outside. In the same scenario, the rooms receiving air from the system are overpressurized. The air can migrate outdoors and cause energy losses or enter wall cavities possibly introducing moisture problems.

Put simply, both negative and positive pressures increase utility costs. A negative house pressure causes leaks, including duct leaks, to draw unconditioned air from the outside, the attic, or the garage. Positive pressure has an opposite effect of pushing expensive conditioned air outside.

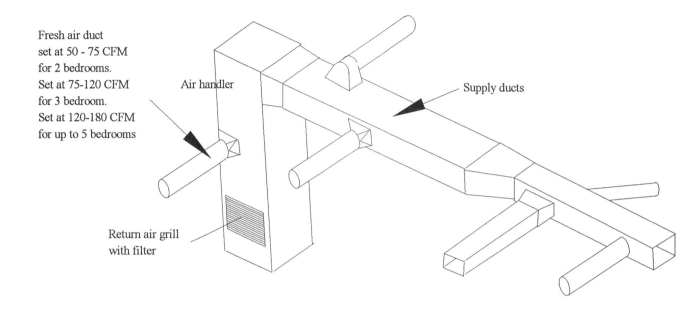

Fresh air duct
set at 50 - 75 CFM
for 2 bedrooms.
Set at 75-120 CFM
for 3 bedroom.
Set at 120-180 CFM
for up to 5 bedrooms

Air handler

Supply ducts

Return air grill
with filter

FIGURE 4.3 Duct system.

There are three basic ways to avoid these problems: (1) by installing ducts within the conditioned space, (2) by properly sealing ducts and, (3) by insulating ducts.

Installing ducts within the conditioned space

According to Mark Ternes of Oak Ridge National Lab's (ORNL) *Energy Division*, "Ducts that pass through unconditioned spaces – attics, garages or crawl spaces – have a good chance of losing energy."

Why is that? Just imagine the energy losses in an air-conditioned home that has leaky ducts in the attic. In the summer, attic temperatures can be as high as 120°F and, in some climates, even 150°F. This extremely hot air is pulled into the return ducts at the same time that supply ducts are fast losing coolness by conduction. Inside, high pressure in the living area is forcing conditioned air outside. Under these conditions, the cooling efficiency of the air conditioner can be cut in half. In fact, DOE says having ducts in the attic wastes millions of dollars of energy every year.

Even if ducts are installed properly, problems can occur if you have ducts in the attic. The return air is generally located in the hall or under the stairs. If the path or the returning air flow is stopped, the house can become pressurized, causing heat losses due to infiltration.

To solve these problems, we *strongly* recommend keeping the ducts within the conditioned space, if at all possible. This can often be accomplished with architectural soffits that add to the design of the house (Figure 4.4).

Since using interior closets and other unsealed parts of the house as a return air duct can draw air from the attic or crawl space due to pressure differences, you should use ductwork for all supply and return ducts. In addition, if the heating unit is gas fired with outside make-up air, the closet containing the furnace should be sealed and insulated from the conditioned space of the house.

FIGURE 4.4 Soffits.

Sealing duct leaks

The proper sealing of plenums, air handlers, and ducts is key to eliminating leaks in a duct system. Some HVAC contractors seal duct joints with duct tape, which is often used improperly and deteriorates in the hot air of attics. Ternes recommends connecting and sealing ducts with mechanical fasteners, mastic, and fiberglass tape. Mastic is a caulking material that can move with the expansion, contraction, and vibration of the duct system components. Fiberglass tape provides rigid structural support and increased strength when used with mastic.

The application process for mastic requires all duct connections be mechanically fastened with screws, rivets, or when using flex duct, with metal bands. To properly seal ducts:

1. Wipe area to be joined clean with a dry rag.
2. Apply the mastic with a trowel or brush (according to its viscosity) and spread 1 in beyond the opening. For ¼- to ½-in openings use fiberglass mesh tape under the mastic. A larger gap needs a rigid material covering.
3. All connections (splices, Ys, Ts, and boots) must be sealed. Additionally, boots should be sealed to the sheetrock. A wire can be used to keep it from pulling loose.
4. Penetrations into the plenum must be sealed. Flex duct inner and outer linings need to be sealed. Do not extend the duct liner through the wall of the plenum to the interior of the plenum.
5. The air handler closet and air handler itself must be sealed, including sealing the air handler to the platform.
6. The return plenum should be lined on the interior with duct board (foil faced in) and sealed. The support platform should be sealed on all sides. Penetrations into the plenum, such as refrigerant lines, must be sealed. The sealing of the support platform can be added to the tasks of insulating/sealing contractor for the house.
7. Return air grills should be sealed at the point of penetration through the walls. Any structural cavities also must be sealed. Place duct boards, cut to fit, on all four sides with the foil sides facing in and seal them in place.
8. Seal boots to sheetrock with polyseal.

Insulating ductwork

Studies show ducts lose about the same amount of energy through conduction as they do through leaks. Losses can be very high if the ducts are uninsulated. Although R-4 insulation is typically used to insulate ducts, use duct that is insulated to *at least* R-6, with a reflective outer surface. It's hard to understand why attics are generally insulated with R-38 insulation, while ducts that contain the cool or warm air are usually insulated with R-4.

We highly recommend wrapping ducts and plenums with R-8 insulation. This translates to a reduction of an average flow-weighted conduction losses of 40 percent.

Airtight Windows

Windows are a primary source of air leakage in a home, allowing up to 17 percent of unwanted air infiltration. Energy-efficient glazing and framing materials are discussed in the following chapter; however, air leakage is an important factor in selecting a window.

In general, the most airtight and least expensive window frames are fixed-paned windows, which costs about 15 percent less than the same-size operable windows. Fixed glazings are available in a full range of sizes and materials, allowing flexibility in the design of architectural details. Keep in mind, because these windows do not open, they cannot be used for ventilation.

Operable windows open and close by either swinging on hinges or sliding within a track. Depending on their quality, hinged windows offer greater potential energy efficiency than sliding windows, primarily due to tighter closure and fairly airtight compression seals. When a window has a good compression seal, it allows about half as much air leakage as windows with sliding seals. Hinged windows include *casement windows* (Figure 4.5), which open sideways with a hand crank, *awning windows* (Figure 4.6), which are similar to casements but have hinges at the top instead of the sides, and *hopper windows,* which are reversed versions of the awning windows with the hinges located at the bottom. Casement, awning, and hopper windows can open fully and encourage ventilation by diverting the air moving along the exterior wall surfaces in the home. However, these windows typically cost more per square foot than other types of openable windows.

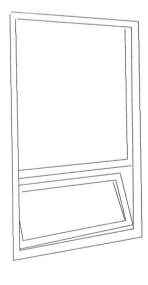

FIGURE 4.5 Twin casement window.   **FIGURE 4.6** Awning and picture window.

Double- and single-hung windows are two of the most common and least expensive types of openable windows. Slider windows (Figure 4.7), a popular type of window, can be inefficient due to a higher rate of air leakage. However, in recent years sliding windows have been improved to meet the requirements of E283-84 of ASTM (American Society of Testing Material) with air leakage of 0.06 to 0.15 cfm/ft (cubic feet of air leakage per minute per foot of window edge). When purchasing windows, look for units with air tightness ratings no higher than 0.2 cfm/ft. The best windows are lower than 0.1 cfm/ft. Sash tilt windows are useful in southern climates because of their ability to direct air flow; however, check to see if the window seals adequately.

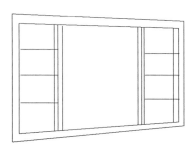

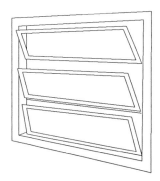

FIGURE 4.7 Slider windows.   **FIGURE 4.8** Sash tilt windows.

Weather-stripping allows the window to operate while providing a tight seal. Window manufacturers supply a variety of weather-stripping with the windows, including rigid vinyl ribs; vinyl covered foam, and rigid vinyl leaf.

Choosing Doors

Doors also contribute to air leakage and therefore you'll want to choose exterior doors that are strong and seal tightly against air leakage. Every time an exterior door is opened during the winter, warm air rushes out and is replaced with cold outside air. During the summer, the opposite occurs. Double door entries help reduce this air leakage. However, even when a hinged door is closed, it may allow considerable air leakage around edges, beneath the sill, past the jambs, and through the lock set. The quality of the doors and the way they are hung affect their tightness when closed.

Solid wood doors have a tendency to warp and change size with different temperatures and humidity levels, making it difficult to maintain a good air seal. Metal doors are a good choice since they are more stable, maintain a better seal, and are unaffected by moisture. The doors remain square when hung, decreasing the air leakage that can occur at the top of a sagging door. Insulated hinged doors with weatherstripping can be energy efficient. A storm door with a primary door also reduces air infiltration.

Sliding glass doors allow considerable air leakage past the sliding sections. Consider using French doors, which are easier to seal. If you choose sliding doors, install doors in protected areas whenever possible. Weather-stripping features are very important, especially along sills and jambs, and at the meeting stile where sliding and fixed sections interlock.

Installation of Windows and Doors

The main objective when installing both windows and doors should be to create an airtight seal between the window or doorframe and the exterior wall. Prior to placing the window into the rough opening, a bead of suitable caulk should be placed on the sheathing or siding around the opening, where the casing will meet the exterior sheathing. The window then can be set in place, leveled, shimmed, and screwed. Some manufacturers provide an anchoring flange and windbreak on vinyl clad windows, others provide a flexible foam bead on the backside of the casing. Swinging doors must be mounted accurately, and hinges should be fastened securely to ensure a tight fit.

Although steel framing permits superior fittings, the window or door won't always fit perfectly into the rough opening. That's why it's important to find a way to seal the cracks that will otherwise cause air leakage. Two easy ways to help ensure that the seal is airtight are to use weather-stripping and caulking, which leads us to our next subject.

Sealing Accessible Air Leaks

Caulking and weather-stripping are the easiest and least expensive weatherization measures and can save more than 10 percent on energy bills. Sealants work much better when applied during construction, rather than as an afterthought. Of course, not only windows and doors should be sealed. Keep in mind, many of the leakage paths through an exterior wall occur at the wall connections. With heat loss in a building reaching as high as 40 percent due to infiltration (National Association of Insulation Manufacturers of America, 1994), overall thermal performance of the completed structure depends on the quality of the construction job.

As the *Thermal Resistance Design Guide* by AISI instructs, "Whether constructed of steel or other framing materials, the overall thermal performance of the completed structure depends on the quality of the construction job. Careful attention to insulation details, caulking around openings, etc. will pay dividends. Steel framing members are inherently straight and true, making it easier to attain quality construction."

With that in mind, let's discuss the different types of caulking and other sealants that are available and determine the areas which need to be sealed to prevent air infiltration.

Caulking

Many caulking compounds can be found on the market today. Pay attention to the manufacturer's description of the material, its performance quality, and directions for application. Use a compound that will adhere to wood, glass, metal, plastic, and masonry, since these materials expand and contract. Also note weathering characteristics, such as cracking, shrinkage, water and mildew resistance. Rope caulking can be used for large openings.

Always use a good caulk gun with an automatic release (Figure 4.9). It's well worth the extra money.

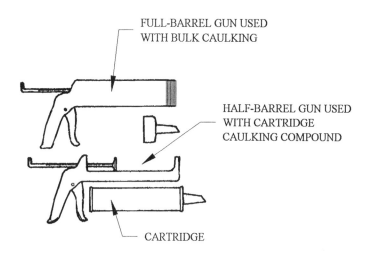

FULL-BARREL GUN USED
WITH BULK CAULKING

HALF-BARREL GUN USED
WITH CARTRIDGE
CAULKING COMPOUND

CARTRIDGE

FIGURE 4.9 Caulk gun.

All areas to be sealed must be cleaned to ensure good adhesion. Start at the bottom of openings to eliminate bubbles. Don't skimp, if caulking shrinks, reapply to form a smooth bead and seal completely. Every possible joint between two materials that penetrates the exterior wall, ceiling, or floor should have a sealant applied. This would include joints between chimney and siding, eaves and gable molding, windowsills and siding, window sashes and siding, windows and masonry, doorframes, masonry and concrete parts, and inside corners formed by siding. Another important area to apply caulking is where the sole plate meets the slab.

On the interior, caulk duct vents, between fireplace and walls, windows, baseboards, furdowns, built-in cabinets, inside closets, and outlet and switch plates. Also look at the areas around ceiling lights and fan fixtures, water lines, the attic access door, and a whole house fan if you have one. Use noncombustible materials and seal the top plates of the wall around wiring and plumbing. You should also look closely at the areas around bath and kitchen vent fans and chimney penetrations. All these are potential sources of air leakage, allowing a lot of hot air into the house in summer and sending all your heated air upward in winter. In addition to the energy benefits, the EPA says that sealing the cracks and joints in walls and floors that contact the soil typically reduces entry of radon from the soil by less than 50 percent.

Weather-stripping

Weather-stripping is a material that can be placed around the moving parts of windows and doors so they will close more tightly. Choose a quality weather-stripping product that you can adjust as the door expands and contracts during the seasons. Spring-plastic weather-stripping is a good choice for double pane windows and is durable. This type of weather-stripping has a self-adhesive backing or can be nailed into place. Metal-backed vinyl weather-stripping is easily installed and very durable. Generally, placing this weather-stripping tightly against the window on the outside will stop air leaks. Adhesive-backed foam weather-stripping is easy to apply but may last only one season.

Weather-stripping should be placed between window and doorframes and the exterior wall, between moving parts of windows, or as a door seal between the bottom of the door and the floor. It is also a good idea to weather-strip along the sides, top and bottom of the sash. Doorframes should also be weather-stripped all around the inside of the frame to ensure a tight seal when closed. Pay special attention to high-traffic doors.

Installing quality door sweeps or thresholds is also a good way to stop air exchange between the interior and exterior of your home. A door sweep can be used on a door with no threshold. Often the sweep, made of metal, has a flexible rubber or plastic edge. The sweep is connected to the door bottom; either inside or out, depending on how the door moves. A gasket threshold replaces an existing threshold and can be attached into the floor directly below the door. This type of gasket wears quickly in high traffic areas.

Last, but not least, install foam gaskets under all exterior wall electric switches and wall plug plates.

Air Quality

As a building becomes "tighter" through reduced air infiltration, air quality becomes a concern. There are several ways to improve the air quality in your home.

Vapor barriers and air barriers

When you make a home airtight, the presence and movement of moisture within it becomes more important. This is due not to an increase in the amount of moisture generated within buildings but to the reduced ability of moisture to diffuse into the air and be carried through the shell and released into the atmosphere.

In the winter, humidity can move through the ceiling as water vapor (Figure 4.10). In the cold attic, the vapor condenses on the steel trusses and drips into the ceiling insulation, which in turn loses most of it efficiency. In extreme cases, icicles can hang from the steel rafters under the roof. This moisture will reveal itself as wet spots in the ceiling and walls. A continuous interior vapor barrier is the most effective way of retarding the amount of moisture in building cavities.

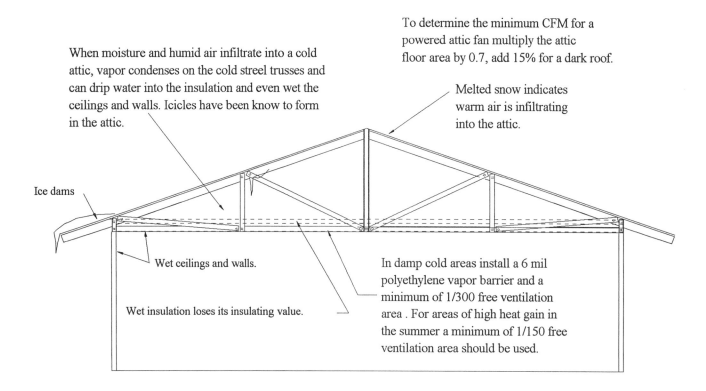

When moisture and humid air infiltrate into a cold attic, vapor condenses on the cold streel trusses and can drip water into the insulation and even wet the ceilings and walls. Icicles have been know to form in the attic.

To determine the minimum CFM for a powered attic fan multiply the attic floor area by 0.7, add 15% for a dark roof.

Melted snow indicates warm air is infiltrating into the attic.

Ice dams

Wet ceilings and walls.

Wet insulation loses its insulating value.

In damp cold areas install a 6 mil polyethylene vapor barrier and a minimum of 1/300 free ventilation area . For areas of high heat gain in the summer a minimum of 1/150 free ventilation area should be used.

Figure 4.10 Problems that occur when humid air infiltrates into a cold attic.

If you're not using exterior insulation, a vapor barrier will help lock out cold winter drafts and keep warm moist air from getting out into the walls of the house. Batt or blanket insulation can be purchased with an attached vapor barrier or asphalt-coated paper or aluminum foil. Other types of insulation should have a separate vapor barrier installed. Keep in mind, vapor barriers must always be installed on the living space side of the insulation.

Continuous air barriers are an important component in any energy-efficient building. Through the use of an air barrier, moisture movement into stud and joist cavities and heat losses can be significantly reduced. Continuous air barriers consist of a material or group of materials that will not let air pass through easily. Materials form a barrier when carefully joined to themselves or each other with a tape, gasket, or sealant. All air barriers must be installed in conjunction with a good vapor barrier. Some construction materials, such as exterior-grade sheathing and extruded polystyrene, combine the properties of both an air barrier and a vapor barrier.

A well-ventilated attic can offer additional protection against moisture problems. Attics with a 6 mil polyethylene air barrier call for a free area of 1/300 for ventilation, assuming that half the vents are in the eaves and half are located at least 3 ft above the attic floor. Installing attic fans is another option. A 1500 ft² attic requires a fan delivering 1050 cfm (or 1200 cfm under a dark roof).

Ventilation systems

When a controlled supply-and-exhaust ventilation system is used in an airtight house, the house can be ventilated when needed to maintain indoor air quality, regardless of outside weather conditions. In this type of system, fresh air is drawn into the house at the same rate as stale air is exhausted. The fresh air is supplied directly to all rooms or moved there by ductwork.

Wide ranges of mechanical ventilation options are available, however, they basically fall into two categories: ventilation only and ventilation with heat recovery. Some of these designs include energy-efficient heat recovery ventilators, also known as air-to-air heat exchangers. Modern, low-volume ventilation systems circulate from 80 to 200 ft³ of air per minute through the house.

Exhaust fans

Exhaust fans provide source-specific ventilation. Fans should have a minimum fan flow rating of 50 cfm at 0.25-in water gauge for bathrooms, laundries, or similar rooms, and 100 cfm at 0.25-in water gauge for the kitchen. Range hoods or down draft exhaust fans used to satisfy specific ventilation requirements for kitchens should not be less than 100 cfm at 0.10-in water gauge. The manufacturer's fan flow rating should be determined, per AMCA 210. Choose exhaust fans with low sone rating to reduce noise.

The minimum duct size for exhaust fans of 50 CFM is 4 in in diameter flex up to 25 ft in length or 4 in in diameter smooth up to 70 ft in length. The minimum duct size for exhaust fans of 100 cfm is 5 in in diameter smooth up to 50 ft in length or 6 in in diameter flex up to 45 ft in length. The maximum of three elbows is permitted for these lengths; for each additional elbow subtract 10 ft from the allowable length shown above.

All exhaust ducts should terminate outside the building and have back-draft dampers. When installed in unconditioned space, insulate with a minimum of R-4.

Outdoor air ventilation

Outdoor air ventilation should be integrated with the forced-air system by installing an outdoor air inlet duct. Connect a terminal element on the outside of the building to the return air plenum of the force-air system, at a point within 4 ft upstream of the air handler. Equip with a damper installed and set to measured flow rates of a minimum of 50 cfm and a maximum of 75 cfm for a two-bedroom house; a minimum of 75 cfm and a maximum of 120 cfm for a three-bedroom house; a minimum of 100 cfm and a maximum of 150 cfm for a four-bedroom house; and a minimum of 120 cfm and a maximum of 180 cfm for a five-bedroom house.

The connection to the return air plenum should be located upstream of the forced-air system blower. To prevent thermal shock to the heat exchanger, do not connect directly into a furnace cabinet. Locate the outdoor air inlet carefully so air is not taken from the following areas: the attic, crawl spaces, garages, or closer than 10 ft from an appliance vent outlet, or the vent opening of a plumbing drainage system (unless such vent outlet is 3 ft above the outdoor inlet).

To provide a return airflow back to the air handler, each door should be undercut to a minimum of 1/2 in above the surface of the finish floor covering. Exhausted air should flow out the exhaust ducts located in the bathroom and kitchen area. This method was used in both my own homes, which are very airtight, and the air quality is excellent.

Radon vent

In areas where radon is present in the soil, the follow information from the Washington State Ventilation and Indoor Air Quality Code may be helpful:

503.2 Floors in Contact with the Earth

503.2.1 General: Concrete slabs that are in direct contact with the building envelope shall comply with the requirements of this section.

Exception: Concrete slabs located under garages or other than Group R occupancies need not comply with this chapter.

503.2.2 Aggregate: A layer of aggregate of four-inch minimum thickness shall be placed beneath concrete slabs. The aggregate shall be continuous to the extent practical.

Gradation: Aggregate shall:

a) Comply with ASTM Standard C-33 Standard Specification for Concrete Aggregate and shall be size No. 67 or larger size aggregate as listed in Table 2, Grading Requirements for Coarse Aggregate; or

b) Meet the 1988 Washington State Department of Transportation Specification 9-03.1 (3) "Coarse Aggregate for Portland Cement Concrete", or any equivalent successor standards. Aggregate size shall be of Grade 5 or larger as listed in Section 9-03.1 (3) C, "Grading"; or

c) Be screened, washed, and free of deleterious substances in a manner consistent with ASTM Standard C-33 with one hundred percent of the gravel passing a one-inch sieve and less than two percent passing a four-inch sieve. Sieve characteristics shall conform to those acceptable under ASTM Standard C-33.

Exception: Aggregate shall not be required if a substitute material or system, with sufficient load bearing characteristics, and having approved capability to provide equal or superior air flow, is installed.

503.2.4 Soil-Gas Retarder Membrane: A soil-gas retarder membrane, consisting of at least one layer of virgin polyethylene with a thickness of at least six mil, or equivalent flexible sheet material, shall be placed directly under all concrete slabs so that the slab is in direct contact with the membrane. The flexible sheet shall extend to the foundation wall or to the outside edge of the monolithic slab. Seams shall overlap at least twelve inches.

Exception: If the membrane is not in direct contact with the bottom of the concrete slab, all overlapping seams shall be sealed with an approved tape or sealant, and the material shall be sealed to the foundation wall in a permanent manner. The membrane shall also be fitted tightly to all pipes, wires, and other penetrations of the membrane and sealed with an approved sealant or tape. All punctures or tears shall be repaired with the same or approved material and similarly lapped and sealed. In no case shall the membrane be installed below the aggregate.

503.2.5 Sealing of Penetrations and Joints: All penetrations and joints in concrete slabs or other floor systems and walls below grade shall be sealed by an approved sealant to create an air barrier to limit the movement of soil-gas into the indoor air.

Sealants shall be approved by the manufacturer for the intended purpose. Sealant joints shall conform to manufacturer's specifications. The sealant shall be placed and tooled in accordance with manufacturer's specifications. There shall be no gaps or voids after the sealant has cured.

503.2.6 Radon Vent: One continuous sealed pipe shall run from a point within the aggregate under each concrete slab to a point outside the building. Joints and connections shall be permanently gas tight. The continuous sealed pipe shall interface with the aggregate in the following manner, or by other approved equal method: The pipe shall be permanently connected to a "T" within the aggregate area so that the two end openings of the "T" lie within the aggregate area. A minimum of five feet of perforated drain pipe of three inches minimum diameter shall join to and extend from the "T." The perforated pipe shall remain in the aggregate area and shall not be capped at the ends. The "T" and it's perforated pipe extensions shall be located at least five feet horizontally from the exterior perimeter of the aggregate area.

The continuous sealed pipe shall terminate no less than twelve inches above the eave, and more than ten horizontal feet from a woodstove or fireplace chimney, or operable window. The continuous sealed pipe shall be labeled "radon vent." The label shall be placed so as to remain visible to an occupant.

The minimum pipe diameter shall be three inches unless otherwise approved. Acceptable sealed plastic pipe shall be smooth walled, and may include either PVC schedule 40 or ABS schedule of equivalent wall thickness.

The entire sealed pipe system shall be sloped to drain to the sub-slab aggregate.

The sealed pipe system may pass through an unconditioned attic before exiting the building; but to the extent practicable, the sealed pipe shall be located inside the thermal envelope of the building in order to enhance passive stack venting.

Exception: A fan forced sub-slab depressurization system includes:

1) Soil-gas retarder membrane as specified in Section 503.2.4;

2) Sealing of penetrations and joints as specified in Section 503.2.5;

3) A three-inch continuous sealed radon pipe shall run from a point within the aggregate under each concrete slab to a point outside the building.

4) Joints and connections may be gas tight, and may be of either PVC schedule 40 or ABS schedule of equivalent in wall thickness;

5) A label of "radon vent" shall be placed on the pipe so as to remain visible to an occupant;

6) Fan circuit and wiring as specified in Section 503.2.7 and a fan.

If the sub-slab depressurization system is exhausted through the concrete foundation wall or rim joist, the exhaust terminus shall be a minimum of six feet from operable windows or outdoor air intake vents and shall be directed away from operable windows and outdoor air intake vents to prevent radon re-entrainment.

503.2.7 Fan Circuit and Wiring and Location: An area for location of an in-line fan shall be provided. The location shall be as close as practicable to the radon vent pipe's point of exit from the building, or shall be outside the building shell; and shall be located so that the fan and all downstream piping is isolated from the indoor air.

Provisions shall be made to allow future activation of an in-line fan on the radon vent pipe without the need to place new wiring. A one hundred ten volt power supply shall be provided at a junction box near the fan location.

503.2.8 Separate Aggregate Areas: If the four-inch aggregate area underneath the concrete slab is not continuous, but is separated into distinct isolated aggregate areas by a footing or other barrier, a minimum of one radon vent pipe shall be installed into each separate aggregate area.

Exception: Separate aggregate areas may be considered a single area if a minimum three-inch diameter connection joining the separate areas is provided for every thirty feet of barrier separating those areas.

Any method of providing fresh air to a building should be designed and installed to meet requirements of the local code and approval by local building officials. As a result, suggestions in this chapter may require modification. As a precaution, review the proposed design of heating and air conditioning systems with local building officials prior to preparation of plans.

5

Windows and Doors

The next three chapters discuss energy-saving components that apply to nearly any material choice for constructing a home; however, we feel these chapters are necessary when considering energy efficiency in steel frame homes. Details, such as choosing efficient windows, utilizing passive solar design, correctly sizing air conditioning units, selecting energy-efficient appliances, etc. can make a difference when compensating for efficiency losses in steel frame buildings.

We'll begin our discussion by showing you how to choose windows with an eye toward energy efficiency and cost effectiveness. As you'll see, selecting windows and doors is more than a matter of appearance and style.

Windows

Windows serve a variety of purposes, bringing light, warmth, and beauty into buildings. Windows also have one of the greatest energy impacts of residential building elements and can be major sources of heat loss in the winter and heat gain in the summer. Windows have significantly lower resistance to heat flow than the wall they replace. In 1990 alone, the energy used to offset unwanted heat losses and gains through windows in residential and commercial buildings cost the United States $20 billion. In an average house, 20 to 30 percent of total heat loss is through the windows.

Reducing these heat losses requires careful selection of energy efficient components. You'll want to consider several factors: the resistance of the material to heat flow, the cost and availability, strength and durability, thermal expansion and contraction, ease of installation, and aesthetics (Figure 5.1).

FIGURE 5.1 Selecting windows involves more than appearance or style.

In the last chapter we discussed ways to reduce infiltration by careful installation. This chapter will help you choose the most efficient and cost-effective glazing and framing materials. The effect of the orientation and shading of windows is discussed in the following chapter on solar design. A final thought before we begin, consider reducing the amount of windows you install. Many homes have far more window area than necessary for adequate view and ventilation. You'll improve indoor comfort and the need for exterior and interior shading. In addition, you'll reduce the number of framing members, which reduces thermal bridging in steel as discussed in Chapter 2.

Windows are rated by either R-value (resistance to the flow of heat) or U-value (ability to transfer heat). The higher the R-value, the more energy efficient the window. U-value is the inverse of the total R, so, the lower the U-value, the higher the insulating value. R-2 is the same as U-0.5, R-4 is the same as U-0.25, etc. R-values typically range from 0.9 to 3.0, U-values range from 1.1 to 0.3, but some highly energy-efficient exceptions exist.

One word of caution. When comparing different windows, some may be rated in terms of the U-value of the center of the glazing, while others state the U-value of the window as a whole. The U-values are different, since the whole window U-value takes into account losses associated with the window frame and the spacers between the panes. Thus, the whole window U-value is more accurate than the center of glazing value and should be used whenever possible.

Several factors that affect the R-value or U-value of a window should be considered:

The type of glazing material

Traditionally, clear glass has been the primary glazing material used in windows. Although glass is durable and allows a high percentage of sunlight to enter buildings, glass by itself is a poor insulator. During the past two decades, though, glazing technology has changed greatly. Several types of glazing are now available that can help control heat loss. When comparing the different kinds of materials, note whether the glazing is most suitable for blocking or allowing solar gain. If you live in a hot climate, you will probably want to block out solar gain. The following are some of the choices:

- Windows equipped with low-emissivity (low-e) coatings allow visible light through but selectively block infrared radiation (heat). The coatings are thin, almost invisible metal oxide or semiconductor films and reflect 40 to 70 percent of the heat that is normally transmitted through clear glass. Heat has a harder time escaping on cold days and entering through the windows on hot ones, which boosts insulating efficiency substantially.
- Spectrally selective (optical) coatings are considered to be the next generation of low-e technologies. These coatings filter out heat normally transmitted through clear glass. Spectrally selective coatings can be applied on various types of tinted glass to produce "customized" glazing systems capable of either increasing or decreasing solar gains according to the aesthetic and climatic effects desired. Because of the energy-saving potential of spectrally selective glass, some utilities now offer rebates to encourage its use.
- Another technology uses heat-absorbing glazings with tinted coatings to absorb solar heat gain. These windows can absorb as much as 45 percent of the incoming solar energy. Some of the absorbed heat, passes through the window by conduction and infrared radiation, but inner layers of clear glass or spectrally selective coatings can be applied with tinted glass to reduce this heat transfer. Heat-absorbing glass reflects only a small percentage of light and therefore does not have the mirrorlike appearance of reflective glass.
- Reflective coatings reduce the transmission of daylight through clear glass. Reflective glazings are commonly applied in hot climates in which solar control is critical and are not typically used in cold climates. However, the reduced cooling energy demands these windows achieve can be offset by the resulting need for additional electrical lighting.

- Plastic and fiberglass glazing materials – acrylic, polycarbonate, polyester, polyvinyl fluoride, and polyethylene are some of the new choices – can be stronger, lighter, cheaper, and easier to cut than glass. Some plastics also have a higher solar transmittance than glass. However, these glazing materials tend to be less durable, more susceptible to the effects of weather, and less scratch-resistant than glass.

Layers of glass and air spaces

Standard single-pane glass has very little insulating value (approximately R-1). It provides only a thin barrier to the outside and can account for considerable heat loss and gain. Increasing the number of glass panes in the unit will increase the window's ability to resist heat flow.

A well-sealed double-pane window has a whole-unit R-value of 2.0, high-performance windows, including triple- and quadruple-glazed windows are now available, providing insulation of up to R-5 or more. However, these multipane windows are considerably more expensive (20 to 50 percent more than standard double-pane units) and limit framing options because of their increased weight. Triple- and quadruple-glazed windows are economically justified only in the most severe heating climates.

Multipane windows have insulating air- or gas-filled spaces between each pane. Each layer of glass and air spaces resist heat flow. Advanced windows are now manufactured with inert gases (argon or krypton) in the spaces between the panes because these gases transfer less heat than air. The insulating value of a low-e window can be improved by 15 to 20 percent with argon or krypton. Though more expensive, krypton insulates better than argon, allowing the window assembly to be thinner. Because of these new technologies, low-e coatings and gas-filled double-glazed windows can achieve many of the benefits of standard triple and quadruple glazing at a lower cost.

Frame and spacer materials

Now that glazings have improved, reducing heat flow through frames has become more important. Windows frames are available in a variety of materials including aluminum, wood, vinyl, and fiberglass. Frames may be composed of one material or a combination.

A wood or vinyl frame will outperform a metal one. In fact, windows with metal frames lose 20 percent more heat than the same window with a wood frame. If customized window designs require that you use aluminum frames, choose windows with "thermal breaks" made of highly insulating material built into the seal and/or the glazing spacer to reduce heat loss.

In general, though, when framing with steel, you'll want to avoid aluminum frames if at all possible. Wood frames have R-values, are unaffected by temperature extremes, and are less prone to condensation, but they require considerable maintenance. Vinyl wood frames are available in a wide range of styles and shapes, have moderate to high R-values, are easily customized, competitively priced, and require low maintenance. Fiberglass frames are relatively new and not yet widely available. They have the highest R-values of all frames and thus are excellent for insulating and will not warp, shrink, swell, rot, or corrode. Some fiberglass frames are filled with fiberglass insulation, an extra bonus.

Spacers are used to separate multiple panes of glass within the windows. Although metal (usually aluminum) spacers are commonly used, again, these are to be avoided because they conduct heat. Several window manufactures now sandwich foam separators, nylon spacers, and insulation materials, such as polystyrene and rockwool between the glass.

Superwindows

State-of-the-art superinsulating windows, or superwindows, combine all the above advanced features: low-e coatings, gas fill, good edge seals, insulated frames, and airtight construction. The low-e coating is applied not only to the glass, but to one or two sheets of a transparent polyester film suspended between the glass panes. The result is windows with whole-unit ratings of up to R-6 (whole unit), about twice as efficient as the same thickness of fiberglass.

Because of their extra cost, superwindows are most likely to be cost-effective in very cold climates. Superwindows generally don't earn their keep in milder climates, where you're better off buying less expensive low-e windows and using shading techniques.

Doors

Although doors are not as critical in the overall heat loss rate as windows, you can reduce unwanted heat transfer by choosing energy-efficient exterior doors.

Many people favor wood doors because of their traditional beauty. However, as mentioned in the last chapter, wood doors may warp, crack, shrink, or cause sealing problems. If you must use wood, choose higher-quality doors that provide better insulation and weather-stripping. As with windows, technology is improving and new choices are becoming available. Some manufacturers are making wood doors with foam-insulated interiors that have R-values around 6. Keep in mind, these newer energy-efficient wood doors cost more.

Another alternative is molded vinyl or fiberglass doors, which are less likely to warp, crack, or shrink than wood. They look like finished wood, withstand abuse, and require little maintenance. Even better, steel doors are made with a steel outer shell filled with insulating material and have an excellent seal against air infiltration. They add the benefits of extra security and are priced comparably to most wood doors. A hollow metal insulated door, with an R-value of 3 to 6 is a wise choice (Figure 5.2).

FIGURE 5.2 Steel doors are a good choice.

Consider an insulated garage door to help reduce the temperature difference across walls common to the home's interior and garage area. Garage doors should be carefully installed and weather-stripped.

Cost versus Benefit

As you can see, several choices exist for windows and doors. When selecting windows, take into consideration the above factors and weigh them against the cost of performance. Nearly half of all residential windows in place are single-pane, which do little more than prevent the wind from blowing through homes. On the other hand, don't feel you have to spend a fortune on the latest superwindows in order to build an energy-efficient steel frame home. In the efficient prototypes, vinyl framed, gas-filled, double-pane windows were used. As we've pointed out before, energy-efficient houses can be cost-effective.

6

Solar Design

Don't let the terms *solar design* or *solar construction* scare you off. Solar construction is simply a practical way to design and build homes that are both comfortable to live in and economical to operate. In the past energy-efficient strategies, such as earth sheltering and complicated solar designs, often dictated the architectural form of the building and overlooked the cost of achieving energy conservation. Today, as this type of housing becomes more commonplace, more emphasis is placed on architectural freedom and cost-effectiveness.

The best techniques have proved to be the simplest and least expensive. Studies show solar high-performance buildings don't need to cost more in order to work well, much of what constitutes a "solar building" is just common sense.

"Passive systems," as described in the following section, are simple and require minimal maintenance and no mechanical systems. Passive solar systems can be integrated to greater or lesser degrees in a building and take many forms. These buildings merely take advantage of energy-savings opportunities that blend natural elements such as sunlight and landscaping with basic energy conservation and efficiency practices, to provide year-round comfort and reduce energy bills. In fact, one way to define solar construction is a well-built house.

The advantages of solar construction are many. Homes stay warmer in the winter and cooler in summer, and, due to their higher heat capacity (thermal mass), remain comfortable even during power failures. To give you an example of the difference solar construction can make, in an average building with no attention paid to window orientation, a house meets 10 to 20 percent of its heating load by solar gains (heat gains from sunlight). With careful attention to selecting the site, orienting windows, and minimizing heat loss, it is possible to have solar gains offset as much as 40 percent of the heating load.

To date, more than 1 million residences and 17,000 commercial buildings in the United States have employed passive solar design. Building occupants are highly satisfied, and studies in commercial buildings document increased worker productivity as a benefit of passive solar design.

Solar Systems

The idea of using solar energy is not new. The comfort-seeking Romans who basked in steamy, glass-walled bathhouses were utilizing the main principle of solar collection: the greenhouse effect.

Sunlight is made up of very short wavelengths of electromagnetic energy (radiation) that passes easily through transparent materials such as glass. If the light strikes a dense, opaque substance, it is absorbed and turned to heat. The heat waves, longer than light waves, cannot readily pass back through the transparent material. Thus, a transparent enclosure becomes a heat trap. This is the same greenhouse effect that turns your car into an oven on a warm day.

There are basically two types of solar systems: active solar systems and passive solar systems. Active solar systems employ mechanical means, such as a fan or pump, to move the collected heat to where it is needed. It is possible, through vast arrays of plumbing and money, to collect, transport, store, and distribute solar heat. The heat is generated in glazed rooftop collectors, transferred in water or air, stored in a basement tank of rock bed, and distributed by a thermostat.

On the other hand, passive systems rely on natural energy characteristics in materials and air created by exposure to the sun to save energy. Unless the occupants are masochists, they also have auxiliary fuel-heat systems. Good southern exposure, proper orientation, proper placement and sizing of windows, and thermal mass are important in passive systems.

When reading about solar construction, you'll also come across the terms *direct-gain systems*, *indirect systems*, and *isolated-gain systems*. Direct-gain occurs when sunlight entering directly through the windows is absorbed, converted to heat, and stored in floors or walls. Typically the south-facing side of the building remains unshaded during the heating season. Sunlight passes through areas of properly sized glazing to collect solar heat. The collected solar heat is absorbed by concrete floors. The heat is then stored in the concrete floor and concrete walls. At night, as the room cools, the heat stored in the concrete radiates into the room.

Indirect-gain systems interpose heat-absorbing mass between the glazing and the living space. The Trombe wall is a primary example of an indirect gain approach. The wall consists of a thick masonry wall on the south side of the house. A layer of glass or plastic is mounted about four inches in front of the wall's surface. Solar heat is collected in the space between the wall and the glazing and distributed over several hours. Heat migrates through a wall mass at a rate of 1 in per hour, so the peak heat release from an 8-in wall occurs about 8 hours after the peak solar gain. When the indoor temperature falls below the wall's surface, heat begins to radiate into the room. Another version of the mass wall is the water wall. Water containers are positioned between the living space and the glazing. Some sunlight passes through the containers for daylighting and immediate heating of the living space, while the remainder is again absorbed for later release to the living space.

Isolated-gain, sun-space systems enlarge the space between the glazing and the mass wall, creating a pleasant living space that can be used for many purposes, including the growing of plants. These living spaces, sometimes called *solar rooms*, *greenhouses*, or *solariums*, have heating characteristics similar to mass wall systems. Solar heat is collected through the sun-space glazing then absorbed and stored in a variety of ways, including masonry and water walls. Heat can be moved into the house with the help of a small fan or blow, or circulated by simply opening connecting windows and doors.

Although more sophisticated systems, such as trombe walls, storage systems, sun spaces, and water roofs, save energy, they probably won't be cost-effective. This doesn't mean these systems should never be used. Often these systems give added amenities and benefits beyond energy savings. For example, sun spaces can be a design feature, as well as an energy system. You can't put a price on the value of sitting in a sun space on a sunny winter day or eating fresh vegetables grown in the winter. Perhaps quick payback periods are not your only concern. For many owners, there is a value that cannot be measured in dollars and cents: the personal satisfaction of conserving energy.

However, since cost effectiveness is a priority in this book, we'll be concentrating on passive, direct-gain systems, which generally don't affect the appearance or the cost of the house and use ordinary building components. Although you won't find all the possible techniques for passive solar design in this chapter, we've selected and illustrated methods we've found to be valuable and applicable to steel frame homes. The details presented will cover most situations, climates, codes, and local building practices.

We'll discuss four key components to passive solar design: (1) building site and orientation, (2) window placement and shading, (3) landscaping, and (4) thermal mass.

Building Site and Orientation

Solar construction requires changes in a builder's approach to siting. Along with concerns for access, drainage, views, privacy, etc., you should consider slope and orientation, wind direction and velocities, vegetation, and the availability of solar radiation. Put simply, sun and wind have the potential to increase or decrease your use of energy. Proper choice of building shape and orientation can often reduce energy costs by a third or more at no extra cost. That's why you'll want to consider the way the sun and wind affect your site and, if possible, orient your home to use them to your advantage (Figure 6.1).

To take advantage of passive solar heating, choose a lot with southern exposure. A south-facing slope, which receives more solar radiation than a level site, is the best choice. In winter, when the sun is low in the southern sky (Northern Hemisphere), solar radiation strikes the earth at a low angle. Sites sloped to the south are actually pitched towards the winter sun and receive more of it's energy, making these sites the warmest of all in winter. Sites that slope to the north receive much less direct solar radiation in winter than south-facing or level building sites and are often considered "cold sites."

In cool and temperate climates, homes will perform best if rectangular in shape, with the length running east to west (Figure 6.2). This exposes long, vertical walls to the south allowing the low winter sun to penetrate the glazed areas and warm the interior. In the summer, the walls can be shaded with overhangs, trellises, or deciduous trees. Keep in mind, buildings of a compact form generally require less energy for heating and cooling than buildings that ramble and sprawl all over the site.

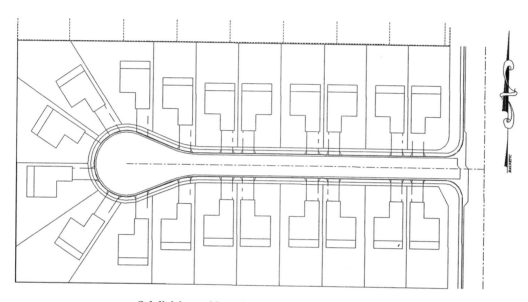

Subdivisions with roads running
east and west offer best solar exposure.

FIGURE 6.1 Subdivision layout for maximum solar exposure.

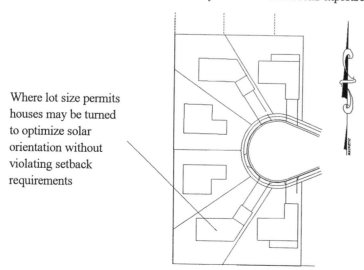

Where lot size permits houses may be turned to optimize solar orientation without violating setback requirements

FIGURE 6.2 Buildings oriented with lengths running east to west.

Since the rate of air infiltration of a house is affected by its orientation to prevailing winds, you'll want to utilize natural and created site features to protect your home. There are several ways to do this. Protective slopes or large trees on the north side can block cold winter winds. The proper location of a garage can also effectively protect the heated spaces in a building. Keep in mind, a building with a corner pointing into the winter wind loses much more heat than one with a wall square to the wind. If you are in a hot region, remember wind deprives you of heat during cold periods but is helpful in removing unwanted heat in the summer. When evaluating the orientation of your home, keep these simple things in mind (Figure 6.3):

*   South-facing buildings are warmer.
*   North-facing buildings are colder.
*   East-facing buildings receive warm, early morning sun.
*   West-facing buildings receive hot afternoon sun.
*   Hill tops, ridges, and higher elevations are more exposed to the wind and will be colder in the winter and cooler in the summer.
*   Sites on flat expanses are open to full wind forces.

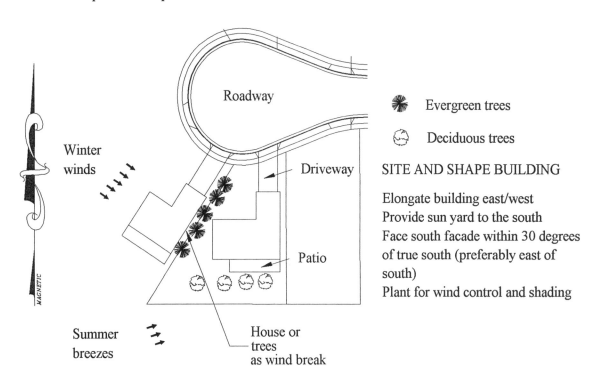

FIGURE 6.3 Orientation of house to prevailing winds.

By properly siting your home you'll not only gain the benefit of offsetting space heating, you'll make the most of natural lighting. Of course, while recognizing energy performance as a priority, use common sense. The positioning of a building on its site must be balanced with the requirements for proper drainage, ease and depth of excavation, orientation toward roads, and views. For example, a well-drained site is more important than a south-facing slope. Facing the building south saves energy but isn't critical to the performance of the house. You wouldn't ignore a spectacular view because solar-heated buildings are designed to minimize glazed areas on the northern exposure. Instead, provide the view in an energy-efficient manner by careful sizing and selecting glazing and perhaps using movable insulation or heavy drapes. As we pointed out before, today's simpler energy-efficient measures do not need to be design-restrictive.

Window Placement and Shading

Windows can optimize the benefit of solar heat and natural ventilation, reducing the amount of energy needed inside the home. The placement and sizing of glazed areas affect both solar heat gained and total heat loss. (See the previous chapter for more information on energy-saving options for windows, such as glazing, framing materials, coatings.)

Window placement

Put simply, to take advantage of solar heating, move as many windows to the south side of the house as practical. Maximize window areas that contribute to winter heating (south, southeast, southwest) and minimize window areas on walls that contribute most to winter heat loss (north) or summer overheating (east, west). See the floor plan in Figure 6.4 for an example of proper window placement.

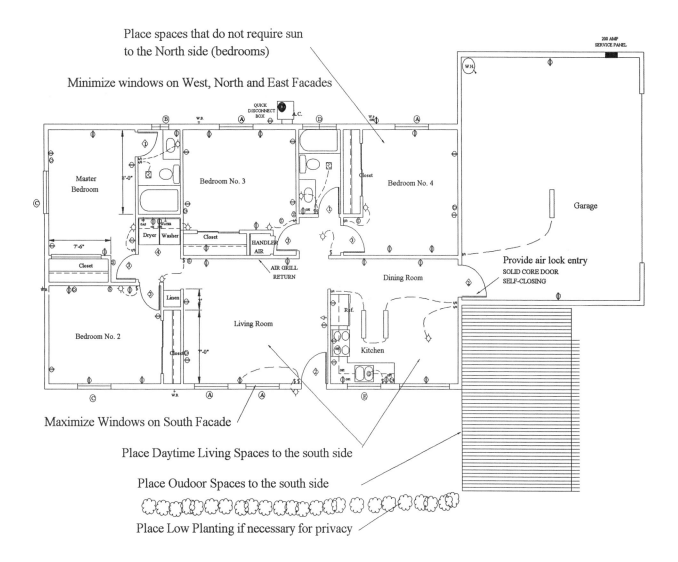

FIGURE 6.4 Window placement for maximum solar heating.

Window shading

When using overhangs for shading, don't forget sun angles. Anywhere outside the tropics, the sun is much lower in the sky during winter and nearly overhead in summer. By designing a building so that low winter sun shines in and high summer sun is blocked by proper placement of window overhangs, any structure will be comfortable (Figure 6.5).

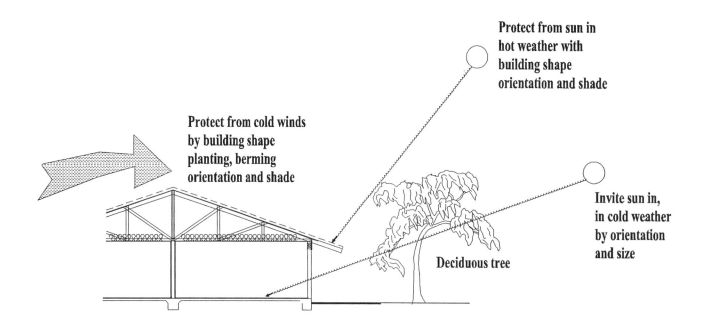

FIGURE 6.5 Shading and overhangs, keeping sun angles in mind.

You can further optimize the effect of direct gain heating in temperate and cool temperate zones by insulating windows with heavy, close fitting curtains that are drawn after sunset. In severely cold climates, double glazing or insulated window shutters may prove economically viable. However, be sure to open drapes and shutters on the south side of your house to let solar heat in during the day.

In warmer climates, the importance of shading windows cannot be overstressed, especially in houses with total south, east, and west glass areas above 10 percent of the floor area. To illustrate, a contractor/builder in Palm Springs, California, insisted on using single-pane windows to save money. The house complied with the energy code, *if* solar window screens were installed. The house had a spectacular view that faced west. The screens blocked the sun, but also blocked the view. Of course, it didn't take long before the owner removed the screens. The room could not be cooled by the calculated air conditioning. To settle a potential lawsuit, the contractor increased the size of the installed air conditioner to provide sufficient cooling. Technically, the house no longer met the minimum requirements for energy-efficiency standards, and the homeowner pays substantially higher electric bills. This example not only stresses the importance of shading but illustrates how cutting corners on energy efficiency can cost in the long run. In this case, installing dual-pane heat mirror glass would have been the best solution.

On south- or near-south-facing windows, a simple horizontal roof overhang can serve as a very effective shading device. East and west windows are harder to shade than south windows because the sun is much lower in the sky when it is in the east or west. This reduces the effectiveness of overhangs. An adjustable awning can be used, or you can provide extra shading by a pergola or movable blinds. Other options include light-colored window shades, white mini-blinds, shutters, and tinted glass; however, shading outside the building is most effective. Shading from the interior reduces the effectiveness because the heat has already entered the house.

A very useful strategy for shading can be the use of natural vegetation, trees, shrubs, and other plants. With proper landscaping, solar gains can be admitted during the winter months when needed and excluded at other times. (For more details on landscaping, see the following section.)

When placing windows, you'll also want to take into consideration cross ventilation. Adding air circulation not only prevents overheating in the summer, but (1) gives instantaneous cooling whenever the inside temperature is higher than the outside one, (2) removes overnight the heat stored in the building during the day, and (3) provides the feeling of cooling on the skin by accelerating its evaporative cooling. Proper attic ventilation will also save energy, provide comfort, and even lengthen the life of roof shingle.

These ideas are not complex at all and were commonly used in ancient China and Greece. We've just used fuel instead of common sense. In summary, in order to use windows for enhancing the energy efficiency of a home, keep the following in mind:

*   Windows facing the west should be smaller to reduce summer heat gain.
*   Windows on the northwest and the north should be avoided or as small as possible, to reduce winter heat loss.
*   Southern, southeastern, and southwestern windows have the best potential for summer solar heat control and, with proper design, can make good use of solar heat during cold months.
*   Openings should be oriented to take advantage of summer breezes. Arrange large windows to provide cross ventilation.
*   Windows closer to the ground allow better cooling.

Landscaping

In the last few decades, the ability of landscaping features to improve the energy efficiency of residential, business, and public buildings has garnered increasing attention, both in the scientific and professional communities.

Landscaping around the home has a large effect on temperature, air movement, and humidity. It's possible to achieve as much as a 30 percent reduction in cooling and heating costs through careful landscape planning. Landscaping can reduce direct sun from striking and heating up building surfaces and prevent reflected light carrying heat into a house from the ground or other surfaces. By reducing wind velocity, an energy conserving landscape slows air leakage and in the winter blocks cold winds that removes heat from the home. Additionally, the shade created by trees and the effect of grass and shrubs will reduce air temperatures adjoining the house and provide evaporative cooling. By selecting the proper vegetation and strategically placing it on your property, you can conserve energy and enhance the beauty of your home at the same time.

We'll consider several ways you can use landscaping to increase energy efficiency, through shading, windbreaks, groundcover, and evaporative cooling. Use plants and trees suited to your particular climate. Native plants will endure temperature extremes better than plants not indigenous to your area and will generally require less maintenance.

Shading

Using landscaping to shade windows and walls reduces solar heat gain that penetrates to the interior of the home. You can shade rooves too, by choosing trees with a large canopy, further reducing cooling costs and increasing comfort. Research indicates that shade has a dramatic effect on ground temperatures. When shaded, ground temperatures were found to drop an average of 36°F in only 5 minutes. That's why temperatures on a forest floor are much cooler than those recorded at the tree tops. You can create similar effects with careful planning and design.

When using trees to provide shade, take into consideration that heat gain from the sun is beneficial in the winter. The ideal situation is to place trees so that they provide shading during the cooling season but not during the heating season.

The best way of doing this is by using large deciduous trees (trees that lose their leaves during the winter) on the west, east and south sides of a house. Since even bare-branched trees can block 30 to 60 percent of the desired sunshine in the winter, you'll want to consider the advantages and disadvantages, depending on your climate zone. Evergreens, while good for windbreaks, are not recommended for shading because they have no canopy to protect a home from the sun. They also block sunlight in the winter when solar heat is desirable.

As pointed out before, to properly use trees for shading, you should have an understanding of the sun's path during the summer and the winter. During the summer, the sun rises far to the northeast, is very high in the sky during the middle of the day, and sets in the northwest for about 15 hours of daylight at the summer solstice. At the winter solstice, the sun rises in the southeast, is much lower in the sky during the middle of the day, and sets in the southwest for about 9 hours of sunlight. Since the sun is so high when on the south side of the home, trees would have to be planted too close to the building to provide shading. Some variety of trees can damage foundations, clog drainage lines, and block gutters with leaves. Instead of trees, use a simple overhang on the south side to keep the south wall shaded.

On the other hand, the sun is much lower in the sky during the morning and afternoon, and an overhang can't protect the east and west walls. Well-placed trees on the east and west will provide summer shading and allow winter gain. The west side is more critical because the skies are usually clear and temperatures warmer in the afternoon, providing the greatest opportunity for heat gain. The use of dense trees on the west and northwest sides of a home will block the summer setting sun.

Planting deciduous vines, such as wisteria, on trellises to shade walls and windows on the south side of your home is another way to reduce air conditioning bills and take advantage of the winter sun. If space is limited, trellised vines can be used to shade windows. To provide shade along the east and west sides of the house, evergreen vines, such as English ivy, are a good choice. Plantings of shrubs may also provide shade to reduce glare and reflected radiation while the sun is low in the sky. When planted a few feet away from the house, shrubs can provide extra shade and control humidity without obstructing air currents.

Since paved surfaces, especially blacktop, absorb huge amounts of heat from the sun and radiate it back into the surrounding air, shading these outdoor areas will help keep air temperature down. Shrubs or small trees can also be used to shade outside split air conditioning or heat pump equipment, improving the performance of the equipment.

Windbreaks

By properly placing a windbreak, you can control the wind, diverting winter winds and directing cool summer breezes toward your home. In an experiment conducted by the Lake State Forest Experimental Station in Nebraska, one house was fully exposed to the wind while an identical house was protected by dense shrubbery. The protected house used 23 percent less fuel to maintain the same environment.

Trees selected for windbreaks should be sturdy and resistant to wind dehydration. Strong trees can weather severe storms better than weak-wooded ones, which may break in high winds. Evergreen plantings on the north side will slow cold winter winds while constructing a natural planted channel to funnel summer cooling breezes into the house.

The effective zone of protection for a windbreak can be 30 times the height of the trees. However, the maximum protection occurs within 5 to 7 times the tree height. For example, if the windbreak will be 25 ft tall, it should be placed 125 to 175 ft from the house (Figure 6.6). The windbreak should extend to the ground, making evergreen trees a good choice. Foliage density on the windward side is optimally 60 percent. The tree heights within the windbreak should be varied.

Care must be taken so winter windbreaks do not prevent desired breezes in the summer, spring, and fall. Make certain that plants used for shade are far enough from the house so they don't restrict air flow. Low branching trees should be avoided, or low branches removed, on the southeastern and/or southwestern exposures because they interfere with motion. Winter wind barriers on the north and northwest sides of the home can help to push breezes from the south back toward the house in the summer. Shrubs placed near the windows can also be effective in directing air into the house.

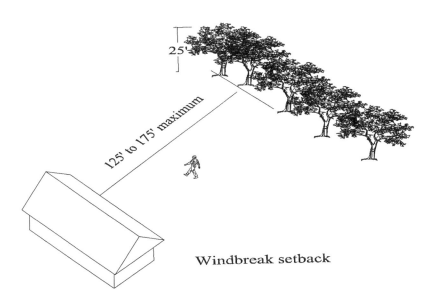

FIGURE 6.6 Providing windbreaks.

Ground cover/turf

Heat waves rippling off a sun-baked parking lot is a familiar sight during the summer. Paved surfaces absorb the sun's heat and radiate it back into the immediate environment. On the other hand, ground cover and/or turf has a cooling effect. The temperature above a ground cover will be 10° to 15°F cooler than above a heat absorbent material, such as asphalt and concrete, or a reflective material, such as light-colored gravel or rock.

Turf grass is undoubtedly the most commonly used ground cover. No other plant material can withstand as much foot traffic. While turf makes an excellent choice for recreational areas, it doesn't grow well in dense shade and is difficult to establish in extremely wet or dry areas. Evidence indicates that taller ground covers with their larger leaf surface can provide even more cooling than shorter ground covers such as mowed grass. There are several alternative ground covers that adapt well to conditions unsuitable for turf. Lily turf (liriope muscari) and mondo grass (Ophiopogon japonicus) are low maintenance ground covers which tolerate dense shade. If damp soils are a problem, day lilies are a good choice.

A heat-absorbent material such as asphalt will continue to radiate heat after the sun has set. For this reason you should minimize the use of heat absorbent and reflective materials, especially near the house. Shade these areas from any direct sun with trees as mentioned above. If possible, driveways should be located on the east or north side of the house to reduce heat buildup during warm afternoons. Brick driveways build up slightly less heat and produce less glare than concrete. Use of mulching materials, treated wood, or brick with interspersing ground cover will produce less heat buildup than large areas of hard, reflective surfaces. If hard surfaces are required, a product called "grasscrete" can be used. It provides a hard surface but allows grass to grow through it to provide a cooler surface.

Evaporative cooling

Ground cover and/or turf also has a cooling effect from evaporative cooling or evapotranspiration. Plants release water through pores in their leaves. When warm winds pass over the leaf surface, surface water absorbs the heat. The warmed water then evaporates into the atmosphere, leaving behind a cooler environment.

A single tree absorbs as much as 600,000 btu of solar radiation every day and evaporates water into the air. This maintains lower temperatures that drop toward the ground and move the warmer temperatures up (by convection) and away from the home. Trees are literally nature's air conditioners.

Thermal Mass

Thermal mass is defined as the ability of a material to absorb and release energy when changing temperature. By including materials with high heat storage capacity, such as concrete slab, tile floors, masonry walls, and stone or brick chimneys built within the insulated shell, the heat storage capacity of the entire house increases. With this increased mass, heat energy will be absorbed, stored and radiated to the home, smoothing temperature swings and utilizing energy more efficiently by retaining it inside the building longer.

The most cost-effective option is usually a concrete-slab floor exposed to direct sunlight. Depending on the floor finish, this could be the most inexpensive floor option and therefore cost nothing. As mentioned before, solar energy enters the home through large glazed areas and strikes the floor. Heat is stored in the finish surface and insulated slab until the slab becomes warmer than the adjacent air, when heat begins to radiate to the space. Concrete slabs used for heat storage must be well insulated along their perimeters. (See Chapter 3 for more information regarding slab-edge insulation.)

Installing the correct amount of thermal mass in relation to glazing is important. An adequate amount of thermal mass will prevent large temperature swings in heated spaces. A simple rule of thumb is that south-facing glazing should have a minimum area of one-fifth the floor area of the rooms to be warmed by the direct gain method. In mild winter climates, the ration of south-facing glass to area of rooms heated by direct gain may be as low as one-eighth.

Foundation walls and masonry partitions are sometimes used as thermal mass in solar homes. Adobe walls used in the southwest function similarly to concrete and help protect buildings from the summer heat. The mass both absorbs excess heat and presents a cool surface that encourages radiative heat loss. Poured concrete, cement block, and adobe walls can also be finished with masonry parge, either stucco or smooth.

Conclusion

Harnessing the sun can be an economical and reliable way to reduce energy losses in steel frame buildings. We've only briefly considered the basic ideas behind passive solar buildings. Technological developments are constantly improving the use of solar energy. If you'd like to learn more, we suggest you consult the many books available on the subject. One book we liked was *The Builder's Guide to Solar Construction* by Rick Schwolsky and James I. Williams.

Remember, the extent to which a passive solar house reduces the use of conventional heating fuels can be enhanced by the careful use of other energy conservation measures as described in this book, such as minimizing steel frame members, proper insulation, caulking and weather-stripping, purchasing energy-efficient heating and cooling equipment, choosing the proper windows and doors, etc.

7

Heating and Cooling Systems

This chapter discusses the selection of HVAC units, heating systems, and water heaters. Improved construction techniques and solar features which reduce heating and cooling loads and allow builders to install less costly, sized-down heating and cooling distribution systems. We'll begin by discussing cooling systems.

Cooling

There are several ways to cool a house. Window fans positioned to blow air out of the heated or sunny side of your house and pull cooler air from the northern or eastern side of the house work well up to a certain temperature. Ceilings fans and whole-house fans centrally positioned in the ceiling under the attic work can cool the house to a comfortable level. Window air conditioners cool small areas and, although once inefficient, have improved their efficiency due to federal energy-efficiency mandates. However, since central air conditioning is no longer considered a luxury item and these methods are seldom used by themselves, we'll be focusing on whole-house HVAC units (Figure 7.1).

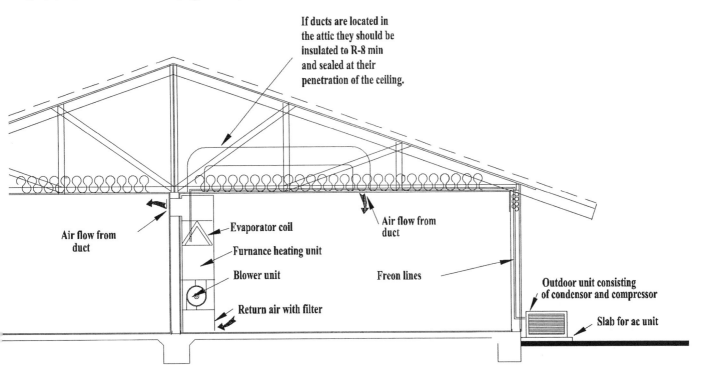

FIGURE 7.1 Whole-house air conditioning units.

What makes a cooling unit work? Simply stated, a cooling system circulates a liquid refrigerant, which attracts heat, collects it and carries it to a point where it can be discharged outside the area being cooled. This circulation is known as the refrigeration cycle. All popular cooling units work by means of a refrigeration cycle, although they do not all collect and discharge heat in the same manner or use the same refrigerant.

Central air conditioners typically use electric vapor compression. Cooling is extracted outside in a condenser and then transferred indoors through a "freon" medium. The cool air is then forced into the house by way of the duct system. As we discussed in Chapter 2, the more efficient the duct system, the more cool air you will receive.

If you want air conditioning, a heat pump makes sense (Figure 7.2). Heat pumps are all-in-one systems that cool in the summer and heat in the winter. To cool a home, the refrigerant absorbs heat from a home's interior and pumps it to the outside. To heat a home, the refrigerant is reversed to absorb heat from the outside air, which is pumped inside. Even on cold days, heat pumps can find and use heat, since refrigerant boils at temperatures below freezing. An air conditioning system alone costs a few thousand dollars, for a few dollars more, a heat pump can provide comfort and savings year-round. Another option for cooling is the new super-efficient mini-split ductless air conditioners and heat pumps, which are quiet and can be less expensive to operate than a central air conditioner. Ductless systems are about the only whole-house cooling option for houses without an air duct system (hot water or electric baseboard heat). The electricity savings are significant. We'll talk more about heat pumps, including ductless heat pumping, in the section under heating.

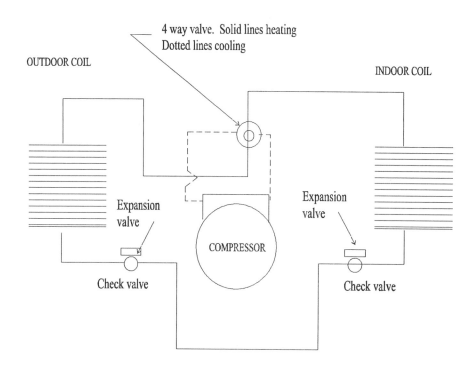

FIGURE 7.2 Schematic diagram of a heat pump.

Cooling capacity

A central air conditioning system's cooling capacity is measured in btu (the British thermal unit) per hour. Technically the btu is the amount of heat needed to raise 1 lb of water 1°F. In other words, the btu measures the amount of heat a cooling system can remove from your home as it cools the air for 1 hour. Consequently, the higher the btu-per-hour rating, the larger the living space the system can cool. The cooling capacity of models available today varies from 10,000 to 60,000 btu per hour.

Sizing air conditioners

Properly sizing your air conditioner is extremely important. The old rules of thumb often used by builders aren't as valid with today's "tighter" homes. Don't estimate or guess when sizing an air conditioner. A common mistake is to simply base the size of the air conditioner on the amount of square footage. Although this approach is rapid and simple, it does not account for orientation of the walls and windows, the difference in surface area between a one-story and a two-story home with the same floor area, the differences in insulation and air leakage, and many other factors.

Do not oversize your air conditioner. Even after correctly calculating the cooling loads, some people buy a 4-ton unit when the calculated cooling loads could be met by a 3 ½-ton air conditioner, so they'll "have enough cooling." In fact, Pacific Gas and Electric Company (PG&E) found in their Model Energy Communities Project that 53 percent of the air conditioners checked were a ton (12,000 btu/h) or more oversized. Unfortunately, many people think bigger is better, but there are several reasons why you should never oversize an air conditioner.

In addition to unnecessary higher costs, wasting energy, and raising utility bills, the air conditioner will short-cycle (run for shorter periods of time than it should) and remove very little moisture from the air, resulting in an uncomfortable house. To illustrate, I have two steel frame homes, one in Washington State and one in the desert area of Palm Springs, California. Because of the low humidity in the desert, 72°F feels cool, however, with the higher humidity in Washington, 68°F feels warm. Humidity greatly affects your comfort level, so when an oversized air conditioner doesn't cycle long enough to remove humidity from the air, you'll need cooler air to feel comfortable. Likewise, when an oversized heating unit dries out the air, you'll need warmer air to feel comfortable. This is an oversimplification, but a fact.

An oversized air conditioner can also mask problems from duct leaks, improper flow across the coils, and improper charge. In addition, oversizing contributes substantially to utility demand peaks.

Calculating cooling and heating loads

How can you avoid oversizing your air conditioner? By designing an efficient and effective air conditioning system using established techniques. (In larger homes, a professional should design your air conditioning system. A professional will save you the fee many times over in energy savings.) To ensure homeowners enjoy the comfort of a year-round environmental system, accurate methods of computing heat gains and losses and duct sizes have been developed. Manual J is a widely accepted method of calculating the sensible (the heat associated with temperature change that occurs when a structure loses or gains heat and is easily detected by a dry bulb thermometer) cooling and heating loads under design conditions. The program was jointly developed by the Air Conditioning Contractors of America (ACCA) and the Air-Conditioning and Refrigeration Institute (ARI) and calculates room-by-room loads for duct design purposes, and whole-house loads for equipment selection purposes.

While Manual J is simplified, it is not simple. Because of the many values and tables, it's easy to make an error. In most cases, either a set of tables specific to your area or a computer program should be used to reduce the likelihood of errors. While individuals who have used Manual J extensively are convinced that it has a substantial oversizing margin, field studies haven't determined the exact size of that margin. The existing duct leakage in homes consumes some, but usually not all, of this safety margin. If duct leakage were brought under control, as we previously discussed, units could be sized smaller. However, even with these setbacks Manual J is useful as a basic load-calculating method as long as one understands and takes into consideration its limitations.

ACCA has also produced Manual S for selecting equipment and Manual D for duct design. Although many find designing a duct system a laborious process, it is essential for proper equipment performance and comfort. The amount of time necessary to design a duct system is certainly warranted in tract construction where the design is used repeatedly. The Manual S ensures that both the sensible capacity and the latent capacity of the selected equipment will be adequate to meet the cooling load. The two main weaknesses of Manual S are that it dictates using 50 percent (or 55 percent) indoor relative humidity and it sets no upper limit for latent capacity. In dry climates the infiltrating air carries less moisture into the house, the indoor relative humidity is lower, and the latent load is lower. With less moisture in the house, the air conditioner runs at a higher sensible capacity. The result: For dry climates Manual S oversizes by approximately 20 percent compared to Manual J load.

The following are some recommendations made in the article, "Bigger Is Not Better: Sizing Air Conditioners Properly" in the May/June 1995 issue of *Home Energy Magazine:*

- Consider both location and level of insulation of ducts.
- When selecting cooling factors for roofs, floors, and walls, consider their R-value and type.
- Always account for the effect of the overhang shading.
- Pay great attention to window type, material, and interior shading. An error in this area can throw off the window heat gain by as much as 100 percent.
- Consider ventilation load if appropriate.
- Select equipment based on the detailed manufacturer's performance data. Do not rely on the nominal tonnage, since different units may have more than 10 percent capacity difference.

Several other energy programs are available to calculate heat loss and gains and properly size your air conditioner. Right J, a short form of the Manual J program and J-Works are two methods we've used and would recommend. For more information on energy programs, see Chapter 9.

If you or your air conditioning contractor do not have access to these energy programs or prefer to have professional help, energy consultants are available at a reasonable fee through the mail or over the internet. See my web page, http://www.cris.com/~insuform, for a listing of these consultants. (This listing is provided for your convenience, without any compensation. We do not assume liability.)

SEER

The capacity and tonnage of an air conditioner is one thing. How efficiently the system does its job is another matter. The seasonal energy efficiency rating (SEER) is a standard method of rating air conditioners and should be considered carefully. To determine the SEER, divide the air conditioner cooling capacity by the amount of electric power it consumes, in watts (SEER = Btu per hour/watts averaged over the cooling season). The SEER is listed on the yellow EnergyGuide label attached to each central air unit.

Put simply, the higher the SEER, the lower your energy bills will be. For example, a central air system with a SEER of 12.0 uses 33 percent less energy than a system with a SEER of 8.0. A few models are available with a SEER of over 15.0. New natural gas central A/C units have an equivalent SEER as high as 27.0. Installing one of these new units can cut your cooling costs by more than 50 percent; however, these units are high in cost.

When choosing an air conditioner, you'll need to answer a few questions such as: What are your budgetary limits? How long can you wait for the unit to pay for itself in energy savings? Are you investing solely to save on energy bills or are other factors also important? If you are environmentally motivated, the air conditioning units with higher SEER ratings might be your choice. If you are trying to keep your costs down, air conditioners with a SEER of 11.0 are adequate. Of course, keep in mind the initial cost is offset by reduced utility bills over the long term.

The SEER should not be confused with the EER, the energy efficiency ratio, the efficiency of the air conditioner. EER changes with the inside and outside conditions, falling as the temperature difference between inside and outside gets larger. For room air conditioners and packaged roof-top units, the EER should be 10 or higher.

Heating

When choosing a heating system you'll want to consider distribution, type of fuel, efficiencies, compatibility with a ventilation system, and cost. Central forced-air systems are the most common heating systems. They have a furnace with an air-distribution system and can easily be integrated with other mechanical systems such as cooling, air filtration, and forced-air systems. In all but a few parts of the country, the cheapest heating fuel is natural gas, followed by either electricity or oil.

Efficiencies are rated by different methods such as heating system performance factor (HSPF) and annual fuel utilization efficiency (AFUE), or thermal efficiency/combustion efficiency factor (Et/Ec). In all cases, the higher the rating number, the more efficient the unit. Ratings of 3.2 for forced air units and 3.8 for water-source heat pumps are considered good. Using the AFUE method, new models may be rated as high as 0.90, which means they deliver over 90 percent of the heat energy contained in the natural gas. Make sure your heater is rated at least 0.70. For combustion furnaces such as gas or oil, an AFUE or Et/Ec rating of 8.1 is considered good.

With these facts in mind, we'll consider the various types of heating units.

Fuel-fired furnaces

As we just mentioned, the most common way to heat homes is forced warm air (FWA) natural gas heating systems. Heat is forced, or supplied, into the home by a blower fan via a duct distribution system. The air then returns to the source through the ductwork. As we've stressed before, the more efficient the duct system, the better the heat is transferred into the home. Natural gas systems have high efficiencies, low fuel costs, low environmental impacts, quick heat response, and moderate equipment costs. More efficient models have an electronic ignition instead of pilot lights. Natural gas burns cleaner than fuel oil and presents few operational difficulties.

Variable-speed, two-stage heat output gas furnaces provide comfort and indoor air quality. You can expect a savings of 40 to 50 percent on your total heating costs. A two-stage furnace operates in the extra quiet low heat output and slow blower speed mode, except in the most severe weather (typically only 10 percent of the time). This reduces indoor temperature swings because the furnace runs longer each cycle in the low-heat stage. These furnaces are extremely quiet. However, variable-speed models will cost several hundred dollars more than single-speed gas furnaces.

Burners that use propane or oil are used in some parts of the country. A propane furnace provides similar efficiency, comfort, and design to a gas furnace but is more expensive to use. Oil furnaces are less efficient, typically in the 80 percent range, and require more maintenance than a gas or propane furnace.

After installation, furnaces should be cleaned and inspected on a regular basis. With proper adjustments, furnaces can run at peak efficiency. The efficiency of the furnace is on a scale of 1 to 100 percent. At 100 percent all the inputted fuel is turned into heat. If $100 is spent on fuel, $100 worth of heat is given to the house. However, if the furnace is running at 70 percent efficiency only $70 of the $100 spent is transferred to the home and $30 is lost up the chimney. As you can see, a 50 percent efficient furnace can cost a lot of money!

Electrical heating equipment

Heating with electrical resistance takes place when a current passes through a resistance load and the power used is converted directly to heat. These systems are relatively easy to size properly, do not require combustion air for operation, and cost less than other central heating systems. Since no combustion takes place, it is a much cleaner system than the fuel-burning types. However, electric resistance heating is inefficient and costly to operate.

Heat pumps

Heat pumps are becoming more popular in new construction. Heat pumps extract heat from the outside air, ground, or water and release it to the house. There are basically four types:

Electric heat pumps: Electric heat pumps are a viable option, especially if you live in a mild climate and use air conditioning in the summer. The energy efficiency of electric heat pumps is rated according to HSPF. New systems should have a HSPF of 7.0 or greater. The one drawback of electric heat pumps is the high initial cost and the relative inefficiencies of air-source systems during cold weather. Electric heat pumps are limited by outdoor temperature. At a certain point they can no longer extract heat from the air and then secondary fuel systems start. However, a heat pump perfectly complements a gas or oil furnace. The pump's clean, comfortable, and efficient cooling and heating back up your furnace until the weather gets really cold. Together, this combination is your most economical, most efficient option. Like all heating systems, heat pumps work best in a fully weatherized and energy efficient home.

Gas-powered heat pumps: The new natural gas-powered heat pumps are the ultimate in efficiency year-round. The heating and cooling outputs are variable, providing greater comfort. Although a gas heat pump is expensive to install, the monthly utility bills savings can pay back its higher cost over its lifetime.

Ductless heat pumps: Ductless mini-spit heating and air conditioning equipment operate in a similar manner as a conventional heat pump. The most significant difference lies in the distribution system. A conventional heat pump passes air over a refrigerant coil, then distributes it to individual rooms or zones through a system of ducts. By contrast, the ductless system, as the name suggests, has no ducts. As the *Technology Brief Series, Cost-Cutting Residential Construction Techniques* newsletter, October 1993 issue points out, "Small diameter-refrigerant lines are run directly from an outdoor compressor to air handlers located in individual rooms or zones. The systems are termed mini-split because of the separation of the compressor from the individual indoor air handlers." Most indoor units are mounted directly to the wall or rest on the floor.

Over the long term, design changes combined with the elimination of duct networks could make the ductless systems more attractive on a first-cost basis than traditional HVAC systems. As we pointed out in Chapter 4, duct leakage and conductive losses to unconditioned space can increase energy consumption by as much as 25 percent. By its very nature, the ductless heat pump virtually eliminates the distribution inefficiencies that plague traditional forced-air systems. Further, refrigerant lines are more easily insulated than ducts.

Geothermal heat pumps: Another heat pump that is available but not commonly used is the geothermal heat pump. These heat pumps extract the constant temperature of the ground soil or ground water, which is used for heating or cooling homes. "There is an increasing demand from homeowners for this environmentally sound, highly efficient option for heating and cooling," says John Manning of *Carrier Corporation*, which recently introduced the WeatherMaker system. Manning points to a recent study by the Electric Power Research Institute showing a high degree of customer satisfaction with water source systems compared to conventional systems and to a report by the EPA acknowledging the energy and environmental benefits of ground source technology.

An added energy- and money-saving advantage is the water heater option. In the summer, heat extracted in the cooling process goes directly into the home's water tank, providing free hot water. In winter, the water is heated at an efficiency that is two and a half to three times greater than a standard electric water heater.

Other heating options

Gravity heating and radiant heating are two other methods of heating that are not commonly used. Gas cooling, absorption cooling, congeneration, electric thermal storage, and infrared heating are nontraditional technologies that can be cost effective in the proper application. These systems require careful study to determine their viability in your situation.

Experts generally agree that wood stoves and fireplaces are not efficient as a household heating source. Wood stove systems have a wide range of ratings depending on efficiency. Catalytic wood stoves with combustion efficiencies near 90 percent are considerably more expensive than standard wood stoves with combustion efficiencies of 20 to 40 percent. While a typical fireplace provides some warmth, in most cases, more warm air escapes up the chimney than enters a room. If you choose to have a fireplace, make sure it has a tightly fitting damper that is kept closed when not in use. Install glass doors over the fireplace opening, which will help control the flow of preheated room air from moving up and out your chimney.

Water heaters

When choosing a water heater, you may be tempted to select a model that is inexpensive to buy, ignoring how much it will add to your energy bill each month. Such a strategy will cost you dearly in the long run. According to the *Consumer Guide to Home Energy Savings,* that "cheap" $425 electric tank heater can cost you $5500 (13 times its purchase price) in energy costs over its typical 13-year life. You have several choices when selecting water heaters. Since water heaters are the third largest home energy users, right behind cooling and heating systems, choose a model that saves energy and money.

Most conventional water heaters are storage tanks. You can greatly reduce water-heating energy consumption by choosing one of the most efficient models on the market. New electric water heaters should have an efficiency factor (EF) of at least 0.90, and new gas water heaters should have an EF of at least 0.57. If possible, choose natural gas- and propane-fired units, which typically cost about 40 percent as much to run as electric units.

Solar water heaters are gaining popularity. Installing a solar water heater can cut your annual utility bills by $200 or more for a typical family of four. Changing technology of solar water heaters are constantly improving the efficiency, appearance, and ease of installation. Passive solar water heaters do not require any external energy and have no mechanical moving parts. The simplest passive solar systems are "batch heaters" consisting of hot water storage tanks in insulated boxes with glass lids and reflective liners. In some climates, these systems can provide most or even all of your hot water needs. In less ideal climates, the system can act as a preheater for conventional water heating methods, reducing energy consumption considerably. The equipment is relatively expensive, but since the energy is free, the systems pay for themselves in a few years. More intricate passive systems, such as thermo-siphoning units, tend to perform better but are more expensive. Another type of water heat is the active solar system. They differ from the passive systems in that they use pumps, sensors, and heat exchangers. This usually means higher initial costs and more maintenance but also maximum efficiency. Solar systems typically have an EF greater than 10.0.

A waste heat recovery system (HRU) operates with your central air conditioner or heat pump by recovering heat normally exhausted to the outdoors. As long as your central system is operating, this "free" heat is used to heat water. The backup electric or gas system provides hot water when the air conditioner is off. If installed on a heat pump, the HRU will produce some hot water in the winter months also. An average family of four could save from $100 to $140 per year or 50 percent of a home's water heating costs with an HRU. The system typically pays for itself within 4 years.

Another option is the demand or tankless heaters. Without a storage tank, these water heaters avoid heat loss through tank walls and pipes, reducing energy use 15 to 20 percent. However, unless the heaters are extremely efficient, they aren't adequate for several major hot water uses at the same time. Combining a demand heater with a solar water heating system can be efficient.

Whatever type of water heater you use, make your hot water system even more efficient by taking the following steps:

- Insulate your water heater tank with at minimum of R-24. An insulation jacket will double the insulation value of most water heaters.
- Insulate hot water pipes wherever they are accessible, especially within 6 feet of the water heater and anywhere they run through unheated or cool areas. Tape the seams with acrylic tape; duct tape won't last. This will reduce energy losses at the tank and as hot water flows to the faucet. An added benefit is you won't have to wait as long for the water to get hot when you turn on the faucet.
- A simple board of rigid insulation under the tank of an electric water heater will prevent heat from leaking into the floor.
- Anticonvection valves and loops can prevent convective heat losses through the inlet and outlet pipes of the water heater. A $5 pair of these simple devises can save 2 to 7 percent of average electric water heating energy.
- Installation of low-flow showerheads can save upward of 10 percent on hot water use, reducing the hot water flow rate from 6 or more gallons a minute to as little as 2 gallons per minute.

- Turn the hot water temperature down to 120°F or around 140°F if you use a dishwasher without a booster heater. You will not only reduce heat loss from your water heater but will extend its life and lessen the chance of scalding.

Take your time when choosing air conditioners, heaters, and water heaters. You'll save energy and money and collect dividends for many years to come.

8

Appliances and Lighting

Although figures vary, appliances generally account for about 20 percent of the home's total energy consumption. While the amount of energy saved through energy efficient appliances is not nearly equal to that gained through other measures, such as reducing steel frame members or weatherproofing the house, you can achieve noticeable cuts in your utility bills by being energy conscious when you shop for new appliances.

Lighting also consumes more energy than you may think. Lighting for buildings, housing, signage, and streets accounts for 25 percent of all electrical energy consumed annually in the United States. However, new, efficient lighting technologies and strategies have the potential to reduce lighting energy use by 50 percent, saving 20 billion dollars annually while increasing productivity and comfort.

Sound impossible? The Solar Lobby calculated that if every home in the United States installed just four compact fluorescent lights, we would save as much electricity as our six large nuclear power plants produce. Even just one bulb will save you as much as $34 over the life of the bulb. Since lighting is consistently used, it can often account for up to 20 percent of your annual household electricity budget depending on usage. Energy-efficient lighting makes good economic sense without compromising comfort or convenience.

To compare energy usage of appliances and lighting, see the chart below (Figure 8.1).

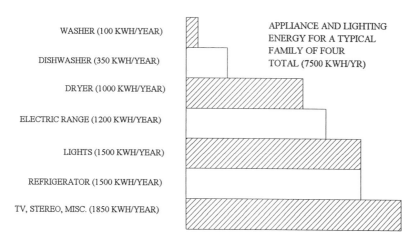

FIGURE 8.1 Family energy use per year.

We'll leave it up to you to control television and stereo and miscellaneous usage, which tops the chart. Instead, this chapter focuses on appliances and lighting. We'll begin our discussion with appliances.

Appliances

There can be a significant difference in the energy consumption of appliances. Determining and comparing the energy efficiency of different models is usually easy, because federal regulations require that Energy Guide labels are present on all major appliances to help select the most efficient models. The Energy Guide label includes:

- The Appliance Type and Capacity information on the top of the label helps you to compare appliances of similar size and features.
- The big dollar figure toward the middle of the form is the Estimated Yearly Energy Costs based on a national average electricity cost. The lower the amount, the more efficient the appliance. Remember to compare similar models to find the most efficient.
- Yearly Cost will give you the energy cost of the appliance for a year based on the energy costs in your area.

This information is extremely valuable, however, the labels don't indicate which *features* are most energy efficient. For example, top loading vertical axis washing machines are not as efficient as front loaders with a horizontal axis. Similarly, refrigerators have different efficiencies according to features such as defrosting characteristics, door style, and size. Remember too, that in most cases, if clothes dryers, cooking equipment, and dishwasher booster heaters are to be used during peak electrical demand periods, gas-fired equipment should be selected. Actual energy use is not much different, however, demand charges make gas-fired equipment far less costly to operate.

Most consumers are looking for the greatest return for the least investment. Keep in mind, this doesn't necessarily mean you should buy the least expensive product on the market. In fact, spending a little more money initially for an energy-efficient product is often more economical in the long run. One way to analyze the cost-effectiveness of a product is to estimate the simple payback period, or the amount of time required for the investment to pay for itself in energy savings. You can estimate the payback period by dividing the total cost of the product by the yearly savings. For example, an energy-efficient dryer that costs $500 and saves $100 a year in energy costs has a simple payback period of five years. If the payback period approaches or exceeds the projected life of a product, the energy savings will not be worthwhile. Also take into consideration installation (if any), operation, and maintenance costs.

The primary appliances that we'll be examining are refrigerators, dishwashers and clothes washers, dryers, ranges/ovens, and microwaves.

Refrigerators

A refrigerator uses more electricity than any other kitchen appliance. New models can be as much as 50 percent more energy efficient than older refrigerators. Because energy-efficient refrigerators offer one of the most effective opportunities for reducing electricity demand and delaying construction of new power plants, some utility companies are now offering rebates. The amount of the rebate often increases with the efficiency.

New technology offers improved insulation qualities and higher efficiencies through design. In the past, European appliances were more energy (and water) efficient, durable, and expensive. However, times are changing. Appliances sold in the United States are going above and beyond regulations dictated by DOE. The Super Efficient Refrigerator Program (SERP) developed in 1994 by 24 public and private electric utility companies, was designed to bring super-efficient refrigerators to market on an accelerated schedule. The prize incentives helped reduce the retail cost of energy-efficient refrigerators in some parts of the country.

The winning SERP entry was a 22-cubit side-by-side refrigerator/freezer manufactured by Whirlpool. The refrigerator is nearly 30 percent more energy efficient than required federal standards, partially due to new CFC-free insulation systems used in the unit. Whirlpool also became the first to use vacuum insulation panels in the United States to achieve high levels of efficiency (Figure 8.2).

You'll want to remember, the initial cost of all the energy efficient appliances is higher than conventional models, but they will pay off in the long run.

The following are some energy-efficient features you will want to look for:

*   Top freezer models outperform the side by side models, although the award winning model had side by side doors.
*   Manual defrost models use half the energy of automatic defrost but must be defrosted periodically to remain energy efficient. Partial automatic defrost are also energy efficient but are harder to find.
*   Automatic icemakers and through-the-door dispensers will increase energy use somewhat; however, they save energy by not having to open the freezer door to get ice.
*   The most energy efficient models are in the 16- to 20-cubit-ft sizes, so consider size carefully. Buying a refrigerator that is larger than you need means spending more money than necessary on your electric bill. Each additional cubic foot of refrigerator space will increase the amount of electricity consumed. On the other hand, it is more energy efficient to have one large refrigerator than two small ones.
*   Chest freezers are more efficient than upright models.
*   Check refrigerator door gaskets to make sure they are tightly sealed. Some refrigerators reduce inside pressure to make a tighter seal. These units are much more efficient. Some 1997 models will follow Whirlpool's example and install advanced vacuum insulation panels in refrigerator walls and doors, which have five times the insulating value of current units.

One last note regarding refrigerators: if at all possible, do not put the refrigerator near the stove or dishwasher. After installation, vacuum the condenser coils of refrigerator and freezers every three months or as needed. Dust-covered coils make compressors work harder and increase energy usage.

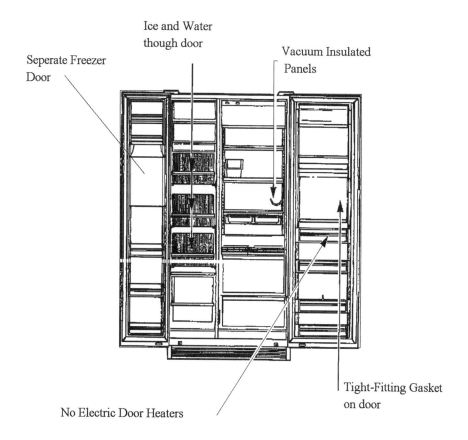

FIGURE 8.2 Details of the award-winning energy-efficient refrigerator.

Clothes washers

Clothes washers in the U.S. are predominantly vertical axis top loaders. However, as mentioned before, top loading vertical axis machines are not as efficient as front loaders with a horizontal axis. The improvement by using a front loader can exceed 50 percent in both energy and water use, resulting in estimated savings of $86 per year. Horizontal axis machines are the primary type used in Europe and Japan. The front loading aspect of many horizontal axis machines may be less desirable due to the bending required to put clothes in and out of the machine. Some horizontal axis machines can be loaded from the top but may be more difficult to find.

Major U.S. manufacturers of clothes washers started reintroducing horizontal axis machines in 1992. White Westinghouse is currently one of the principal U.S. manufacturers producing horizontal axis machines. They also supply machines for Frigidaire, Gibson, and Sears. Maytag has introduced their Neptune washer, a horizontal-axis unit that holds 20 percent more laundry then traditional machines. Other features include a front-loading door tilted at a 15 degree angle for easy access.

Various northwest utilities, convinced people will prefer tumble action washing machines, are giving $130 instant rebates for purchasing horizontal washing machines that meet "WashWise" high efficiency standards. The washing machine drum is turned on its side, using only enough water to fill the first few inches of the tub. The result is less impact on water and sewer systems. Less water also means less energy needed to heat the water.

Machines with the WashWise symbol will:

- Reduce water use by up to 40 percent.
- Reduce energy use by up to 60 percent.
- Reduce detergent use, and reduce wear and tear on clothing.

Features and options that affect the amount of hot water used play a primary role in the overall efficiency of a clothes washer. You'll want to consider:

- A model with many wash and rinse cycle choices. Warm wash cycles clean very well. Only oily stains may require hot washes. Cold water washing is adequate with proper detergents and presoaking. Cold rinses are effective. "Suds-saver" (reusing slightly soiled wash water) and presoaking are energy conserving options.
- Generally, washing a full load is more efficient. However, if a small load is necessary, the clothes washer should have the option of using a smaller amount of water.
- Higher spin speeds will reduce drying times.

Clothes dryers

Today's dryer's offer a variety of features, including temperature and moisture sensors, that can save you time and energy costs. These features eliminate the inefficiency of overdrying and the inconvenience of underdrying. Of the sensing controls, moisture sensors offer the greatest savings. Here are some energy-saving tips:

- Be sure your dryer is vented properly. If you vent its exhaust to the outside, use the straightest, shortest duct available.
- After installation, make sure you clean the lint filter after every load.
- Drying clothes one load after another will take advantage of the dryer's retained heat.

Dishwashers

Major U.S. and European manufacturers make efficient dishwashers. Look for dishwashers that highlight energy and water savings. Use the following guidelines:

- Use models that require the least amount of water.
- Booster heaters conserve energy by allowing the primary water heater to operate at a lower 120°F temperature setting. Without a booster heater, in most case dishwashers need 140°F water to liquefy some fatty soil and dissolve detergent fully. This feature will reduce your overall water heating energy costs.
- Again, washing a full load is best, but wash cycles options that can do smaller loads will conserve energy.
- A no-heat drying cycle is readily available and is worthwhile. These cycles use a forced-air unit to provide satisfactory drying results without the energy needed to generate heat.
- Maytag's IntelliSense dishwasher uses a metered fill rather than a timed fill device (a float measures water for a more precise water fill each time regardless of water pressure). Another environmental plus: better wash action means prewashing of dishes isn't required, saving even more water. The dishwasher, which also uses a no-heat drying cycle, is said to save up to 21 percent in energy, 27 percent in water, and up to 19 percent in wash time, compared to dishwashers that meet the current DOE standards. It became the first dishwasher in North America to earn the International Power Smart Seal of Energy Efficiency.

Ranges/Ovens

Some ranges and wall ovens are now designed with a thick blanket of fiberglass insulation covering the sides, back and top of the oven cavity. The design is said to allow less heat leakage than the separate batts of insulation typically used. Some other features you may want to look for:

- Large windows allow the user to see inside, so the door doesn't have to be opened during the cooking, resulting in less heat loss.
- Convection cooking means the oven cavity has a fan to circulate heated air over, under, and around food. Baking temperatures can be reduced by 25°F and roasting times are reduced by 25 percent, for more energy savings.
- A pilotless ignition reduces gas use by 30 percent over the use of a constantly burning pilot light.

Microwaves

There is no doubt about it, the microwave oven is one of the most energy-efficient appliances money can buy today. For example, did you know that it takes 18 times the electricity to bake a potato in a regular oven than in a microwave? Similar savings can be achieved with cooking meat, vegetables, sauces, soups, and puddings. Microwave ovens are about 30 percent more efficient than conventional ovens. The microwave is also a great time saver and will add value to your house. The following are some ways to make microwave cooking even more efficient:

- Food cooks faster in small quantities.
- Warm food in the microwave rather than keeping it warm in your range oven.

For more information

As you can see, technology is providing more and more choices for energy efficient appliances. As we've said before, you don't have to buy the most expensive appliances to have an energy efficient house. Because the refrigerator uses the most electricity, you'll want to consider the energy efficiency of this appliance more carefully.

For more information about energy-efficient appliances, contact: *Most Energy-Efficient Appliances*, American Council for an Energy-Efficient Economy (ACEEE), 2140 Shattuck Avenue, Suite 202, Berkeley, CA 94704 for a current listing of the highest efficiency appliances on the market. ACEEE also provides general and technical information on energy efficiency, including a publication titled *The Consumer Guide to Home Energy Savings.*

The Association of Home Appliance Manufacturers (AHAM) provides energy-efficiency information for specific brands of major appliances and runs a certification program for certain types of appliances. The address of their information center is: 20 North Wacker Drive, Chicago, IL 60606.

Lighting

How much energy you need to get the amount of light you want basically depends on five things: room characteristics, light sources, light fixtures, and controls.

Room characteristics

A room's shape, size, and color affect how much light you'll need. Spaces with dark surfaces and high ceilings are harder to light efficiently. In contrast, low ceilings and light-colored surfaces work well when the light source is at ceiling height. In any lighting design, it's important to use light-colored surfaces on furniture, wall coverings, ceilings, etc. to minimize artificial lighting.

Consider a lighting design that uses natural daylight to supplement artificial lighting. Daylight is the least expensive and most pleasing light source. Proper window placement can make the most of natural light. Keep windows and blinds open unless heat absorption is a problem. Although skylights can admit natural light, take into consideration the heat loss that typically occurs with skylights.

Light sources

The first step in an efficient lighting design is to select correct illumination levels for the tasks being performed. Most lighting systems are designed with higher light levels than necessary, resulting in higher cost and operating expenses. Once the proper illumination levels are determined for each space, the most efficient light source should be used.

When installing a new bulb, look at its wattage. Wattage is a measure of the energy the bulb uses. Most people look at wattage as a means of measuring the amount of light, but the true measure of the amount of light a bulb produces is its lumens. A bulb that provides greater lumens while using fewer watts is, therefore, an energy saver. The lumens measurement may be on the bulb itself but usually are found on the bulb's container.

An incandescent light bulb is the conventional lamp used in most residential and many commercial and industrial lighting applications. Incandescent lamps have changed little since they were invented by Thomas Edison in the late 1800s and are not energy efficient. Reduced-wattage bulbs are available and may make sense if a slight decrease in light output is acceptable, such as in closets and hallways. Halogen lamps have a slightly greater efficiency than conventional incandescent lamps. However, there are many other choices:

Compact Fluorescent Lamps (CFLs): Compact fluorescent lamps have the potential to make a tremendous difference, using one-fourth as much power as incandescent. In addition to higher energy efficiency, fluorescent lamps last about ten times longer than incandescent bulbs, lowering lifetime costs and adding the convenience of changing bulbs less often.

When people think of fluorescent lighting, generally they think of flickering, bluish, humming lights. While this was true one or two decades ago, modern fluorescent lighting has come a long way. Fluorescent systems are now available that don't produce visible flicker or an audible hum and do provide higher-quality light (including some that closely approximate the light produced by incandescent lamps). New options include "warm" or "deluxe" tubes instead of "cool white" or "standard" for more natural light. We've personally used fluorescent lighting in our homes and would highly recommend their use.

Compact fluorescent require one of two types of ballast – electronic or magnetic (core and coil). Electronic ballasts cost twice as much but have several advantages. They start without flickering, prevent electromagnetic interference (EMI), start in very cold weather, and produce more light while using less power.

According to the Building Technologies Program 1994 *Annual Report*, tests have shown lamp position and geometry can have a significant effect on the light output, light distribution and shade losses and thus fixture efficiencies. The data suggest that a circular lamp is more efficient than the A-lamp (which is symmetrically oriented). By tilting lamps or changing the shade geometry, it may be possible to fully utilize CFLs.

High-Intensity Discharge Lamps: Another type of high-efficiency bulb is the HID (high intensity discharge). While HID fixtures are not applicable to all lighting situations, they can be used in small areas requiring highly intensified light beams. The HID bulb provides 2 to 5 times the light of an incandescent bulb of the same wattage and can last from 10 to 25 times longer. Presently, HIDs are primarily used in commercial applications and include mercury vapor, metal halide, and sodium lamps. Because the color quality of HID lamps is generally poor (although many improved color models are available), they are primarily used for outdoor lighting and illuminating large spaces such as warehouses or supermarkets. Lamp technology research is concentrating on making HIDs more efficient.

Sulfur Lamps: In October 1994, the DOE officially announced the new sulfur lamp. These lamps use microwave energy to excite sulfur gas, producing an efficient, bright, white light. Sulfur lighting has the potential to be not only the most efficient light source but also the most environmentally benign. Sulfur lamps do not contain mercury, used in all other conventional light sources, and are expected to maintain their efficiency and light output over their entire lifetimes. Although no electrodes or phosphor coatings are necessary, current microwave lamps require the globe of gas to spin in order to distribute energy and produce uniform light output, adding mechanical complexity. If sulfur lighting is to bring about significant reductions in energy usage, many challenges still remain, but this lighting source is definitely something to watch for in the future.

Radio Frequency Lamps: Radio frequency (RF) lamps use high-frequency electromagnetic fields and have penetrated the market only a tiny bit, but new RF technologies are being unveiled. Their cost is somewhat more than compact fluorescent lamps. There are some unanswered questions about the RF emissions, but manufacturers have taken steps to incorporate RF shielding in their designs.

Light fixtures

Efficient light fixtures typically are more open and use highly reflective surfaces. Instead of directing light to the floor, they direct light to surfaces and tasks.

In the kitchen, fluorescent or surface-mounted fixtures are good choices. Commonly used are wood-trimmed and "cloud" or "floating" ceiling fixtures with inexpensive four-foot fluorescent lamps. These lamps last about 20,000 hours, compared to 10,000 for compact fluorescents and only 1000 hours for typical incandescents. If cabinets are located above counters, you can provide efficient working light with shallow under-cabinet fluorescent fixtures. Avoid using recessed lighting fixtures, especially in upper ceilings where they protrude into the attic; these fixtures are limited in their range of illumination and are sources of heat loss and air and moisture leakage.

In the bathroom, small mirrors are commonly surrounded by bar fixtures. Compact fluorescent bar fixtures that use 5-watt lamps (without medium-base sockets) are now available. Compact fluorescent surface drums and recessed downlights over the sink are other options. Mirrors larger than 3 feet may be lit by long-tube wood-trim or cloud fluorescent ceiling fixtures. If fluorescent fixtures around the mirror provide adequate light for the bathroom, you don't need additional lighting. Fluorescent ceiling- or wall-mounted fixtures are good choices for large bathrooms or bathrooms with compartments.

One more hint: replace multiple bulbs in a fixture with a single bulb. For example, replacing two 60-watt bulbs with a single 100-watt bulb allows you to receive the same amount of light output with 17 percent less energy use.

Controls

Make certain light switches are convenient and easily accessible. Install motion sensor controls where the lights are likely to be left on in unoccupied spaces, such as bathrooms, kitchens, etc. Use spring-wound or switch timers in areas that are used infrequently.

Put outdoor security lighting on a timer, photosensor, or motion sensor, so the light will be on only when needed. Solid-state dimmers or high-low switches reduce lighting intensity and save energy. Dimmers should not be used with most fluorescent lamps as they can damage the units. Rheostat-type dimmers do not reduce power to the lamp, so they do not save energy. Solar-powered lights are another landscape lighting option. These lights have built-in photo sensors and, best of all, don't require the burning of fossil fuels.

Of course, lights not in use should always be turned off. Contrary to popular belief, turning off florescent lights even for a short time saves energy. Frequent switching does shorten lamp life, but energy savings will more than compensate for replacement costs.

Lighting tips

To summarize, the following are ways to save energy when it comes to artificial lighting:

- Turn off all lights when leaving a room.
- Convert incandescent to fluorescent fixtures where practical.
- Install solid state dimmer switches. Rheostat-type dimmers do not cut your energy use.
- Use natural sources of light whenever possible.
- Reduce the size, or wattage, of bulbs where only fill-in or general lighting is needed.
- Use spotlight bulbs over areas where you need light for a specific task.
- Put outdoor security lighting on a timer or photosensor.
- Use light colors to help reflect light.

PART **2**

Materials, Details, and Techniques for Energy-Efficient Steel Framing

9

Conducting Your Own Energy Analysis

Most states have enacted an energy code based on the Energy Policy Act of 1992. In a recent survey conducted by Pacific Northwest National Laboratory, code officials indicated that lack of understanding by the building community was the number one obstacle to residential building energy code compliance. Many builders don't understand how to provide the necessary data and give inadequate information to code officials. Code complexity and lack of simple tools to verify compliance have presented a significant obstacle to builders. However, in this chapter we'll detail simple methods of providing data as it relates to steel framing. Included are several names of energy programs you can use to conduct your own energy analysis. We'll also take you step-by-step through one of the programs, MECcheck™. Let's begin by discussing this self-explanatory and user-friendly program.

MECcheck™ Model Energy Code Compliance Materials

MECcheck™ is a set of compliance materials that includes: MECcheck™ software, the MECcheck™ Manual, and MECcheck™ Prescriptive Packages. The U.S. Department of Energy (DOE) Building Standards and Guidelines Program at Pacific Northwest National Laboratory developed the MECcheck™ Compliance materials. (See Appendix A for a complete copy of the 1995 MECcheck™ Manual.) You can obtain the manual and software free by down-loading off the Internet (http://www.oikos.com/library/mec), or they can be purchased for $20 by calling 1-800-270-CODE.

DOE and the authors are to be congratulated, this is one of the best and easiest programs we've used to provide energy audits. The "trade-off approach," explained in Chapter 4 of the manual, enables you to trade off insulation, window-efficiency levels, slab-edge insulation, etc. to compensate for the lower R-values that result from thermal bridging of metal studs.

The Model Energy Code (MEC) contains energy-related building requirements applying to many new U.S. residences. Recently, the DOE issued a notice that states revise their residential building codes to meet or exceed the Model Energy Code or explain why they choose to use different standards. Several states have adopted the MEC as their residential energy code; states that don't have energy codes are under pressure to comply. Any residential building that is to be financed by the government by their various programs will require MEC documentation. The 1997 Uniform Building Code states in its Chapter 13 Appendix, "To comply with the purpose of this appendix, building shall be designed to comply with the requirements of the Model Energy Code." Therefore, most building departments should accept calculations for residential steel buildings using the MECcheck™ compliance process.

Recently, a contractor we know took his plans to a local building department. The plan checker said, " I see you have a metal front door. That's good, but you also have metal stud walls. That's bad." A typical response, perhaps. However, by submitting calculations using MECcheck™, the builder was able to convince the building officials that a steel frame building can be just as energy efficient as a standard wood frame house.

The EZFRAME Program

In addition to the MECcheck™ program, we'd highly recommend the easy-to-use EZFRAME software, which allows computer users to experiment with innovative steel stud wall alternatives and quickly calculate the results. EZFRAME 2.0 (P400-94-002R) is available for $14 per copy. To receive the program diskette and a user's manual, mail a check or money order payable to the California Energy Commission, to:

California Energy Commission
Publications Office, MS-13
P.O. Box 944295,Sacramento, CA 94244-2950
Phone: (916) 654-5200

Or you can download the EZFRAME program, the materials library, and the user's manual off the internet, ftp://energy.ca.gov/pub/efftech//ezframe. The Energy Commission also offers the nonresidential ENV-3 Form, which can be used for both residential and nonresidential buildings. The ENV-3 Form uses precalculated metal framing factors developed and tabulated for limited framing sizes and insulation levels. A copy of the user's manual is found in Appendix D.

Other Energy Programs

Over the years, we've calculated thousands of homes for heat and loss gains for compliance with the Title 24 Energy Requirements of the State of California, one of the most comprehensive energy requirements in the country. We've used Comply 24, Micropas, and Calres2 computer programs to produce the required compliance certification forms. These programs are quite involved and require training to use. The most reasonably priced is the Calres2 program produced by the State of California, which costs $125. If you're building in California, you'll probably have an energy consultant prepare the required Title 24 calculations. Nevertheless, it will help the consultant if you supply a copy of your MEC documentation.

Other software programs are available based on Manual J and can be used in other states, such as Florida. Right-J, a short form of the Air Conditioning Contractors of America (ACCA) Manual-J program, and the only residential load-sizing program approved by ACCA, allows you to enter the R-value of the wall section as computed by the EZFRAME, a nice feature. The program also correctly sizes heating and air conditioning units, which prevents oversizing. For more information, contact Wrightsoft at 1-800-225-8697, or write Right J, Wright Associates, Inc., 394 Lowell St., Lexington, MA 02173. The price for the software is $395. A videotape showing you how to use the program is available. For a discount, mention this book.

J-Works runs on Windows 3.1 and Windows 95. Weather data for over 750 cities is easily accessed from the program, as are the common wall, ceiling, door window, floor factors, and other data needed to perform an accurate loan calculation. J-Works includes a pop-up calculator with a feet/inch-to-square conversion function, which makes blueprint take-off's easy. The price of $149.50 for the Windows version is affordable, and they will send a free disk on a trial basis upon request. For more information, contact: J-Works, Micro Works, Inc., 907 Buck Daniels Road, Culleoka, TN 38451.

Preparing your own energy audit

We strongly suggest you make the effort to learn how to prepare your own energy audit using the MECcheck™ program. To help you, we're going to take you step-by-step through the MECcheck™ Software Approach. If at all possible, use this approach, which is simpler than the prescriptive package approach. If you don't own a computer, some of the office supply companies rent computers. The software is designed to run on most DOS-based computers and allows trade-offs between all building envelope components and heating and cooling equipment efficiencies. With minimal input, you can quickly compare different insulation levels to select a package that works best for your proposed building.

For this demonstration, we'll be using a sample two-bedroom steel-framed house, located in St. Louis, Missouri. We've selected the following materials (Figure 9.1):

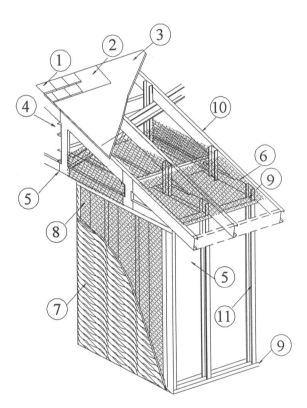

Building Components
1. Asphalt shingles
2. Asphalt felt
3. 15/32" plywood or 7/16" OSB
4. Air space in attic
5. 1/2" gypsum board
6. 10" blown in insulation
7. 7/16" OSB panel sheathing
8. 6" thick R-19 insulation
9. 1-1/2" deep leg track 20-ga
10. Metal trusses at 24" O.C.
11. 6" x 1 5/8" 20-ga studs at 24" O.C.

Walls are 6-in, 20-ga studs at 24-in on center, insulated with R-19 batts and sheathed with OSB (without exterior rigid foam insulation), with your choice of exterior finish.
The roof is a steel truss system with plywood sheeting and loose fill R-30 insulation.
Windows are vinyl-framed, dual-pane, single-hung windows.
Doors are steel foam-core.
The heating unit is a high-efficiency heat pump with a 7.3 HSPF and a 10.5 SEER.
The floor is a concrete slab with perimeter insulation.

FIGURE 9.1 Building components.

The following list of building components are used in the MEC energy analysis

Building component	Area	Insulation Level
Ceilings with attic	816 ft²	R-30 (metal trusses) *22
Walls	836 ft²	R-19 (metal studs @ 24-in on center)
Windows	92 ft²	U=0.56 vinyl (dual pane)
Doors (entrance and garage)	40 ft²	U=0.35 (metal foam core)
Floors (slab unheated)	116 lf.	R-7 (24-in depth) perimeter

Use the above information to specify component types and their area, insulation R-values, and glazing and door U-values.

MECcheck™ can be used with either the keyboard or a mouse. If using the keyboard, use Tab to move the cursor to the next field, Shift + Tab to move to the previous field, Alt + the underlined letter to choose a button or menu option (for example, Alt + F to selects the File menu), and Enter to select the current button, menu option, or list item. When opening the program, you'll see the Required Information Box first.

Required information

Your building's location and construction type (single-family or multifamily) are entered in this screen. From the scroll down menu, you'll select a state and a city. If you cannot find the city in which your house will be located, choose a city close to your building site that has similar weather conditions.

For our sample house, select **Missouri , St. Louis.** Next, select **single family.** Click **OK**. You're now in the main screen, or the Building Description screen (Figure 9.2).

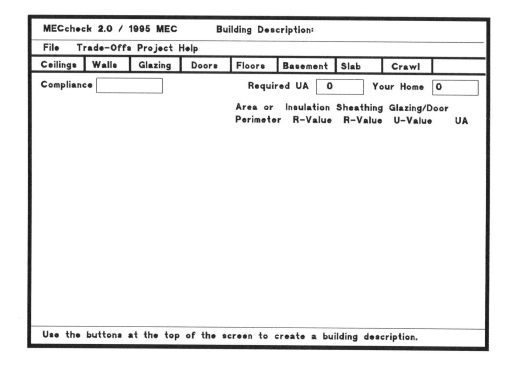

FIGURE 9.2 Building description screen (before entering data).

You'll be using the eight buttons located in the tool bar at the top of the screen to choose building components describing your home. As you choose building components, a list of your choices is created and displayed on the screen. Almost all components in this list require an area or perimeter entry. Some components require insulation R-value, sheathing R-value, and/or an assembly U-value. Figure 9.3 shows what the Building screen will look like after you've entered all your data.

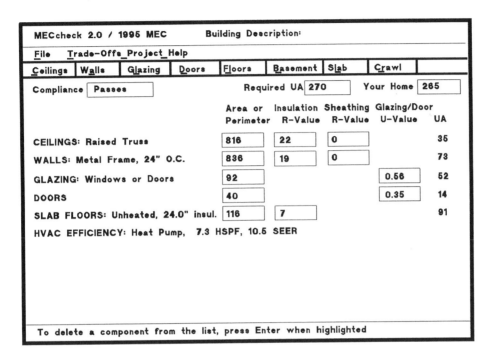

FIGURE 9.3 Building description screen (after entering data).

Ceilings

Next, select **Ceilings** (Figure 9.4).

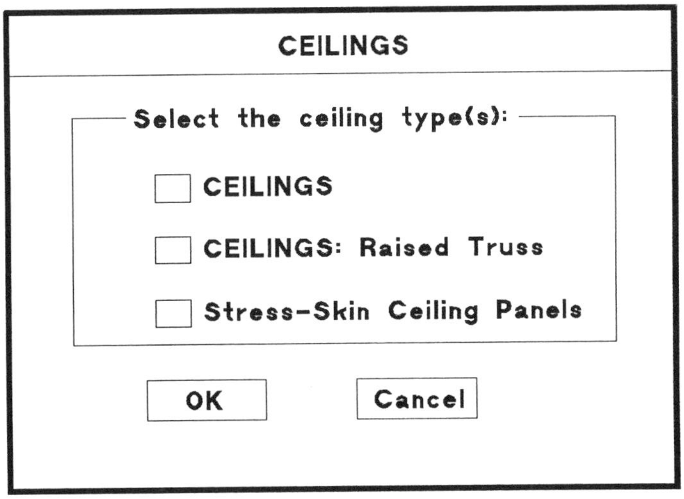

FIGURE 9.4 Ceilings screen.

For the sample house, select **Ceiling, Raised Truss.** Click **OK**. The program will transfer your ceiling types to your list of building components. You may notice that the Compliance field, which tells you whether your building complies with the MEC, currently reads "invalid area." While inputting information, the Compliance field may read fails, passes, or invalid area, changing as you add information. Ignore the message until you've entered all the necessary information.

Next, provide the net ceiling area; do not include the area of skylights. Enter **816** ft² for the sample house.

Provide the R-value of the cavity insulation to be installed. The program doesn't adjust the value to compensate for the thermal bridging of steel, so you won't enter 30 as the R-value. The calculation below shows that a steel truss with R-30 has an actual U-value of 0.045 (Figure 9.5). (See example 6 in Appendix D.)

Table 4.1 of the MECcheck™ Manual shows a U-value of O.045 for a wood truss with R-22. Therefore, for the sample house, enter **22** for your insulation R-value and **0** for sheathing R-value.

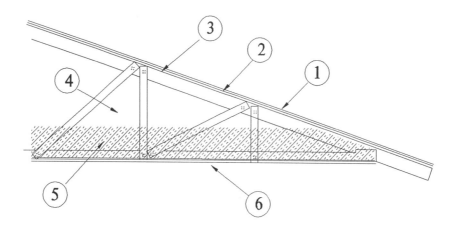

FIGURE 9.5 Steel truss.

Assembly name R-30 steel truss
Framing material, metal; framing spacing, 24"O.C.; framing percentage, 4.0%; absorptivity, 0.70; roughness, Asphalt shingles.

The below calculations are based on the Fr * (if the component is in the framing member), Th (thickness of the assembly)

ASSEMBLY U-VALUE *

Construction components	Fr	Th (in)	R-Value Cavity	Frame
Outside air film			0.17	0.17
1. Roofing, asphalt shingles		0.250	0.44	0.44
2. Membrane, vapor-permeable felt		0.010	0.06	0.06
3. Plywood		0.500	0.62	0.62
4. Air space	*	24.000	0.80	0.80
5. Insulation, mineral fiber, loose fill	*	9.25	30.00	30.00
6. Gypsum or plaster board		0.500	0.45	0.45
Inside air film			0.61	
Unadjusted R-values			33.15	0.00

ADJUSTED TOTAL U-VALUE = 0.045   **TOTAL R-VALUE =22.20**

Note: This calculation is based on the Comply 24 software, developed by Gabel Dodd/Energy Soft, LLC.

Walls

Next, select **Walls** (Figure 9.6**).**

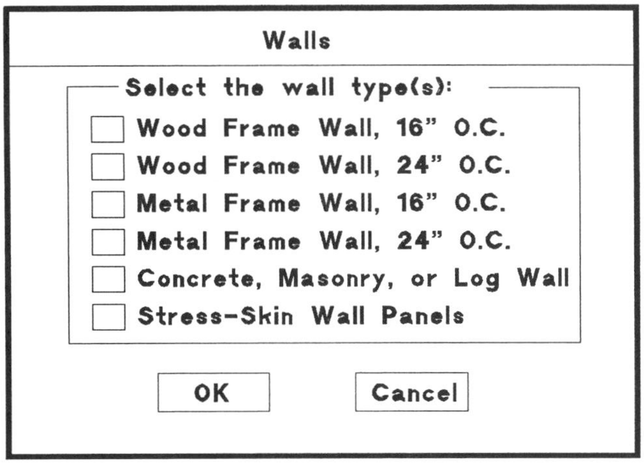

```
┌──────────────────────────────────────┐
│                Walls                  │
├──────────────────────────────────────┤
│  ┌ Select the wall type(s): ──────┐   │
│  │ ☐ Wood  Frame  Wall, 16" O.C.   │   │
│  │ ☐ Wood  Frame  Wall, 24" O.C.   │   │
│  │ ☐ Metal Frame  Wall, 16" O.C.   │   │
│  │ ☐ Metal Frame  Wall, 24" O.C.   │   │
│  │ ☐ Concrete, Masonry, or Log Wall│   │
│  │ ☐ Stress-Skin Wall Panels       │   │
│  └─────────────────────────────────┘   │
│                                        │
│      ┌──────┐      ┌────────┐          │
│      │  OK  │      │ Cancel │          │
│      └──────┘      └────────┘          │
└──────────────────────────────────────┘
```

FIGURE 9.6 Walls screen.

Select **Metal Frame Wall 24" o.c.** for the sample house. Click **OK**. Provide the net wall area excluding the area of the windows and doors, which is **836** ft². Enter **19** for the insulation R-value, and **0** for sheathing R-value.

This program automatically adjusts the R-value to compensate for the thermal bridging of 20-ga (33mils) studs. If using another gauge, for example 18-ga (43 mils), adjust the value by comparing Table 4.4 in the Manual (Appendix A) with the actual U-value you're using. This U-value can be determined using the EZFRAME program, or download the "Thermal Resistance Design Guide" by AISI, from their web page, http://www.steel.org/markets/construction/techctr.htm, and compare values.

When using the EZFRAME program, adjust the web punchout percentage so that the R-19, 20-ga stud wall has a U-value of 0.088 to match Table 4-4. Changing the web thickness will provide the following results:

Gauge	U-Value	Adjusted Cavity R-Value
18-ga (43 mils)	0.094	R-15
16-ga (54 mils)	0.099	R-13
14-ga (68 mils)	0.103	R-11

You can see why it's important not to increase the gauge of the stud unnecessarily. As pointed out in Chapter 2, this proves to be self-defeating in two ways: by increasing the cost of the house because thicker studs cost more and by reducing the energy efficiency of the wall.

If using an insulated foam concrete system, select the **Concrete, Masonry, or Log wall** option, chose **Log wall , 7" thick**, which has similar thermal properties. (Concrete, by itself, has little R-value.) Next, enter the R-value of your wall system. For example, if using the OTW Building System, you'll enter **10.5** for the R-value.

Glazing

Select **Glazing** (Figure 9.7).

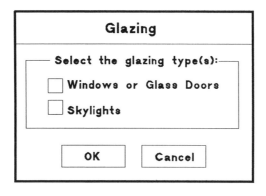

FIGURE 9.7 Glazing screen.

Glazing is any translucent or transparent material in exterior openings, including windows, sliding glass doors, skylights, and glass block. For our sample house, select **Windows or Glass Doors**. Click **OK** to transfer your list of building components.

Enter the area, which is **92** for the sample house. Enter the U-value from the manufacturer's data or use the default values. In this case, we used **0.56** from Table 4-9 of the MECcheck™ Manual.

Doors

Select **Doors.** Enter the nominal door area or rough opening area of all exterior doors. For the sample house, enter **40.** Enter the U-value from Table 4-10 in the manual. In this case, we used steel doors with foam core for a U=**0.35.**

Slab floors

Select **Slab** (Figure 9.8).

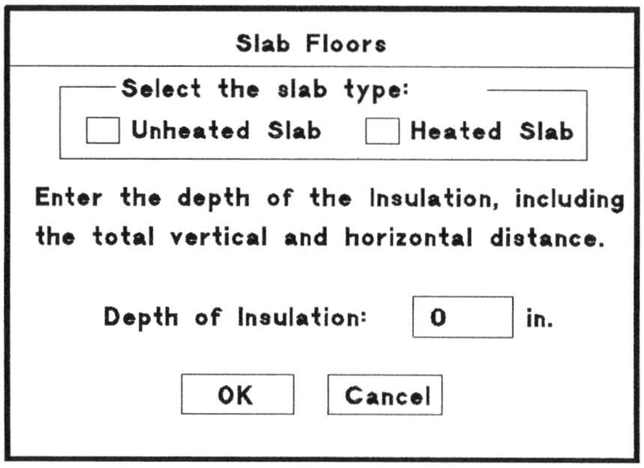

FIGURE 9.8 Slab floors screen.

Next, you'll select the type of slab you'll be using. For the sample house, select **unheated slab**. Provide the depth (in) of the insulation you intend to install around your slab perimeter, the top of slab to the bottom of the insulation, which is **24**. Click **OK**.

After adding a slab component, enter the perimeter of the slab floor, which is **116**. Enter **7** for the R-value of the perimeter insulation.

If your house doesn't pass

At this point, the software calculates the results and tells you whether your building complies with the MEC. You'll notice that the sample house didn't pass. Don't worry, we purposely chose a building design that doesn't comply with the MEC to demonstrate the several options you have.

If your house doesn't pass, you may want to change the glazing of the windows or sheathing value. For example, for our sample house, options included using windows with a U-value of 0.42, or adding rigid foam to the walls. In the next section, we'll discuss how to select materials and equipment and compare the benefits both in cost and energy efficiency. What if you don't want to change the building materials you've chosen? You may want to use the Trade-Offs Menu.

Trade-offs menu

This option gives you credit if you're using high-efficiency heating and/or cooling units. Select **Trade-Offs** (Figure 9.9).

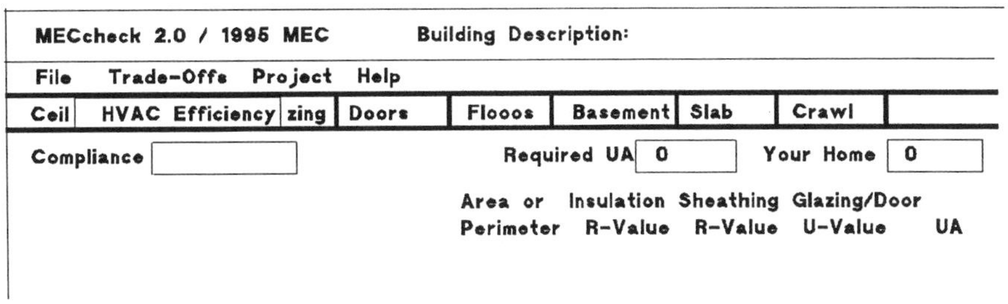

| MECcheck 2.0 / 1995 MEC | | Building Description: | | | | | |

| File | Trade-Offs | Project | Help |

| Ceil | HVAC Efficiency | zing | Doors | | Flooos | Basement | Slab | Crawl | |

Compliance [   ]   Required UA [0]   Your Home [0]

Area or   Insulation Sheathing Glazing/Door
Perimeter R-Value   R-Value   U-Value   UA

FIGURE 9.9 Trade offs menu.

To adjust the heating and/or air conditioner efficiencies, select **HVAC Efficiency**. Next, you'll enter the annual fuel utilization efficiency (AFUE), heating seasonal performance factor (HSPF), or seasonal energy efficiency ration (SEER) in the appropriate spaces. This information can be obtained from the manufacturer's data sheets.

For our sample house, we're going to use the most economical alternative, a heat pump with a 7.3 HSPF and a 10.5 SEER. Under HEAT PUMP: Heating Mode, HSPF, enter **7.3** and under Cooling Mode, SEER, enter **10.5**. Click **OK**.

You'll notice the Compliance box now reads "pass" for our sample home. Select **File**, then **Create Report**, then **Send Report to Printer**. Click **OK**. Print the report (Figure 9.10) and save the file.

TYPICAL COMPLIANCE REPORT

MECcheck™ COMPLIANCE REPORT
1995 Model Energy Code
MEC CHECK Software Version 2.0

CITY: St. Louis
STATE: Missouri
HDD: 4948
CONSTRUCTION TYPE: Single Family

DATE OF PLANS: 1997

TITLE: EXAMPLE HOUSE FLOOR PLANS

COMPLIANCE: PASSES
Required UA = 278
Your Home = 265

	Area or Perimeter	Insul R-Value	Sheath R-Value	Glazing/Door U-Value	UA
CEILINGS: Raised Truss	816	22.0	0.0		35
WALLS: Metal Frame, 24" O.C.	836	19.0	0.0		76
GLAZING: Windows or Doors	92			0.560	52
DOORS	40			0.340	14
SLAB FLOORS:Unheatred, 24" insul.	116	7.0			91
HVAC EFFICIENCY: Heat Pump, 7.3 HSPF, 10.5 SEER					

COMPLIANCE STATEMENT: The proposed building design represented in these
Documents is consistent with the building plans, specifications, and other
Calculations submitted with the permit application. The proposed building
Has been designed to meet the requirements of the 1995 CABO Model Energy Code.

Builder/ Designer_____Date_____

FIGURE 9.10 Compliance report.

On page 2 of the report under Ceilings, you should indicate in the comment section that the actual insulation is R-30, to compensate for the thermal bridging of the steel. The insulation is a full 10-in depth at the exterior wall.

For those of you who don't have a computer, follow the directions in the MECcheck™ Manual located in the Appendix of this book, beginning in Chapter 4 to prepare a trade-off worksheet step-by-step. Just remember to adjust the roof framing values for thermal bridging, and adjust the wall framing values when using studs heavier than 20-ga (33 mils).

Selecting Material and Equipment for Cost and Energy Efficiency

While completing the above energy audit, you may have wondered why various components were selected for the sample house. The following questions will give you reasons for our choices. Take into consideration, the sample house is built to meet the minimum energy requirements for a cost-effective entry-level home.

Why was a metal truss roof used instead of a wood truss, when the metal truss reduces the R-value from 30 to 22?

Building your own trusses on site, especially when you're experienced, may cost less than buying trusses. As we pointed out earlier, building your own metal trusses can compensate for energy losses by reducing infiltration and minimizing steel frame members. Blow-in insulation is cost and energy efficient since it fills the voids around the flanges better than batt insulation and in a 10 inch depth gives a slightly higher R-value. In addition, the truss is fire resistant and you'll eliminate any problems of scheduling for delivery.

Why were 6- in studs used instead of 3 5/8-in studs with foam insulation on the exterior?

Because we're taking into account the cost effectiveness of a home, it's important to note the actual costs for the various insulation systems. We've listed several insulation products along with their cost below. Depending on your area, these costs may vary.

Insulation Product	**Cost**
Polystyrene (white bead-board)	Approximately 4.5 cents per ft²/ R/-value.
Extruded Polystyrene (blue or pink board)	Approximately 5.5 cents per ft² R/-value.
Urethane (brand name Celotex)	Approximately 6.5 cents per ft² R/-value.
Fiberglass batts	Approximately 1.5 cents per ft² R/-value.

Since cost is the criteria, the following was taken into consideration: the material cost for a 3 5/8-in wall with R-11 and R-5 sheathing costs about the same as a 6-inch wall insulated with R-19. Although the wall with sheathing has a higher R-value, when we added the labor cost of installing the rigid foam to the studs, a difficult and time consuming procedure, we found the 6-inch stud with R-19 preferable. In addition, installation of finish exterior is easier when using OSB.

Why did you use vinyl framed dual pane windows instead of aluminum frame windows?

Once again, we compared cost effectiveness to energy efficiency. The cost of the vinyl windows is only about 25 to 30 percent more than the aluminum frame windows. The aluminum dual pane windows have a U-value of 0.87; the vinyl windows have a U-value of 0.56. Thus, the vinyl windows are 50 percent more efficient than the aluminum windows and well worth the extra cost.

Likewise, steel foam core doors and upgraded heating/air conditioning units speak for themselves. Most experts agree these items are both energy and cost efficient.

Why did you use an insulated concrete slab instead of a raised floor?

Insulating the concrete slab is one of the most cost effective ways of saving energy in a home. As discussed in Chapter 3, insulating the slab provides thermal mass and is both easy and reasonable to construct.

In addition, raised steel frame floors tend to vibrate when the upper flange flexes. Many builders find that a single ply of ¾-inch sub-floor is unsatisfactory and add a second layer of ½-inch plywood, at a combined cost of about a dollar per square foot. Some experts suggest using heavier gauge steel to eliminate the problem, which not only raises the cost of construction, but increases energy losses. When you compare the cost of an insulated cement floor to a raised steel frame floor, the savings are considerable and the energy efficiency is improved. Chapter 11 details the construction cost of both methods.

10

Materials and Tools

Materials

Steel framing is made from light-gauge galvanized steel, cold-formed into C-shaped components. The main structural components, the steel studs (Figure 10.1), usually are channel-type, roll-formed from corrosion-resistant coated steel, and designed so that the facing materials can be screw-attached quickly.

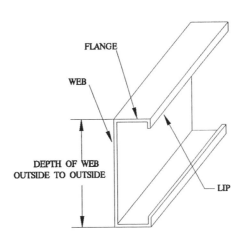

FIGURE 10.1 Cold-formed steel studs.

Light steel-framing members and accessories should meet ASTM specifications as listed in section 603.2.1 of the Material, CABO code; 43 mils (18-ga), 33 mils (20-ga), 27 mils (22-ga), and 18 mils (25-ga) materials are normally formed from steel with a minimum yield of 33,000 PSI or 25,000 PSI; 97 mils (12-ga), 68 mils (14-ga), and 54 mils (16-ga) materials are formed from steel with a minimum yield of 50,000 PSI. Metal building components usually are produced from galvanized steel that meets ASTM specifications as listed in section 603.2.3 of the Corrosion Protection CABO Code, or other rust resistant coating equal to these standards.

Selecting the suitable framing materials depends on a number of factors, including size, thickness, configuration, and grade of steel. To help you choose the appropriate materials and understand the purpose of the framing components, the following describes the main structural components of a lightweight steel frame home.

Wall studs

Wall studs come in a wide range of sizes and thicknesses. The studs are easily and quickly installed either on the construction site with self-drilling screws or as shop-fabricated wall panels. Basically, there are two kinds of studs: drywall studs and structural studs.

Drywall studs, sometimes called standard screw studs, metal studs or non-load-bearing studs (Figure 10.2), are strong, non-load-bearing components most commonly used for interior partitions. The web of these studs vary in size, from 1- 5/8-in to 6-in. Most builders use 3-5/8-in and 6-in sizes, however they are also available in 3½-in and 5½-in sizes. They are made in numerous gauges, from 25-ga to 12-ga. The most commonly used for residential construction are 25-ga and 26-ga. When ordering material for web stiffeners, specify 1¼-in flange drywall studs to eliminate cutting 1-5/8-in flanges to fit into the structural stud.

Structural studs, or load-bearing studs, are steel members with a heavier thickness, used in load-bearing construction (Figure 10.2). Again, these studs come in various sizes, from 2½ in to 12 in, and 20-ga to 12-ga; the most common thickness used for load-bearing joists, studs, headers and roof framing are 18-ga and 20-ga. Studs typically have a 1-5/8-in flange, but can be ordered with a 1-3/8-in, 2-in, and 2½-in flange.

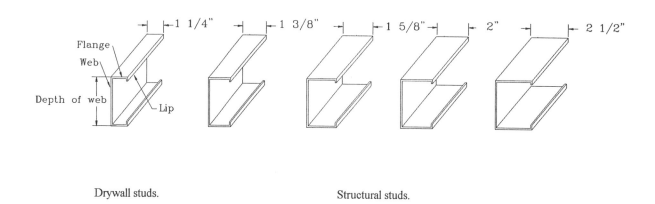

Drywall studs.   Structural studs.

Figure 10.2 Studs

The 1996 CABO One and Two Family Dwelling Code requires a legible label, stamp, or embossment with the following information:

1. Manufacturer's identification
2. Minimum uncoated steel thickness in inches (mm)
3. Minimum coating designation
4. Minimum yield strength, in KIPs per square inch (ksi) (kPa)

If using steel for the first time, rely on these manufacturers' stamps or labels to identify the size, gauge, and length of a piece. Studs and tracks (tracks are discussed later in this chapter) are often color-coded according to the gauge. Typically, 25-ga has no color, 20-ga is red, 18-ga is green, 16-ga is blue, 14-ga is yellow, and 12-ga is black. However, not all stud manufacturers use the same color code, so verify codes with your local supplier.

When selecting a stud gauge and size, in addition to determining whether the assembly is for a load-bearing or non-load-bearing wall application, consider anticipated wall height, insulation requirements, and anticipated wind loads. Equally important are frame spacing and the maximum span of the surfacing material. Tables in Appendix B as well as tables and charts in steel manufacturer's catalogs identify the physical and structural properties of the steel-framing materials, wind-loads, floor and roof applications, limiting heights, etc. Be sure to consider local and state building codes when selecting framing materials. Structural factors such as limiting height and span, required number of screws, metal thickness, and spacing should not be exceeded, since they affect the flexural properties and strength of an assembly.

Punchouts

Generally, studs are manufactured with punch holes in the web to accommodate plumbing and electrical installations. See Figure 10.3 below for examples of typical punchout patterns.

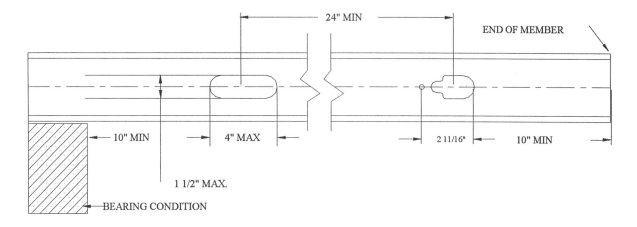

FIGURE 10.3 Punchouts.

For optimum effectiveness, a punch hole is usually provided at 12-in from each end. Intermediate holes are generally placed at 24-in intervals along the web, with no less than 24-in between the last intermediate and end hole. Studs can be ordered without the punchouts when used as forming material and for trusses.

By installing studs correctly, you'll align the punchouts for the plumbing and wiring that run through them. Unlike wood, steel studs have a top and bottom. Look at the holes punched in the sides of the studs; if the holes are key-shaped, the notch goes toward the bottom. When studs have oval-shaped knockouts, the painted ends are the bottoms. If cutting studs, be sure to measure from the bottom. When cutting load-bearing studs, avoid leaving a knockout within 10-in of the ends.

Track

Studs fit into light-gauge steel tracks (sometimes called *channel runners*) that serve as top and bottom plates (Figure 10.4). The track has a web and flanges, but no lips.

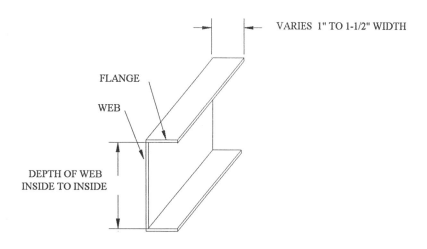

FIGURE 10.4 Track.

The metal track has 1-, 1¼-, and 1½-in flanges. The 1½-in track is typically called a *deep leg track*. Tracks are designed to accompany all stud/joist designs. For instance, they serve as channel runners at the top and base of all load-bearing wall construction and work as end caps for joist construction. Tracks also help unitize headers and other structural components that combine members for added strength.

The 1997 Uniform Building Code specifies a minimum of 1¼-in flange, when used in a shear wall configuration. The track is made without punchouts and is available in 25, 20, 18, 16, 14, and 12 ga.

Other steel-framing components

Several other steel-framing components aid in the installation of lightweight steel framing. The following are descriptions of the more common items:

Flat straps: Flat steel strapping reinforces joists and braces steel studs laterally. The straps are available in various lengths and various widths for diagonal bracing.

Lateral braces or bridging: Horizontal, lateral braces or bridging (Figure 10.5) prevent the stud from bending under wind and axial loads and provide resistance to stud rotation. Install horizontal bracing at proper intervals to fully utilize the stud's load-carrying capacity.

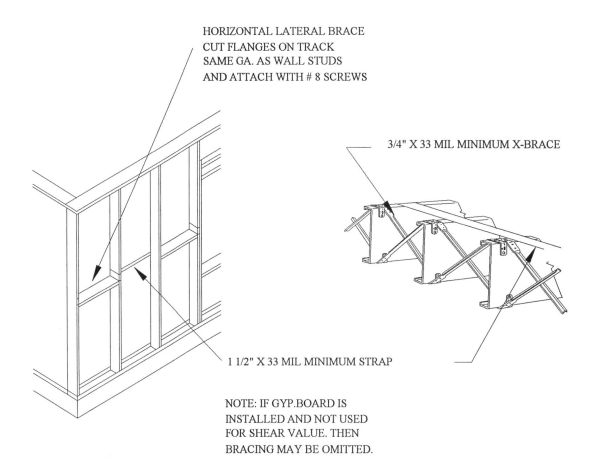

HORIZONTAL LATERAL BRACE
CUT FLANGES ON TRACK
SAME GA. AS WALL STUDS
AND ATTACH WITH # 8 SCREWS

3/4" X 33 MIL MINIMUM X-BRACE

1 1/2" X 33 MIL MINIMUM STRAP

NOTE: IF GYP.BOARD IS
INSTALLED AND NOT USED
FOR SHEAR VALUE. THEN
BRACING MAY BE OMITTED.

FIGURE 10.5 Lateral braces or bridging.

Z furring and drywall furring channels (Figure 10.6): Z furring channels are used to mechanically attach insulation blankets, rigid insulation, or gypsum panels or base to the interior side of monolithic concrete or masonry walls. They are also used for sound control in metal floor joists and provide additional space for pipes, conduits, or ducts. Drywall furring channels, also called *hat channels*, are used to attach gypsum board or plywood to concrete, masonry, and steel studs.

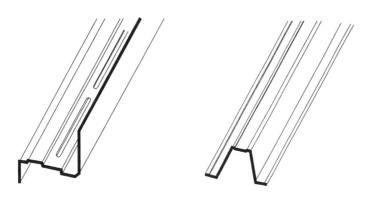

FIGURE 10.6 Z furring and drywall furring channels.

Web stiffeners: Web stiffeners are attached to the joist web when increased load-carrying capacity is required. Web stiffeners prevent web crippling at points of reaction and concentrations of axial loads.

Clip angles: A clip angle is a L-shaped short piece of metal typically used for connections.

Ordering materials

With both stick-by-stick and panelized construction, you'll need to prepare a cut list to order the required steel materials. Steel framers typically make a cut list while preparing the material take-off for estimates. The necessary data can usually be found on the blueprints for the home, including information such as the manufacturer's part number, gauge, component, and the dimension of the piece. (If you're using a preengineered system, the manufacturer will prepare the steel shipping list. When the parts arrive, make sure the parts received match the manufacturer's shipping list.)

To determine the approximate required steel in a building, substitute 3-5/8-in x 33mils (20-ga) studs for 2 x 4 wood studs, and substitute 6-in x 33 mils (20-ga) studs for 2 x 6 wood studs. A typical steel frame house uses from 6 to 10 lb of steel per square foot. However, these substitutions may vary in two- or three-story structures. The weight of the truss also has an effect on the design, so review the details of the plans with a structural engineer to ensure structural integrity.

To get the necessary data for ordering, use the manufacturer's catalog. Steel ordered through a supplier can be cut and delivered to the job site, saving considerable time. However, the cut list will probably not be 100 percent accurate, especially when building model homes. If you're building production houses and repeating the same model, expect the cut list to be about 95 percent accurate. If you're new to steel construction, add approximately 10 percent to your material list estimates and have an account with a local steel fabricator or commercial building supplier.

When ordering material, you'll find that prices vary from region to region and from manufacturer to manufacturer. Most manufacturers give substantial discounts when buying in truckload quantities. A truckload of materials is approximately 40,000 lb.

Screws

The most common steel-framing fastener is the self-drilling screw. In one operation, it can drill the hole and securely fasten just about any material to steel framing. These screws come in a variety of styles to fit a vast range of requirements.

Generally, screws should be approximately 3/8-in to ½-in longer than the thickness of connected materials. At least three exposed threads should extend through the steel to assure an adequate connection. These are minimum guidelines. Select the correct fastener, as described in the following section, for each application. Consult the manufacturer's specifications for the correct screw type and length.

The following screws (Figure 10.7) are the most common and useful varieties of head configurations:

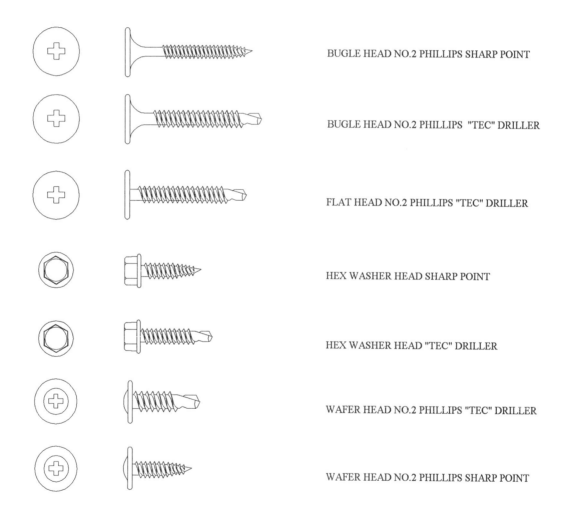

BUGLE HEAD NO.2 PHILLIPS SHARP POINT

BUGLE HEAD NO.2 PHILLIPS "TEC" DRILLER

FLAT HEAD NO.2 PHILLIPS "TEC" DRILLER

HEX WASHER HEAD SHARP POINT

HEX WASHER HEAD "TEC" DRILLER

WAFER HEAD NO.2 PHILLIPS "TEC" DRILLER

WAFER HEAD NO.2 PHILLIPS SHARP POINT

FIGURE 10.7 Varieties of head configurations.

Bugle head #2 phillips sharp point: This screw is used to attach ½-in or 5/8-in drywall to 25-ga metal studs. Generally the screws are #6 x 1-in to 1-1/8- in. The 1997 UBC specifies, "Drywall screws shall be a minimum #6 by 1 inch (25mm)."

Bugle head #2 phillips "tec" driller: This screw is used to attach ½-in or 5/8-in drywall or plywood in up to 68 mils (14-ga) metal studs. Generally the screws are #6 x 1-in to 1-1/8-in.

Flat head #2 phillips "tec" driller: The 1997 UBC has the following additional requirements for screws in shear panels: (1) "Plywood and OSB screws shall be approved and shall be a minimum #8 by 1 inch (25mm) flat head with a minimum head diameter of 0.292 inch (7.4mm). (2) "Fasteners along the edges in shear panels shall be placed not less than 3/8 inch (9.5mm) in from panel edges. Screws shall be of sufficient length to ensure penetration into the steel stud by at least two full diameter threads."

Hex washer head sharp point: This screw is used for framing in 18 mils (25-ga) steel studs and track. The screws are normally #8 x ½-in.

Hex washer head "tec" driller: The hex head is generally used for trusses and exterior applications. Generally the screws are #8 x ½-in or #10 x ¾-in.

Wafer head #2 phillips "tec" self-drilling: The 1997 UBC specifies "Framing screws shall be #8 by 5/8 inch (16mm) wafer head self-drilling." The screws are used in up to 68 mils (14-ga) material.

Wafer head #2 phillips sharp point: These #7 or #8 x 7/16-in screws are used for framing in 18 mils (25-ga) interior walls and are the most economical.

Choosing a screw point

The two common screw points used in cold-formed steel framing are: The sharp point for steel up to 0.035-in thick. The point may be a sharp needle or piercing point. The Tec or self-drilling is normally fluted and designed for steel up to 0.112-in thick. Screws should conform to the requirements of section 603.2.4 fastening requirements, CABO code.

Depending on the thickness of the material, the thread along the screw's shank should be held back from the point of the screw to prevent the threads from engaging the steel until the drilling process is complete. This prevents overdrilling or stripping the threads after partial penetration.

Applying screws

Screws positively and mechanically attach the metal base to the steel framing. Do not predrill holes in the receiving material when using the self-drilling/self-tapping screws. The screws are normally installed with a screw gun having a minimum speed of 2500 rpm for screws up to #10. The battery-power screw guns with 1400 rpm take longer to install the screws. The allowable shear for #8 and #10 screws in 33 mils (20-ga) varies and can be from 164 lbs up to 340 lbs with approximately 100# tension value (Appendix C, page 10). Verify these values with the manufacturer.

Screws should be installed and tightened in accordance with manufacturer's recommendations. Install screws as close to the web of the stud as possible to eliminate flexing of the flange.

The self-drilling Phillips head screws are more difficult to use than the hex head. The screws have a tendency to flip sideways when pressure is applied, especially the longer lengths. As a result, these screws generally take longer to penetrate the first layer of steel. In addition, tips often burn before completely fastening the steel. Some imported screws perform better than others, so if you're having problems you may want to try a different brand. However, even superior screws will have some burned tips.

One final note: Don't try to save money like the homeowner who saw screws all over the floor. He picked up the screws, sorted them and put them back in the boxes with the new ones. Needless to say, the framers were very unhappy the next day.

Tools

Only a few essential tools are required to perform two basic operations: cutting and fastening. Get rid of your chisels, cats-paw, nail punches, and assorted nails and reload your tool pouch with aviation snips, vice-grip clamps, self-drilling screws, and screw-gun replacement tips.

Cutting tools

We'll begin by discussing tools used for cutting. As mentioned before, you should avoid cutting in the field whenever possible by using precut stock. When completing the material take-off for estimates, make a cut list for the steel suppliers, and they will do the cutting. However, even with precut stock, some onsite cutting will be necessary for most projects, and you'll want the following tools.

Chop saw: A steel cutting blade in your skill saw will cut steel; however, we recommend purchasing a chop saw (Figure 10.8). The cuts are square and accurate, which is important when framing with steel. Chop saws cost approximately $200 and are safe to use. The saws normally use a 14- x 3/32- x 1-in abrasive cut-off wheel specifically designed for use on portable cut-off machines. Use premium quality, formulated for cutting applications in metal only. Do not install worn blades for use in hand-held cut-off machines or circular saws. The saw should have a minimum of 15 amp (3 hp) and a quick release locking vice.

FIGURE 10.8 Chop saw.

Hand circular saws: A 4 ½- to 5-in hand grinder (Figure 10.9) equipped with an inner flange, dry cutting diamond blade designed for the grinder makes a handy tool for trimming and hand cutting steel studs, runners, and joists. The tool is light and able to fit into tight places. In fact, other people must find this tool very useful, since I've had two stolen and one borrowed and never returned. The saws sell from about $60. The diamond blade costs $30 to $80 (shop around). If you're using a saw with ¾ hp, be careful. The built-in breaker may trip if you cut too fast. Skill saws are economical; the saws with over 2 hp and a dry cutting diamond blade (costing $40 to $80) are convenient tools.

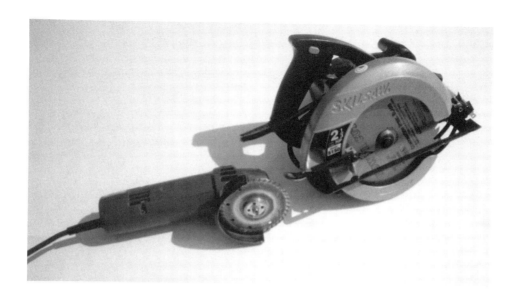

FIGURE 10.9 Hand circular saw.

Plasma cutter: An item that's not essential, but is extremely helpful, is a plasma cutter. These tools cut steel up to ½-in thick quickly and cleanly, making cutting steel feel like slicing butter. Plasma cutters are invaluable for special cuts such as holes in studs for ducts, plumbing, and angle cuts. A plasma cutter costs approximately $400 to over $2000. The less expensive 110-volt models require a separate air compressor; the more expensive 230-volt models have a built-in air compressor. I purchased a plasma cutter from a mail order house for about $400 and thought it was an excellent tool. Unfortunately, this tool was also stolen. Take my advice, if you find a tool that works well, watch it like a hawk.

Aviation snips: Aviation snips (Figure 10.10), sometimes called *metal snips* or *hand tin snips*, make straight and curved cuts in steel framing components. These tools are used for small cuts on the flanges of studs. Aviation snips are available in several sizes and styles. Do not attempt to cut heavier materials than those for which the snips are designed. Use a good brand, such as WISS. The cheaper brands dull quickly and make hand cuts difficult. The red handle is for right hand cuts; the green handle for left hand cuts, and the yellow handle for straight cuts.

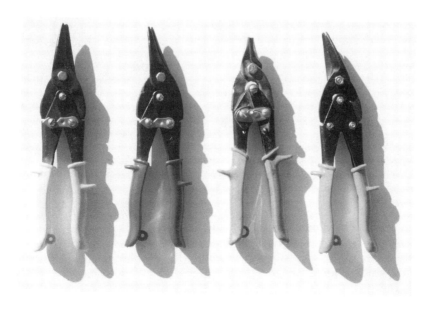

FIGURE 10.10 Aviation snips.

Electric shears: Electric shears are quieter than chop saws and faster than aviation snips. However, they must be used correctly. Many workers struggle when learning to use them. If you apply too much pressure, the shears do not cut properly. If the shears are not held at a correct angle, they have a tendency to stall, making it difficult to cut over the flange of the stud. Some models become dull, and their blades have a tendency to break. Personally, I prefer using a hand grinder with a dry cutting diamond blade for cutting steel on the job site.

On the other hand, heavy duty electric shears with a retail list price of $400 to $500 performed well in a Thailand building project. Local workers were trained to use the shears effectively in a short time.

Table saw: Table saws (Figure 10.11) are used to cut the rigid insulation and the plywood strip for the top of the track, as described in the following chapter.

FIGURE 10.11 Table saw.

Fastening tools

Residential steel-framing studs, runners, and joists can be fastened together with screws or welds. Although manuals and books on steel framing typically have a section on welding, instructions for welding are not included in this book. Welding is not widely used in residential construction today because of uncertainties of how the welds will be inspected, and the galvanized coatings at and near the weld have to be recoated where destroyed from heat. In any case, welding should not be done on any material lighter than 43 mil (18-ga).

You may have noticed, a spot welder wasn't included in our list of tools. Unless workers are specifically trained in welding, spot welding is not recommended. If not used correctly, the arc will burn a hole through the metal and destroy the connection. When using a portable spot welder for trusses on two residential projects, joints were inconsistent in strength. In addition, portable spot welders are heavy and awkward to use.

Other fastening techniques such as clinching, crimping, and press jointing are being used in other countries but have not been widely accepted in the United States. Many of these newer techniques may have the greatest potential for panelized construction.

For these reasons, this section focuses on fastening with screws. Selecting the best fastener for a particular job enables fast, economical, and efficient connections. To help you choose the best tools, we've listed various screw guns, screwdrivers, screw nailers, and driver's tips that are available. (Screws were described in the "Material" section earlier in this chapter.)

Power screwdrivers: Screws are generally applied with a positive-clutch electric power tool, commonly called an *electric screwgun* or *power screwdriver*. When using a screwgun to drive a screw, set the adjustment for the proper screw depth, place the screw in the bit tip, and start the screw straight. Battery-powered models (Figure 10.12) have a clutch that prevents "spinning the screws" (particularly on hex head screws) or stripping out the metal where the screws enter. These screwdrivers also help eliminate the problem of Phillips bit tips that frequently break off in the screw.

Use a minimum 9-volt model. A 12- to 18-volt model makes the work easier. Most new models come with two fast charge batteries. Battery-powered screwdrivers cost from $100 to over $200. These screwguns usually don't have a depth sensitive nosepiece; however, an adjustable depth driver called a *dimpler* or a *dimpler screw tip* can be used for attaching drywall and roof sheathing.

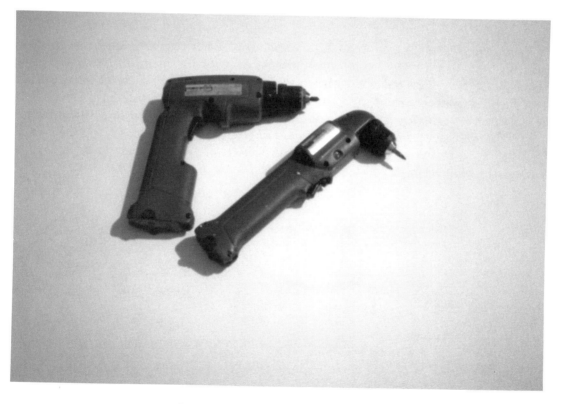

FIGURE 10.12 Battery powered screwgun.

Electric drywall/framing screwdriver: An electric drywall/framing screwdriver with a depth-sensitive nosepiece is typically used for attaching drywall and roof sheeting. These tools cost from about $80 to $160. Figure 10.13 shows the screwdriver with the nosepiece removed and with a long screw tip installed directly into the gun with the tip holder removed.

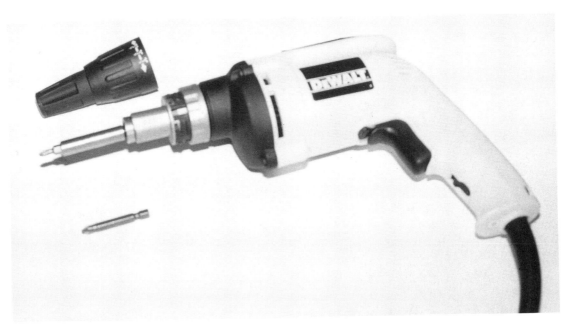

FIGURE 10.13 Electric drywall/framing screwdriver.

Screwguns with automatic screw feeds and screw-nailers: Screwguns are available with automatic screw feeds. Some screw feeds attach to the end of certain brands of screwguns (Figure 10.14), or you can purchase a pneumatic screw-nailer. Screwguns and screw-nailers vary in price from about $100 for the screw gun attachment to over $1000 for the pneumatic screw-nailer. While these tools increase the speed of attaching screws, the screws cost considerably more than typical screws in bulk. In addition, the special tips for the screwgun attachments have a tendency to break and take more time to replace than ordinary tips for screw guns. The pneumatic screw-nailer weighs about 9-lbs. Screw-nailers are best used when large numbers of fasteners are installed in a repetitive basis, such as making trusses. When cutting and fitting of material is required, a screw gun works better.

FIGURE 10.14 Screw nailers.

Coil nailers: Developed in 1990, coil nailers are fast and cost-effective tools that allow rapid attachment of wall, roof, and floor sheathing to light-gauge steel studs, trusses, and joists. Coil nailers are durable and are one of the most popular Hitachi tools. Model VH650 weighs about 7-lbs and has a load capacity of 385 nails. Depending on the air compressor size, these tools can nail 10 to well over 100 nails per minute. The nail sizes range from 1½- to 2 ½-inch x 0.083/0.099, and are available with smooth, ring, and screw shank styles, in 4M carton packaging. Verify the value with the manufacturer prior to use if used for shear application.

Brad nailers: These tools work quite well in attaching the base board and door trim in the interior of the house with metal studs. The 1-in and 1¼-in brads bend like a staple behind the gypsum board when they hit the metal stud.

Driving tips

Driver tips come in various lengths and sizes (Figure 10.15). You'll need a good supply of Phillips tips (see 4 and 7 in Figure 12.15), since they break. Purchase quality tips, marketed as antislip tips, for tips that last longer. Using the longer Phillips tip (see 4) without a separate holder (see 3) eliminates the problem of the short tip (see 7) slipping out of the holder and remaining in the screw. If you use holders, replace them whenever the slip anchor wears out and the holders no longer grip enough to hold the tip in place.

Many framers complain that it's difficult to drive screws through steel, especially when the framers are in an awkward position or have to screw upward or toward themselves. Learning to apply the correct amount of pressure to start the screw and penetrate the steel takes practice. If the end of the bit wobbles when using a holder, the application is even more difficult. Try keeping the tip as close to the screw gun as possible. Removing the depth-sensitive nosepiece and inserting the driver tip close to the drill will reduce tip movement.

The square head tip (see 2) is used for finish screws. Use the dimpler head (see 5) for drywall and plywood installation for battery screwguns or regular drills; a Philips no.2 power bit (see 7) for size 6, 8, 10 screws; and a no.3 tip for size 12 screws. Use magnetic hex socket drivers (see 1 and 6) as required for hex head size. The magnetized tip has a tendency to fill with steel shavings; a strong magnet or a piece of putty will help remove the shavings quickly.

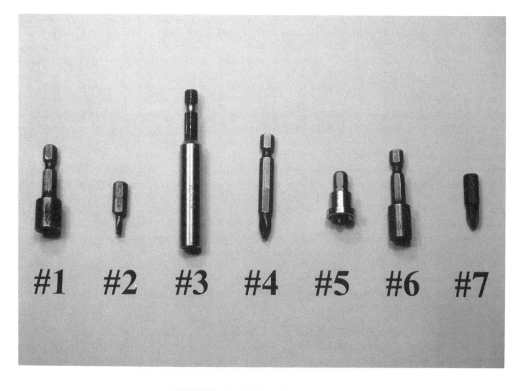

FIGURE 10.15 Driver tips.

Other tools

Aside from the tools listed above, you'll find the following tools useful:

C-clamps: The locking C–clamp comes in various sizes and shapes (Figure 10.16). You'll need various sizes with round tips and several with swivel pads. C-Clamps are used to clamp the steel together to prevent the steel from separating during attachment. At first you may find it difficult to adjust the clamp while your hands are full, however, this problem will be overcome as you work with the clamps. Eventually, the clamps will become like second nature. In the meantime, a clamp manufacturer is working on easier, lighter C-clamps to use in the field.

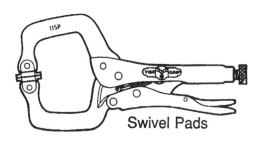

FIGURE 10.16 C-clamps.

Metal stud punch: A metal stud punch punches round holes in sheet metal up to 33 mils (20-ga), and is useful for installing plumbing and electrical conduits.

Miscellaneous tools: You'll also want to have a small square, a level with a magnet for plumbing the walls, safety glasses, a pouch for your screws, and leather gloves to protect you from frequent cuts that can occur while working with steel.

Safety guidelines

Safety procedures are important when working with steel. Sharp, hard edges, along with workers who may be inexperienced with tools, equipment, and procedures, make it necessary to review some safety guidelines to reduce the potential for job site injuries.

- When drilling, grinding, or sawing, operators should wear goggles and use guards and other safety devices.
- Use heavy leather gloves that seem a little large, with long gauntlets that protect the wrist.
- Exercise care when using all cutting equipment. Shield blades to prevent accidents, contain sparks, and protect the worker in case the blade fractures. Never torch-cut a load-bearing steel member.
- When using metal snips, keep hands safely above the work. Do not attempt to cut heavier materials than those for which the snips are designed, typically up to 43 mils (18-ga) material.
- Keep tools in good working order at all times and regularly inspect tools.
- When working with steel shapes, touch them with your gloved hand as you pass to ensure clearance and save on clothing and skin injury.
- When guiding a load of steel being moved by a crane, do not touch the steel directly. Use fiber rope to steer the material.

11

Step-by-Step Framing Details

The steel framing industry has faced an issue common to many new industries and materials, a lack of standardization that limits broad acceptance. However, some common components are beginning to become standards, or at least common practice, for steel frame homes. Many of these form the basis for the *Prescriptive Method for Residential Cold-Formed Steel Framing,* prepared for the U.S. Department of Housing and Urban Development and cosponsored by the American Iron and Steel Institute and the National Association of Home Builders. This prescriptive method was recently developed for home construction, significant parts of which were approved by the *CABO One and Two Family Dwelling Code* in October 1996.

In Appendix B of this book, we've included a copy of the *prescriptive method* and will endeavor to indicate any changes between the *prescriptive method* and the CABO code. The second edition of the *prescriptive method* will be updated to meet the approved code and will soon be available from the American Iron and Steel Institute. To obtain a copy of the CABO Code, contact the Council of American Building Officials, 5203 Leesburg Pike, Suite 708, Falls Church, Virginia 22041.

The limitations to using this code are as follows: The size of the building is limited to 60 ft (18288 mm) in length perpendicular to the joist span, and less than 36 ft (10970 mm) in width. The building is limited to two stories in height, with each floor not greater than 10 ft (3048 mm) high. The building is limited to seismic zones 0, 1, and 2, with maximum wind exposures of 90 mph (144.8 kmh) exposure C or 100 mph (1777 kmh) exposure B. The maximum ground snow load is 50 lb/ft² (2.395 knm²). States such as California, Florida, the western portions of Nevada, Oregon, Washington, and many coastal regions will require the use of a professional engineer or architect to prepare the plans. Consult with your local building officials for requirements.

It is our hope that this chapter will help builders, building designers, and others utilize the CABO code to prepare plans and build steel frame residences, in some cases without hiring a licensed engineer or architect. Since many building officials and inspectors have little experience with plan-checking steel frame buildings, the information in this chapter, along with the standardized load and span tables, details, and fastening requirements found in Appendix B, can be used to show code officials that your plans meet the CABO code. The Boca National Building Code, the Standard Building Code, and the Uniform Building Code accept the CABO code. However, as indicated in the disclaimer, the use of the *Prescriptive Method for Residential Cold-Formed Steel Framing* details for roof framing are subject to approval by the local code authority.

Workers with little or no experience can usually learn how to build with steel studs in a short period of time. The anatomy and sequence of events of steel frame homes are basically the same as wood-frame. The basic framing function remains the same, although the attachment methods are different. To help you make the transition and understand the purpose of the framing components, we'll take you step-by-step through framing a steel frame house, utilizing the CABO code and techniques I've used in my award-winning homes.

The Framing Crew

Before the framing begins, consider the skill of the framing crew. Builders, constructors, and framers who are not familiar with steel framing construction may require some time, practice, and supervision. Ideally, your steel framing crew should consist of one knowledgeable steel framer or lead worker for the layout and cutting and at least two apprentices who can join together the framing components.

Assembly Methods

Basically, there are three assembly methods for residential construction:

Stick-built construction: This type of steel frame system is similar to wood construction. The members are delivered to the construction site as pre-cut pieces or purchased in longer or standard length and cut to desired lengths at the site. The assembly is the same as lumber, except the components are screwed together rather than nailed. Sheathing and finishing materials are fastened with screws or pneumatic pins. As explained in Chapter 2, do not replace steel stick-by-stick for wood when using this method. Heat transfer and thermal bridging will add unnecessary energy penalties to the house.

Panelized construction: Panelization is a system that prefabricates walls, floors, and/or roof components into sections. Panels can be made in the shop or in the field. A jig is developed for each type of panel. Steel studs and joists are ordered in cut-to-length sizes for most panel work. They are placed into the jig, then fastened using screws. This type of construction lends itself to mass production and construction speed and efficiency. In fact, a job can usually be framed in about one-fourth the time required to stick-build a structure. Although the panels are typically transported from the shop to the job site, most often the cost advantages offset the transportation costs.

Preengineered construction: Preengineered systems take advantage of the extra strength and design flexibility of steel. These systems typically space the primary load-carrying members on more than 24-in centers sometimes as much as 8-ft centers. In most preengineered systems the pieces come with prepunched holes and other provisions that make it possible to assemble the structure in a very short time. Furring channels that support sheathing materials provide thermal breaks.

The three phases of steel framing, just as in wood construction, include the foundation, wall framing, and roof framing. We'll outline the stick-built method using standard construction practices. Let's start with the foundation.

Foundation

The following instructions on forming the foundation details methods for:

1. A frost-protected shallow foundation insulated concrete slab (FPSF)
2. A standard insulated slab
3. A raised foundation and finished floor

Since the primary purpose of this book is to provide details for energy-efficient and cost-effective steel-framed homes, this section focuses on details for an insulated slab. However, a standard concrete slab can be used, with a word of a caution: THE SLAB MUST BE SQUARE AND LEVEL. Unlike wood, the steel track won't bridge low spots. The top plate follows the contour of the slab, making it necessary to cut studs or use longer studs. The CABO code specifies that "steel studs shall not be spliced."

Keep in mind, a concrete floor is by far more cost and energy efficient than building a raised foundation floor system. (See section, "Cost Comparison of a Raised Foundation and a Slab Floor" later in this chapter.)

A frost protected shallow foundation insulated concrete slab (FPSF)

As pointed out in Chapter 3, frost–protected shallow foundation technology (Figure 11.1) eliminates the need for basements and deep foundations with crawl spaces that permit infiltration in the home in many cold areas of the country. In addition, the foundation provides thermal mass in the slab to moderate the temperature of the home.

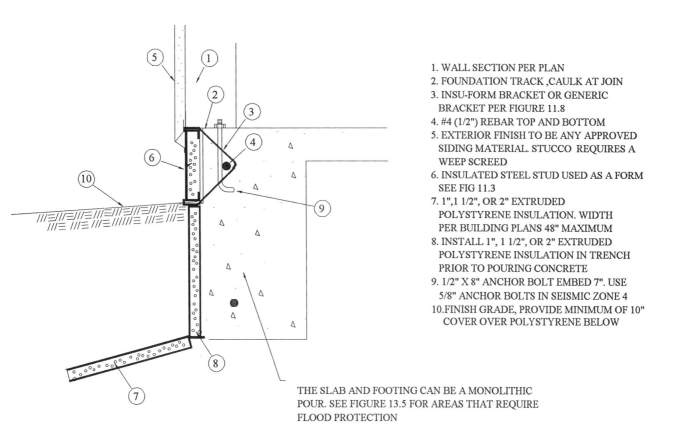

1. WALL SECTION PER PLAN
2. FOUNDATION TRACK ,CAULK AT JOIN
3. INSU-FORM BRACKET OR GENERIC
   BRACKET PER FIGURE 11.8
4. #4 (1/2") REBAR TOP AND BOTTOM
5. EXTERIOR FINISH TO BE ANY APPROVED
   SIDING MATERIAL. STUCCO REQUIRES A
   WEEP SCREED
6. INSULATED STEEL STUD USED AS A FORM
   SEE FIG 11.3
7. 1",1 1/2", OR 2" EXTRUDED
   POLYSTYRENE INSULATION. WIDTH
   PER BUILDING PLANS 48" MAXIMUM
8. INSTALL 1", 1 1/2", OR 2" EXTRUDED
   POLYSTYRENE INSULATION IN TRENCH
   PRIOR TO POURING CONCRETE
9. 1/2" X 8" ANCHOR BOLT EMBED 7". USE
   5/8" ANCHOR BOLTS IN SEISMIC ZONE 4
10. FINISH GRADE, PROVIDE MINIMUM OF 10"
    COVER OVER POLYSTYRENE BELOW

THE SLAB AND FOOTING CAN BE A MONOLITHIC
POUR. SEE FIGURE 13.5 FOR AREAS THAT REQUIRE
FLOOD PROTECTION

FIGURE 11.1 Insu-Form frost-protected shallow foundation (FPSF) detail.

In areas requiring deep footing placement to provide frost protection, the above detail (which may allow a foundation depth as shallow as 12 in (305 mm) can provide immediate savings due to reduced labor and material costs. Since the insulated foundation system serves as the concrete formwork, only one concrete pour is required. This method complies with the requirements of the Model Energy Code (MEC).

The foundation system should comply with section 1806, Footings, as outlined in the 1997 UBC and all local building requirements. The contractor should verify the application of these footing details to his or her individual project.

The ground surfaces should have a minimum slope of 1 unit vertically for every 100 units horizontally (1 percent slope) for a minimum of 3 ft from the foundation walls. Figure 4.3 in the MECcheck Manual (Appendix A) illustrates this method of insulating the slab, with the following instructions: "The slab insulation extends from the top of the slab downward to the bottom of the slab and then horizontally away from the slab for a minimum total distance equal to or greater than the required depth. The horizontal insulation must be covered by pavement or at least 10 inch of soil."

A standard insulated slab

Since insulation must extend from the top of the slab, the question of protecting the insulation from the elements comes into play. Some designers cover the insulation with a flashing, extended below grade; however, this method presents problems. First, moisture can work its way up the floor level by capillary action, since a loose connection provides an inadequate seal. Second, the exposed insulation up to the slab allows insects and termites a passage into the wall section.

The Insu-Form foundation system was developed to solve these problems. As discussed in Chapter 3, in the 1980s I developed and patented a bracket that connects the foundation to an insulated steel stud, which surrounds a concrete slab. The foundation includes styrene foam panels that recess into the ground, helping the slab retain heat. The Insu-Form brackets can be purchased from OTW Building Systems (1-800-541-1575) at a reasonable price. However, standard materials can also be used to make the brackets. This book will detail how to make the brackets and grant each purchaser of this book a license to make brackets. (This license is limited to personal use and not for resale without express written permission.)

The foundation system has been used on several hundred homes and proved to increase energy efficiency substantially. As discussed in Chapter 1, a steel frame duplex using this foundation system won an award in the Leading Edge Contest for construction of one of the most energy-efficient houses, even though most of the homes were wood frame.

The foundation system can be installed flush with the building line, as shown in Figure 11.1, with a stucco finish and a weep screed. Other options include installing the system with approved siding, normally installed flush with the foundation system, or with rigid exterior insulation as shown in Figure 11.2. If rigid insulation is installed, then the outside edge of the bottom track should be installed approximately 1½ in inside the face of the metal form. The metal form should be installed 1½ in outside the building line.

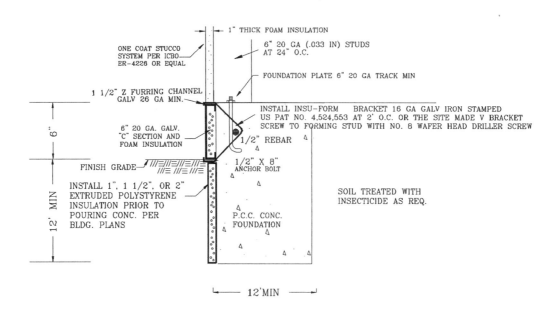

THE EXTERIOR FACE OF THE METAL FORM IS TO BE COATED BY ONE OF THE FOLLOWING METHODS:

1. COAT EXPOSED FACE WITH ELASTOMERIC COPOLYMER BRUSHABLE SEALANT AND PAINT TO MATCH EXTERIOR.

2. SCREW EXPANDED METAL LATH TO EXPOSED FACE AND COVER WITH A FIBER-REINFORCED STUCCO SYSTEM. ICBO ER-4226 OR EQUAL

FIGURE 11.2 The Insu-Form insulated foundation system.

In areas of high wind and moderate temperatures, the foundation's form (see Figures 11.1 and 11.2) may be installed without insulation and filled with concrete. This technique allows straps to be screwed or welded to the form and connected to the studs, which typically gives an excellent uplift capacity. However, a professional engineer or architect should design this type of installation.

Also keep in mind, as discussed in Chapter 4, you should install a conduit under the slab coming from the electric meter to a subpanel located in the hall or other interior location to reduce infiltration.

Forming the slab

A level and square slab for steel framing is extremely important, since studs must fit flush with the bottom and top plates. Use 16- ft long, 6-in by 33 mils (20 ga) unpunched steel studs for forming the slab instead of wood, which is not straight. If rigid insulation is thicker than one inch, use unpunched structural studs.

Support studs with round or flat steel stakes or wood stakes to form the slab as shown below in Figure 11.3. If using round metal stakes, remove stakes easily using a pipe wrench and unscrew stakes out of the hard ground. Use #6 bugle head drywall screws to hold studs in place.

FIGURE 11.3 Supporting studs with stakes to form the slab.

Join metal studs by cutting the web with a 1½- in notch and slide the studs together (Figure 11.4). Cut the flange and bend at 90 degrees at corners.

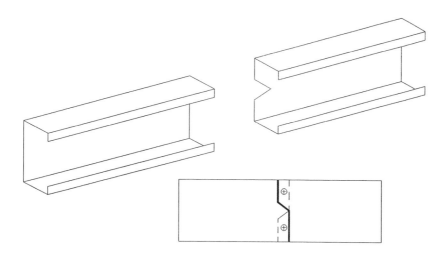

FIGURE 11.4 Joining metal studs by cutting the web and sliding studs together.

One method of supporting the steel stud form is with a bracket (Figure 11.5). The bracket can be made by welding a standard 6-in tee or a 6-in long 6-in x 97 mils (12-ga) track to a standard foundation screed bracket. Use a round metal pin to support the form and #8 hex drillers to attach the forming stud. The bracket makes it easy to adjust forms to a level grade after ground plumbing is installed. Several hundred of these brackets were loaned to contractors using the Insu-Form foundation system. Within a year the braces were gone; however, they are still used by cement contractors for supporting wood forms.

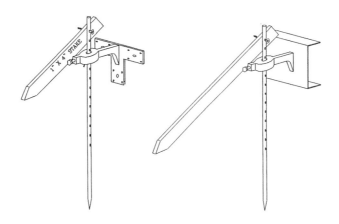

FIGURE 11.5 Supporting the steel stud form with a bracket.

If the foundation is in unstable soil, you may need to use a 6-in wood or metal stud with a 6-in long 6-in x 97 mils (12-ga) track screwed to its end, supported by round metal stakes (Figure 11.6). Use #6 bugle head screws to hold pins to forms.

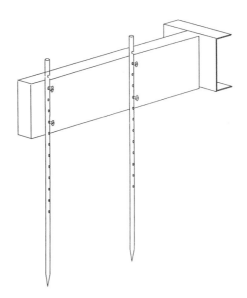

FIGURE 11.6 Form supports in unstable ground.

When insulating the slab with rigid insulation, use a 1-5/8-in flange structural stud with 1½-in thick isocynarate rigid insulation cut into 5¾ in strips for an R-value of 10. One inch of extruded polystyrene in a drywall stud gives an R-5 insulation value, 1½-in extruded polystyrene gives an R-7 insulation value (Figure 13.7). Note: Use isocynarate insulation only when protected by the stud. In applications below grade, always use extruded polystyrene.

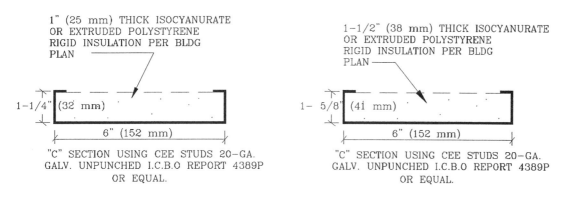

FIGURE 11.7 Sections of insulated form.

Making and installing the Insu-Form bracket

As mentioned earlier in this chapter, you can purchase the Insu-Form brackets, or make your own brackets using the details below (Figure 11.8).

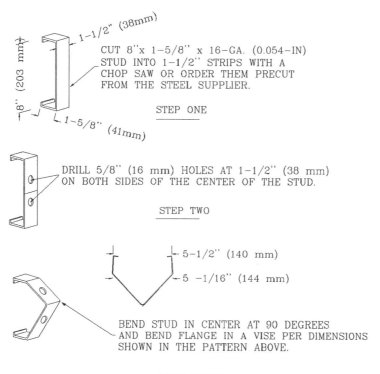

Note: For 8-in forms, make the bracket out of a 10-in stud, for 10-in forms, make the bracket out of a 12-in stud.

FIGURE 11.8 Making your own Insu-Form brackets.

To insulate the slab with rigid insulation, the following steps should be taken:

1. Insert the strip of rigid insulation; 1½-in extruded polystyrene in an unpunched 1-5/8-in flange stud was used for the sample house. Expand the V bracket and install over the flange of the 6-in stud (Figure 11.9).

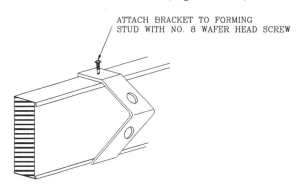

FIGURE 11.9 Expanding the V bracket and installing it over the flange of the stud.

2. Attach to the form stud with a #8 wafer head screw (Figure 11.10).

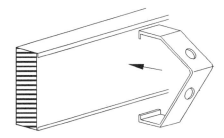

FIGURE 11.10 Attaching to the form stud.

3. Install rebar. Insert and twist an 8-in long anchor bolt into the holes. Provide 7-in embedment into the concrete. If you have difficulty installing the anchor bolts through the holes, it may be necessary to bend the V bracket, as shown below in Figure 11.11.

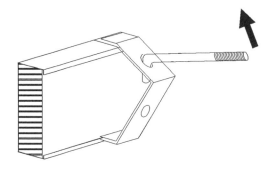

FIGURE 11.11 Installing rebar.

After completing these steps, the insulated form (Figure 11.12) is ready for concrete.

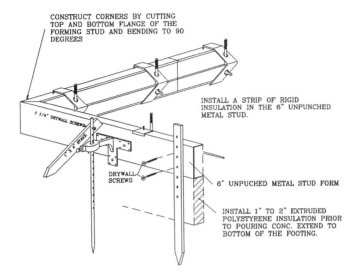

CONSTRUCT CORNERS BY CUTTING TOP AND BOTTOM FLANGE OF THE FORMING STUD AND BENDING TO 90 DEGREES

INSTALL A STRIP OF RIGID INSULATION IN THE 6" UNPUNCHED METAL STUD.

1 1/4" DRYWALL SCREWS

DRYWALL SCREWS

6" UNPUCHED METAL STUD FORM

INSTALL 1" TO 2" EXTRUDED POLYSTYRENE INSULATION PRIOR TO POURING CONC. EXTEND TO BOTTOM OF THE FOOTING.

FIGURE 11.12 Insulated forms.

Prior to pouring concrete

Check all required ground plumbing. If necessary, adjust forms for square and elevation. Insu-Form anchors should be spaced less than 6 ft (1829 mm) apart. Use a minimum of two anchors per piece of wall section, with one located not more than 12-in (305 mm) or less than 7 bolt diameters from each end of the piece. For steel framing, I recommend that the anchor bolts be spaced no more than 24-in on center. After pouring the slab, the forming stakes may be removed the same day.

The exterior face of the metal form is to be coated by one of the following methods:

1. Coat the exposed face with Elastomeric Copolymer Brushable Sealant and paint to match the exterior of the building, or
2. Screw expanded metal lath to the exposed face and cover with a fiber-reinforced stucco system, ICBO ER-4226 or equal.

Installing the bottom track

Snap out the interior wall lines using a chalk line. Use a 4-in x 4-in section of wood to align the track with the anchor bolts. Install the bottom plate by using a ¾-in iron pipe and a hammer to punch through the anchor bolt (Figure 11.13).

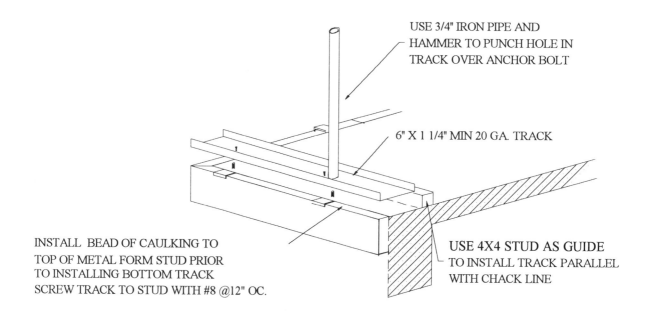

USE 3/4" IRON PIPE AND HAMMER TO PUNCH HOLE IN TRACK OVER ANCHOR BOLT

6" X 1 1/4" MIN 20 GA. TRACK

INSTALL BEAD OF CAULKING TO TOP OF METAL FORM STUD PRIOR TO INSTALLING BOTTOM TRACK SCREW TRACK TO STUD WITH #8 @12" OC.

USE 4X4 STUD AS GUIDE TO INSTALL TRACK PARALLEL WITH CHACK LINE

FIGURE 11.13 Installing a bottom track.

A raised foundation and floor framing

A raised foundation is a common method used in steel framing (Figure 11.14). Joists are the main horizontal structural members. Use 16-in or 24-in centers for layout; 24-in centers are optimum.

The ends of the C-shaped joists fit into track material that replaces wood rim joists. Fasten to the track by driving screws through the flanges and installing a screwed clip that attaches the joist web to the track. Web stiffeners (vertical pieces of stud material screwed to the web of the joist) installed at the ends of the joists reinforce this area (Figure 11.16). Solid blocking and flat steel strapping reinforce joists midspan between carrying beams and partitions. If the span is more than 12-ft long, section 506.3.3 of the CABO code requires bracing of the bottom flanges.

Install the blocking between joists, and pull the flat strapping tight across the tops and bottoms of the joists and fasten the blocking to joists. Use solid blocking to reinforce floor joists that carry walls and other loads (Figure 11.16). Glue and screw-attach plywood subflooring to steel framing.

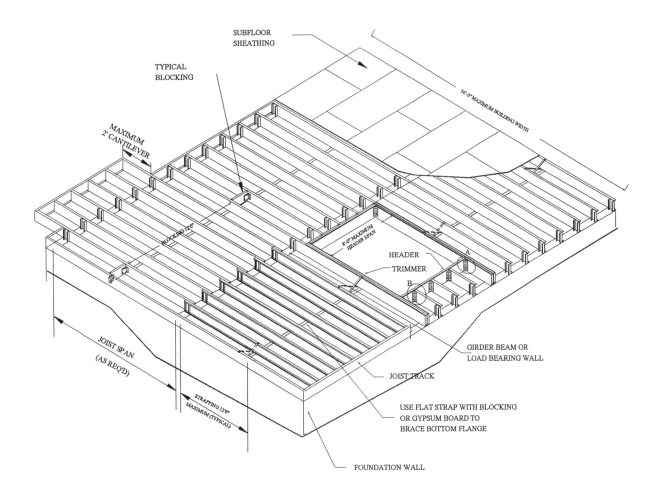

FLOOR ASSEMBLY

FIGURE 11.14 Typical floor framing detail.

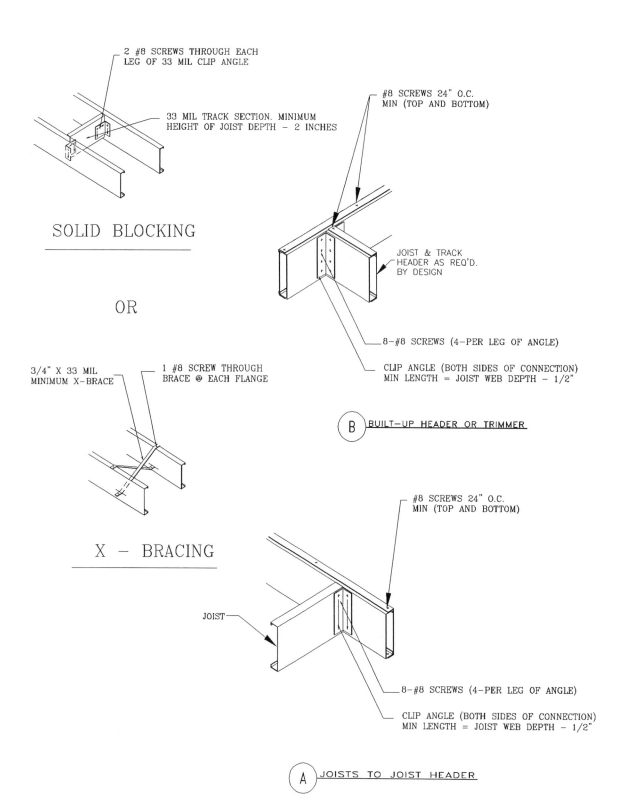

2 #8 SCREWS THROUGH EACH
LEG OF 33 MIL CLIP ANGLE

33 MIL TRACK SECTION. MINIMUM
HEIGHT OF JOIST DEPTH – 2 INCHES

#8 SCREWS 24" O.C.
MIN (TOP AND BOTTOM)

SOLID BLOCKING

JOIST & TRACK
HEADER AS REQ'D.
BY DESIGN

OR

8–#8 SCREWS (4–PER LEG OF ANGLE)

CLIP ANGLE (BOTH SIDES OF CONNECTION)
MIN LENGTH = JOIST WEB DEPTH – 1/2"

3/4" X 33 MIL
MINIMUM X–BRACE

1 #8 SCREW THROUGH
BRACE @ EACH FLANGE

B BUILT-UP HEADER OR TRIMMER

X – BRACING

#8 SCREWS 24" O.C.
MIN (TOP AND BOTTOM)

JOIST

8–#8 SCREWS (4–PER LEG OF ANGLE)

CLIP ANGLE (BOTH SIDES OF CONNECTION)
MIN LENGTH = JOIST WEB DEPTH – 1/2"

A JOISTS TO JOIST HEADER

FIGURE 11.15 Typical floor blocking details.

NOTE: In the details in Figure 11.16, the connection to the raised footing is made by using a 6-in x 6-in x 54 mils (16-ga) clip angle and anchor bolt. The clip angle is screwed to the exterior track with eight #8 screws.

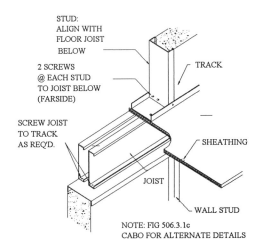

STUD:
ALIGN WITH
FLOOR JOIST
BELOW

TRACK

2 SCREWS
@ EACH STUD
TO JOIST BELOW
(FARSIDE)

SCREW JOIST
TO TRACK
AS REQ'D.

SHEATHING

JOIST

WALL STUD

NOTE: FIG 506.3.1c
CABO FOR ALTERNATE DETAILS

FLOOR JOISTS TO EXTERIOR WALL

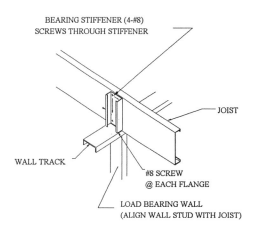

BEARING STIFFENER (4-#8)
SCREWS THROUGH STIFFENER

JOIST

WALL TRACK

#8 SCREW
@ EACH FLANGE

LOAD BEARING WALL
(ALIGN WALL STUD WITH JOIST)

FLOOR FRAMING AT INTERIOR WALL

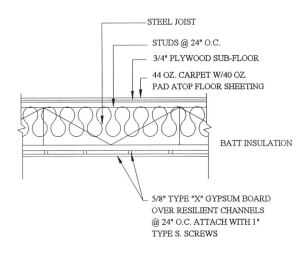

STEEL JOIST

STUDS @ 24" O.C.

3/4" PLYWOOD SUB-FLOOR

44 OZ. CARPET W/40 OZ.
PAD ATOP FLOOR SHEETING

BATT INSULATION

5/8" TYPE "X" GYPSUM BOARD
OVER RESILIENT CHANNELS
@ 24" O.C. ATTACH WITH 1"
TYPE S. SCREWS

CEILING DETAIL

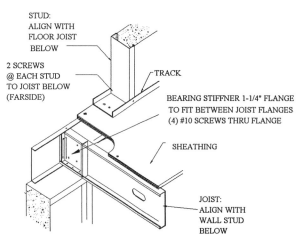

STUD:
ALIGN WITH
FLOOR JOIST
BELOW

2 SCREWS
@ EACH STUD
TO JOIST BELOW
(FARSIDE)

TRACK

BEARING STIFFNER 1-1/4" FLANGE
TO FIT BETWEEN JOIST FLANGES
(4) #10 SCREWS THRU FLANGE

SHEATHING

JOIST:
ALIGN WITH
WALL STUD
BELOW

FLOOR JOISTS TO EXTERIOR WALL

FIGURE 11.16 Foundation and top of wall details.

In Figure 11.16, a ceiling detail is included with resilient channels. In my experience, installing a floor over a living area without insulation or channels allows the metal floor joist to act as a sounding board. Since the transmission of noise can be annoying, this detail is recommended for multifamily units.

Cost comparison of a raised foundation and a slab floor

Steel framing can be used on any type of foundation: slab-on-grade, crawl space, or basement. However, as stated before, a slab floor is more cost effective, since a raised foundation requires an extensive amount of material and labor.

The forming and installation of the slab, including the insulating materials, costs between $3 and $4 per square foot in most locations. The total cost of the slab for the 816 ft² foot sample house described in Chapter 9 (see floor plan in following section) is approximately $3000.

The cost of a raised foundation includes 120 lineal feet of 12-in x 8-in thick footing, plus 120 lineal feet of block four courses high, at a cost of $5 per lineal foot, or a total of $600. Table 6 of Appendix B shows that a 1- 5/8-in x 8-in x 43 mil (18-ga) stud at 24-in on center will span 12-ft, 3-in. To frame the floor you will use 432 lineal feet at a cost of approximately $1 per foot, or a total of $432. Other supplies include:

- 54 lineal feet of 1-5/8-in x 6-in x 43 mil (18-ga) drywall stud bearing stiffeners at a cost of $0.80 per lineal foot, or about $44
- 116 lineal feet of 1-5/8-in x 8-in x 43 mil (18-ga) track at $1 per foot, or about $116
- 816 ft² of ¾-in plywood subfloor at $0.80 / ft², or about $650
- Screws and blocking material, for a total of about $50
- Insulation, for a total of about $260

The cost of the material alone for a raised foundation for the sample house is about $2100. Add another $2000 for labor, and your total cost is approximately $4100. As you can see, using a slab floor can save about $1000 with the extra benefit of providing thermal mass.

Wall framing

After the foundation work is completed, you'll begin framing the walls. As pointed out in the last chapter, 3-5/8-in and 6-in x 33mil (20-ga) exterior walls are common in stick-built residential construction, 3-5/8-in or 3½-in x 1¼ in x 18 mil (25-ga) studs are used for interior walls. Framers use layouts of 16-in and 24-in centers, but the optimum spacing is 24-in on center. The studs fit into light-gauge steel tracks that serve as top and bottom plates. Screw through the track flanges and into the studs to fasten them together. Openings in walls require headers designed to transfer joist and truss loads to jamb assemblies. Depending on the type of house being built, steel framers either stick-build the walls and attach the exterior sheathing or assemble the walls and attach the sheathing before tilting up the wall sections.

Bracing and shear walls commonly call for flat steel strapping that runs diagonally across the interior or exterior walls and fastens to the studs. The strapping is secured to the foundation or slab and run under tension to the top track. You can also use steel sheets secured to studs to create shear walls, however the CABO code specifies plywood or OSB shear panels.

The CABO code requires that structural sheeting be applied to each end (corners) of each exterior wall with a minimum 48-in (1220-mm) wide panel. The minimum percentage of full height structural sheathing on exterior walls is outlined in Table 603.6 of the CABO code (Table 23 of Appendix B) for seismic zone 0, 1, and 2. *The CABO One and Two Family Dwelling Code (1996/1997 Amendments)*, states that the minimum length of full-height structural sheathing per Table 603.6 should be multiplied by 1.10 for 9-ft (2743.2-mm) high walls and multiplied by 1.20 for 10-ft (3048-mm) high walls. In addition, structural sheathing should be installed with the long dimension parallel to the stud framing and applied to each end (corners) of each exterior wall with a minimum 48-in (1220-mm) wide panel. For seismic zones 3 and 4, as well as other applications, use Tables-22-VII-A, B, and C of the 1997 UBC. Figure 11.25 is a table based on the values given in the UBC and has shear values for both gypsum and OSB shear panel applications.

    To help take you through these steps, we'll be using a sample floor plan (Figure 11.17) of a simple 816 ft² two-bedroom home (based on the sample house described in Chapter 9). The sample floor plan below can be expanded with the addition of two bedrooms, a bath, and laundry area. (See Figure 6.4 in Chapter 6.)

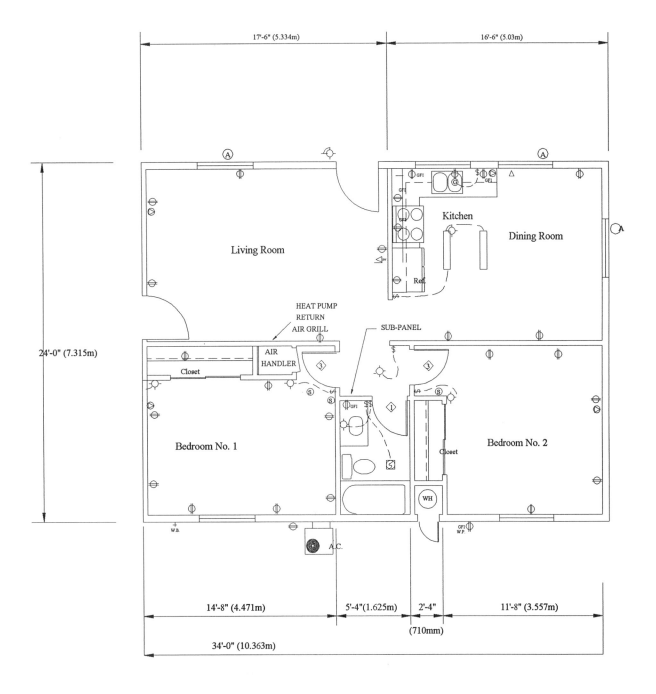

FIGURE 11.17 Floor plan.

Before framing the load-bearing walls, mark the exterior stud locations. Check the plans for all doors and windows and mark their locations on the floor. In the sample house, windows include: three 3-ft, 6-in x 5-ft windows in the living and dining room; one 3-ft, 6-in x 3-ft, 6-in window in the kitchen, and two 3-ft, 6-in x 4-ft windows in the bedrooms. A window was not installed in the bathroom since a fan is necessary to remove stale air from the house. (The fan should be wired into the light switch, so the fan will automatically run every time the room is used. A 10 percent fresh air supply should be provided in the air handler.)

Let's begin framing.

Constructing the corner shear panels

Temporarily clamp studs with vise grip clamps to a sheet of 4- x 8-ft OSB or plywood. Attach the metal stud flush with the edge of the OSB panel on one corner, attach the next stud 24 inches from the end, and attach the end stud centered on the edge of the OSB panel. On the adjacent corner, attach the end stud centered on the edge of the OSB panel, attach the next stud 24-in from the edge of the OSB panel, and attach the face of the last stud 6-in from the edge of the OSB panel. Screw the OSB to the stud with #8 screws at 24-in centers. (Figure 11.18). When 5½-in or 3½-in studs are used with 7/16" OSB, the entire panel will fit into a 6-in or 4-in track. If 6" studs are used then the OSB must slip outside the flange of the top and bottom track, so don't screw the ends of these studs at this time.

The details below are to be constructed for each corner of the building. For the sample house, construct four sets of corner shear panels.

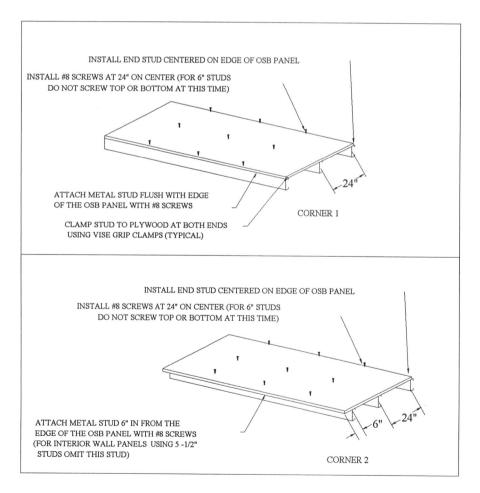

FIGURE 11.18 Typical corner shear panels.

Installing corner shear panels

Table 603.6 of the CABO code (Table 23 Appendix B) provides the minimum percentage of full height structural sheathing required for each wall.   For the sample house, we used a wind speed and exposure of "up to 70C or 90B," and a roof pitch of less than 9:12, which requires a minimum of 30 percent of full-height structural sheathing for both walls.   In conditions where design wind speeds are in excess of 90 mph and in seismic zone 3 or greater, the amount and type of structural sheathing should be determined by accepted engineering practices.

Percentages are given as a percentage of the total wall length.   For example, the 24-ft end wall of the sample house requires 30 percent of full height sheathing, resulting in (0.30 x 24-ft = 7.2-ft) of wall with full-height sheathing.   Using these percentages, the two end corners for the sample house are adequate.   However, the 34-ft side walls, which also require 30 percent of full height sheathing (0.30 x 34-ft =10.2-ft) are not adequate, requiring one more sheet of plywood.   Table 603.3.2a allows 2 x 6 x 33 (20-ga) studs at 24-in on center, 8-ft high, for building widths up to 36-ft for one-story buildings.

Use the details below to install the corner shear panels (Figure 11.19).

Install one each of corner 1 and of corner 2. Plumb wall and screw with #8 screws at 6" O.C. to bottom track, and along corner. If wall uses 6-in or 4-in studs use a shim between stud and plywood sheet to slip stud into bottom track and leave screws out of the top 24-in of the panel, until the top track is installed

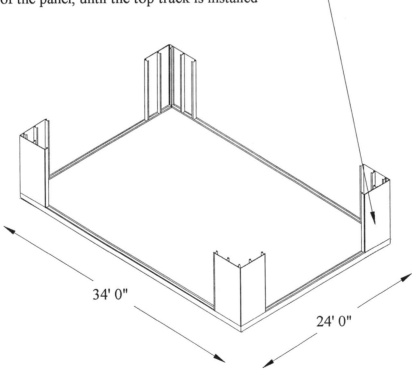

34' 0"

24' 0"

FIGURE 11.19 Installing corner shear panels.

Headers

Steel accepts any kind of door or window on the market. Doors and borrowed light openings should be rough-framed with steel studs and runners. Check the rough-opening dimensions provided by the manufacturer for windows. Measure from the top to find the proper location for the bottom sill. Unlike wood construction, extra headers are not needed over windows and doors.

Size the headers using Table 16 through 22 of Appendix B, *The Prescriptive Method for Residential Cold-formed Steel Framing*. If utilizing the CABO code, use Table 603.5a for a one-story house and Table 603.5b for a two-story house. For the sample house, we used 30 psf (pounds per square foot) ground snow load, 24-ft building width, and 24-in on center spacing.

Two 2 x 4 x 43 (18-ga) headers span up to 4-ft, 2 in, and two 2 x 6 x 43 (18-ga) headers span up to 6-ft. Table 603.5c of the CABO code requires one jack stud and one king stud for openings up to 3-ft, 6-in, and one jack stud and two king studs for openings that are 3-ft, 6-in to 8-ft. For the sample house, the single hung vinyl windows are 3-ft, 6-in x 4 or 5-ft and the steel foam doors are 3-ft x 6-ft,8-in.

The headers used in Figure 13.20 are two 4 x 1 5/8-in x 43 mils (18-ga) with one jack stud and one king stud. Use a sheet of plywood or OSB as a jig to hold the studs in place and screw clip angles to the king stud, as shown below (Figure 11.20). The clip angle should be a minimum of 2-in x 2-in and have a minimum thickness of the header members or the wall studs, whichever is thicker.

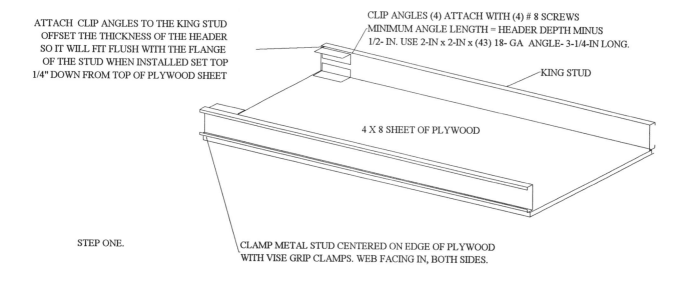

Note: If 5½-in or 3½-in studs are used for the wall framing then the OSB should be screwed to the studs as shown per Figure 11.18

FIGURE 11.20 Installing clip angles to king studs.

Install 1½ in of rigid insulation in the 3-5/8-in 18-ga header in the front and back and attach headers to clip angles with four #8 screws, or in accordance with Table 603.5d of the CABO code or Table 22 in Appendix B. The CABO code permits the use of #8 screws, Table 22 specifies #10 screws. We recommend #8 screws. Install jack studs and space to fit the size of the rough height opening dimensions according to manufacturer's specifications (Figure 11.21a). In most cases, windows sizes are the same as the "called out size," which is 3-ft,6-in x 4-ft for the sample house. Install track for all required door and window sections (Figure 11.21b).

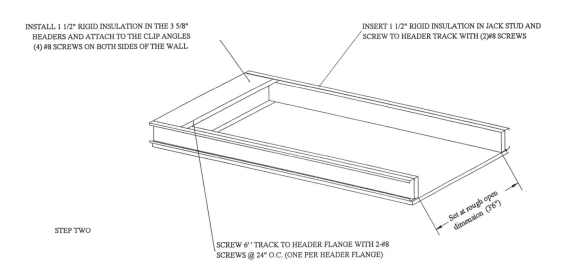

INSTALL 1 1/2" RIGID INSULATION IN THE 3 5/8" HEADERS AND ATTACH TO THE CLIP ANGLES (4) #8 SCREWS ON BOTH SIDES OF THE WALL

INSERT 1 1/2" RIGID INSULATION IN JACK STUD AND SCREW TO HEADER TRACK WITH (2)#8 SCREWS

Set at rough open dimension (3'6")

STEP TWO

SCREW 6'' TRACK TO HEADER FLANGE WITH 2-#8 SCREWS @ 24" O.C. (ONE PER HEADER FLANGE)

FIGURE 11.21a Installing header and jack studs.

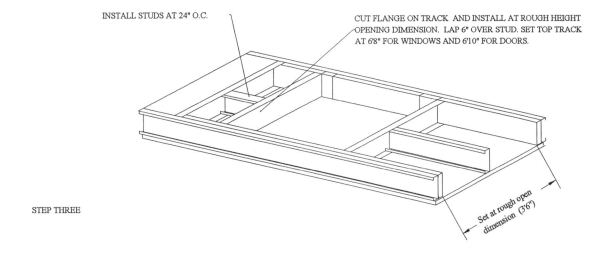

INSTALL STUDS AT 24" O.C.

CUT FLANGE ON TRACK AND INSTALL AT ROUGH HEIGHT OPENING DIMENSION. LAP 6" OVER STUD. SET TOP TRACK AT 6'8" FOR WINDOWS AND 6'10" FOR DOORS.

Set at rough open dimension (3'6")

STEP THREE

FIGURE 11.21b Installing track for window framing and wall studs.

Erecting the wall sections

To erect the wall section, take the following steps (Figure 11.22):

STEP 1:   Install the window and door sections in the wall section per building plan layout.

STEP 2:   Plumb these sections using 3-5/8-in studs as shown below.

STEP 3:   Screw the stud securely to a metal pin firmly driven in the ground.

STEP 4:   Install a level line from each corner at 8-ft, 7/16-in above the floor. You can use a piece of OSB screwed to the corner to hold in place. Dental floss works quite well.

STEP 5:   Install the top track. Size the pieces correctly and you can splice the sections at the headers. Provide two #8 screws at 24-in on center on the top to each flange of the header.

STEP 6:   Install the wall studs (or interior wall panels, see Figure 11.18), slipping them into the top and bottom track. Make sure you align the studs at 24-in on center, measuring from the corner. Attach with #8 screws, top and bottom on both sides.

STEP 7:   Cut a 6-in wide strip of OSB and screw it to the top track with two #8 screws at 24-in on center. This strip raises the ceiling height, eliminating the need to trim drywall in interior walls and allowing 8-ft interior studs without binding. The strip also serves as a thermal break between the trusses and wall section and permits the installation of the wiring without punching a large hole in the top plate.

STEP 8:   If you used 6-in or 4-in wall studs, you now can finish sheeting the exterior with OSB using #8 screws, 6-in on center edges, and 12- in on center in the field.

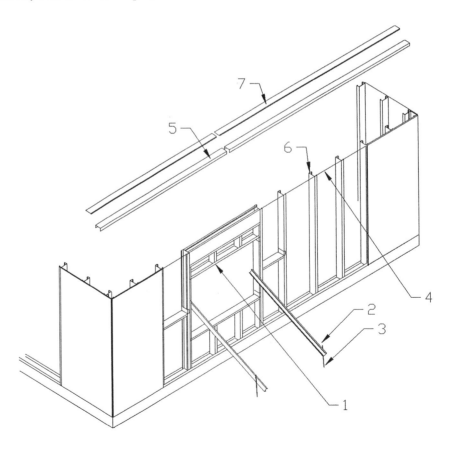

FIGURE 11.22 Erected wall section.

Strap bracing

Bracing and shear walls commonly call for flat steel strapping running diagonally across the interior or exterior walls and fastened to the studs. Secure the strapping to the foundation or slab and run it under tension to the top track (Figure 11.23). Use two pretension tools, which can be easily constructed with standard materials (Figure 11.24), to rack and align the building. After tightening the straps and installing the screws, remove the tool.

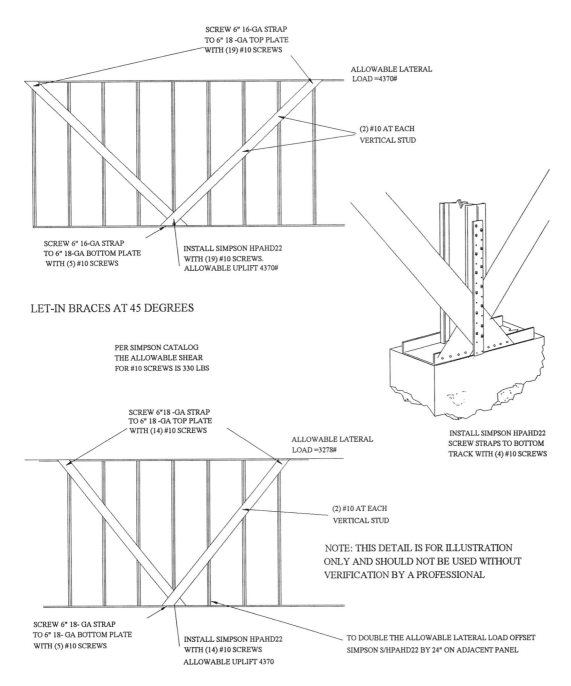

FIGURE 11.23 Strapping.

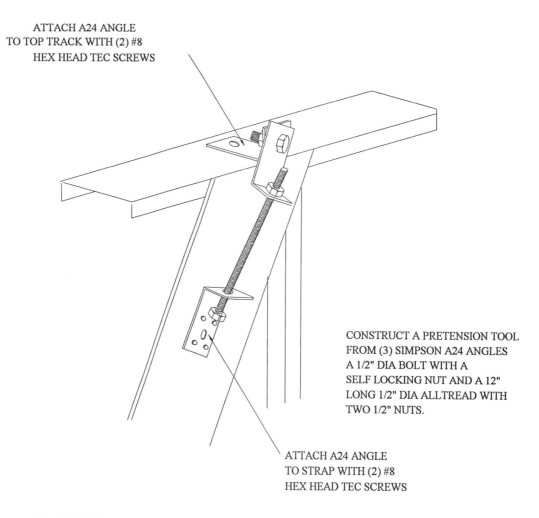

ATTACH A24 ANGLE
TO TOP TRACK WITH (2) #8
HEX HEAD TEC SCREWS

CONSTRUCT A PRETENSION TOOL
FROM (3) SIMPSON A24 ANGLES
A 1/2" DIA BOLT WITH A
SELF LOCKING NUT AND A 12"
LONG 1/2" DIA ALLTREAD WITH
TWO 1/2" NUTS.

ATTACH A24 ANGLE
TO STRAP WITH (2) #8
HEX HEAD TEC SCREWS

NOTE: THE 1997 UBC STATES " PROVISION SHALL BE MADE FOR PRETENSIONING
OR OTHER METHODS OF INSTALLATION OF TENSION-ONLY BRACING TO GUARD
AGAINST LOOSE DIAGONAL STRAPS.

FIGURE 11.24 Making your own pretension tool for strap bracing.

At the present time, shear panels are the only bracing method permitted by the CABO code. A professional engineer or architect must design any method of strapping.

The code requires seismic lateral forces be increased when you use light-steel-framed bearing walls with tension-only bracing. In seismic zone 4, where the calculated lateral force for shear panels might be 0.183 of the vertical load; the lateral force for tension-only bracing would be 0.275 of the vertical load. Consideration should also be given to the vertical load imposed by the strapping at the top track; an additional stud is usually necessary at that location. In addition, a modification of the foundation system may be required to support this added load. Figure 11.23 illustrates one method of providing lateral strapping. Do not use without verification by a professional engineer or architect.

The following sample schedule in Figure 11.25 provides typical allowable shear loads for gypsum and OSB for metal-framed walls. While it has been submitted and used on various projects, it is included here only as an example of how to prepare a schedule. All shear values and anchor bolt spacing should be verified by a professional engineer for your particular project and loading conditions.

METAL STUD SHEAR WALL SCHEDULE

SHEAR WALL (1) (2)					PLATE SCREWING		
1 MARK	TYPE & MATERIAL	EDGE SCREWING O.C.	A-BOLTS AT FND. FOR SHEAR WALL ON ONE SIDE (4)	BOTTOM (FLOOR PLATE)	TOP (BLOCKING OR TRUSS)	ALLOW SHEAR
G1	1/2" GYP.BD.	#6 BY 1 INCH @7"	1/2" DIA. X 8" @72" O.C.	#6-S @16"	#6-S @ 16" O.C.	97#/ft.
G2	1/2" GYP.BD. TWO	#6 BY 1 INCH @7"	1/2" DIA. X 8" @24" O.C.	#6-S @8"	#6-S @ 8" O.C.	195#/ft.
G3	1/2" GYP.BD.	#6 BY 1 INCH @4"	1/2" DIA. X 8" @48" O.C.	#6-S @8"	#6-S @ 8" O.C.	142#/ft.
G4	1/2" GYP.BD. TWO	#6 BY 1 INCH @4"	1/2" DIA. X 8" @24" O.C.	#6-S @4"	#6-S @ 4" O.C.	284#/ft.
NOTE: THE ABOVE VALUES ARE FROM TABLE 22-VII-B 1997 UBC (WIND FORCES) SAFETY FACTOR OF 3						
P1	7/16" OSB (ONE SIDE)	#8 BY 1 INCH @6"	1/2" DIA. X 8" @24" O.C.	#8-S @6"	#8-S @ 6" O.C.	303#/ft.
P2	7/16" OSB (ONE SIDE)	#8 BY 1 INCH @4"	1/2" DIA. X 8" @18" O.C.	#8-S @4"	#8-S @ 4" O.C.	470#/ft.
P3	7/16" OSB (ONE SIDE)	#8 BY 1 INCH @3"	1/2" DIA. X 8" @12" O.C.	#8-S @3"	#8-S @ 3" O.C.	578#/ft.
P4	7/16" OSB (ONE SIDE)	#8 BY 1 INCH @2"	1/2" DIA. X 8" @12" O.C.	#8-S @2"	#8-S @ 2" O.C.	637#/ft.
NOTE: THE ABOVE VALUES ARE FROM TABLE 22-VII-A 1997 UBC (WIND FORCES) SAFETY FACTOR OF 3						
S1	7/16" OSB (ONE SIDE)	#8 BY 1 INCH @6"	1/2" DIA. X 8" @24" O.C.	#8-S @6"	#8-S @ 6" O.C.	280#/ft.
S2	7/16" OSB (ONE SIDE)	#8 BY 1 INCH @4"	1/2" DIA. X 8" @18" O.C.	#8-S @4"	#8-S @ 4" O.C.	366#/ft.
S3	7/16" OSB (ONE SIDE)	#8 BY 1 INCH @3"	1/2" DIA. X 8" @12" O.C.	#8-S @3"	#8-S @ 3" O.C.	510#/ft.
S4	7/16" OSB (ONE SIDE)	#8 BY 1 INCH @2"	1/2" DIA. X 8" @12" O.C.	#8-S @2"	#8-S @ 2" O.C.	650#/ft.
NOTE: THE ABOVE VALUES ARE FROM TABLE 22-VII-C 1997 UBC (SEISMIC FORCES) SAFETY FACTOR OF 2.5						

NOTE: WHERE OSB IS SPECIFIED, 15/32-INCH (12mm) STRUCTURAL 1 SHEATHING (PLYWOOD) MAY BE SUBSTITUTED

NOTES:

(1)   GYPSUM BOARD TO BE APPLIED PERPENDICULAR TO FRAMING WITH 1 1/2" WIDE 20 GA (0.33 IN) STRAP BLOCKING BEHIND THE HORIZONTAL JOINT AND WITH SOLID BLOCKING BETWEEN THE FIRST TWO END STUDS. END JOINTS OF ADJACENT COURSES OF GYPSUM BOARD SHALL NOT OCCUR OVER THE SAME STUD.

(2)   SCREWS IN FIELD OF THE PANEL SHALL BE INSTALLED 12 INCHES (305MM) O.C. UNLESS OTHERWISE SHOWN.

(3)   UNLESS OTHERWISE SHOWN, STUDS SHALL BE A MINIMUM 1 5/8" (41MM) BY 3 1/2 INCHES (89MM) WITH A 3/8 -INCH (9.5MM) RETURN LIP. TRACK SHALL BE A MINIMUM 1 1/4"(32MM) BY 3 1/2INCHES (89MM). (OTW PANELS TO HAVE MIN. BACK TO BACK 1 1/4" FLANGE STUDS)

(4)   MINIMUM (3) 1/2" DIA. ANCHORS PER SHEARWALL. FOR SHEAR PANELS ON TWO SIDES OF WALL USE ONE-HALF THE SPACING GIVEN IN THE SCHEDULE FOR ANCHOR BOLT, SILL NAILING and TOE NAILING.

(5)   S1 - S4 ...USE 5/8" ANCHOR BOLTS IN SEISMIC ZONE 4

(6)   FRAMING SCREWS SHALL BE NO. 8 X 5/8" (16MM) WAFER HEAD SELF-DRILLING.

(7)   PLYWOOD AND OSB SCREWS SHALL BE APPROVED AND SHALL BE A MIN. OF #8 X 1 INCH (25MM) FLAT HEAD WITH A MINIMUM HEAD DIAMETER OF 0.292 INCH (7.4 MM)

(8)   DRYWALL SCREWS SHALL BE A MINIMUM NO. 6 BY 1 INCH (25MM)

FIGURE 11.25 Shear panel specifications.

Shear panel details

The ratio of the plywood shear panel (Figure 11.26) for steel stud walls must not exceed two vertical to one horizontal , requiring a minimum 4-ft width for an 8-ft foot high wall. A professional should design shear panels. Consideration must be given to loads that are transferred to the footing, which should be sized accordingly.

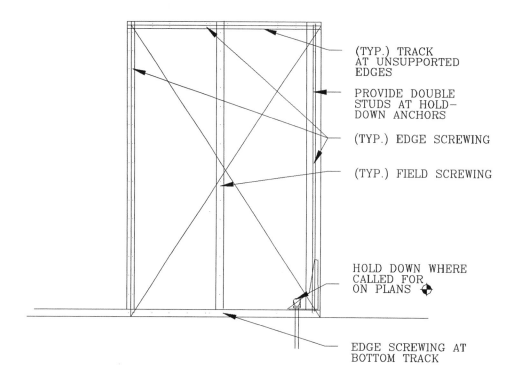

(TYP.) TRACK
AT UNSUPPORTED
EDGES

PROVIDE DOUBLE
STUDS AT HOLD-
DOWN ANCHORS

(TYP.) EDGE SCREWING

(TYP.) FIELD SCREWING

HOLD DOWN WHERE
CALLED FOR
ON PLANS

EDGE SCREWING AT
BOTTOM TRACK

NOTES

1. ALL SCREWS SHALL HAVE A MINIMUM 3/8"
   EDGE DISTANCE.

2. PROVIDE 1-5/8-IN FLANGE STUDS AND BLOCKING AT
   ALL PLYWOOD JOINTS UNLESS NOTED
   OTHERWISE.

3. EDGE SCREW PLYWOOD TO ALL DOUBLE STUDS.

4. LONG DIRECTION OF PLYWOOD SHEETS
   TO RUN PARALLEL TO STUDS, UNLESS
   AN ALTERNATE PROPOSAL IS SUBMITTED
   TO AND APPROVED BY THE ARCHITECT
   OR ENGINEER.

FIGURE 11.26 Shear wall detail.

Hold down details

The design of plywood or OSB shear walls in seismic zones 3 and 4 and in high wind areas normally require the use of hold downs and strapping between the first and second story. The details below (Figure 11.27) are from the Simpson Strong-Tie Catalog. The catalog gives allowable values for use in light gauge steel stud construction. For the nearest dealer call 1-800-999-5099.

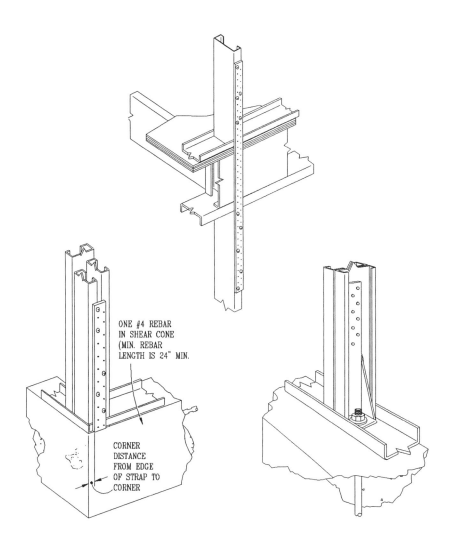

ONE #4 REBAR
IN SHEAR CONE
(MIN. REBAR
LENGTH IS 24" MIN.

CORNER
DISTANCE
FROM EDGE
OF STRAP TO
CORNER

FIGURE 11.27 Typical hold down details.

Roof Framing

Figure 11.28 details how to stack a steel stud roof. Install a footing under the center wall for bearing. According to Table 29 of Appendix A, the maximum span for the sample house, which has a 2 x 6 x 33 (1-5/8-in x 6-in x 20-ga) roof rafter with a 30# snow load is 9 ft, 7 in. Table 26 shows the maximum span for a 2 x 4x 43 (1- 5/8-in x 3-5/8-in x 18-ga) ceiling joist without attic storage is 12-ft, 1-in. The CABO code at this time does not contain span tables for joists. If the building department doesn't accept the span tables in Appendix A, they should accept approved span tables in ICBO ES Report #4389P. Contact Steeler Inc., 10023 Martin Luther King Jr. Way South, Seattle, Washington, 98178, to obtain a copy.

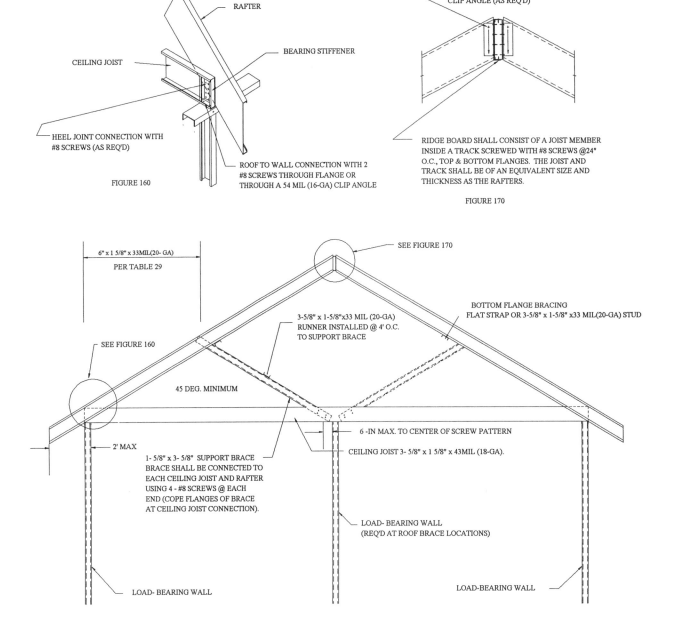

FIGURE 11.28 Roof framing details.

Trusses

In steel construction today, an engineer retained by the builder generally designs roof trusses. The builder submits a cutlist to the manufacturer and assembles the trusses at the job site. However, other options are becoming available. Several steel manufacturers have developed metal trusses sold in a kit. Individual pieces are delivered to the job and assembled in the field. In addition, some wood truss manufacturers are developing steel trusses that will be marketed soon. Until then, fabricating your own trusses on site can be an easy process if a jig is built and members are precut.

Although some builders complain steel trusses are too heavy or difficult to frame, simplifying the design can solve these problems. For example, a Dutch hip roof results in lighter trusses and less detailed connections. As mentioned in Chapter 2, this type of roof use less steel and reduce thermal bridging.

Build a solid jig and lay out the working points carefully. Even a small-discrepancy jig can cause hours of heartache in assembly. Once a jig is built, many builders find two or three workers can assemble a truss in 3 to 5 minutes. Don't use camber, steel trusses do not deflect, and lay out the bottom chord flat.

If you're set on a complicated or cut-up roof, use trusses where practical, then fill in with pieces. Whether you choose trusses or stick-frame, focus on the connection to the walls. Use gussets to reinforce the connection and hurricane clips to secure it. Square-cut stock should be used on all roof framing. Cutting angles and compound miters wastes time and material.

The truss illustrated below (Figure 11.29) can be made out of standard metal studs. The design of this truss uses a total load of 40# per square foot. The materials needed to construct the truss for the sample house are listed below.

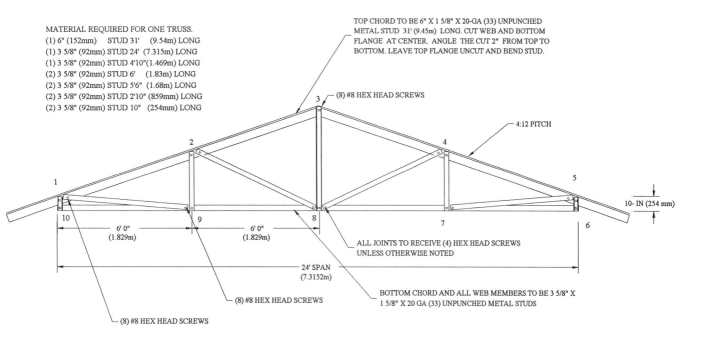

MATERIAL REQUIRED FOR ONE TRUSS.
(1) 6" (152mm)   STUD 31'   (9.54m) LONG
(1) 3 5/8" (92mm) STUD 24' (7.315m) LONG
(1) 3 5/8" (92mm) STUD 4'10"(1.469m) LONG
(2) 3 5/8" (92mm) STUD 6'   (1.83m) LONG
(2) 3 5/8" (92mm) STUD 5'6" (1.68m) LONG
(2) 3 5/8" (92mm) STUD 2'10" (859mm) LONG
(2) 3 5/8" (92mm) STUD 10" (254mm) LONG

TOP CHORD TO BE 6" X 1 5/8" X 20-GA (33) UNPUNCHED
METAL STUD 31' (9.45m) LONG. CUT WEB AND BOTTOM
FLANGE AT CENTER. ANGLE THE CUT 2" FROM TOP TO
BOTTOM. LEAVE TOP FLANGE UNCUT AND BEND STUD.

(8) #8 HEX HEAD SCREWS

4:12 PITCH

10- IN (254 mm)

6' 0"
(1.829m)

6' 0"
(1.829m)

ALL JOINTS TO RECEIVE (4) HEX HEAD SCREWS
UNLESS OTHERWISE NOTED

24' SPAN
(7.3152m)

BOTTOM CHORD AND ALL WEB MEMBERS TO BE 3 5/8" X
1 5/8" X 20 GA (33) UNPUNCHED METAL STUDS

(8) #8 HEX HEAD SCREWS

(8) #8 HEX HEAD SCREWS

FIGURE 11.29 Typical metal truss.

Making a truss with a jig

Bend the top chord of the truss by cutting it at an angle, as shown in Figure 11.30. This method allows the top chord to fit inside the adjacent piece without removing any of the bottom flange. For 6-in top chords, the cut offset is half of the roof pitch.

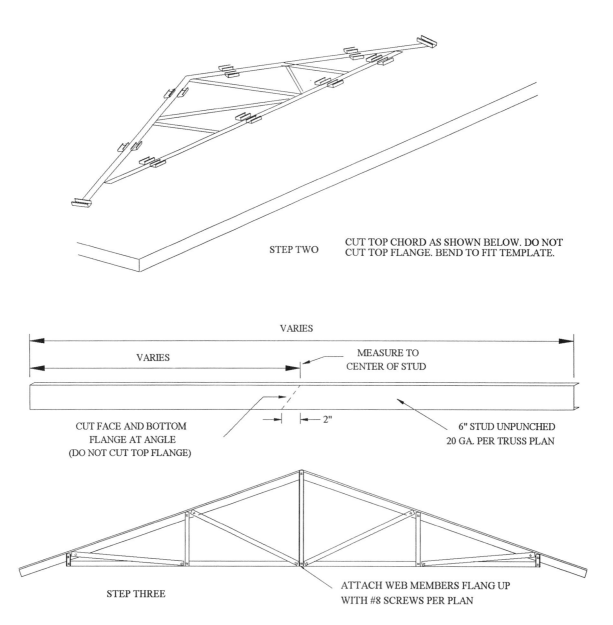

STEP ONE   ON SLAB SNAP OUT WITH RED CHALK
                  LINE THE TRUSS LAYOUT AND INSTALL
                  6" LONG METAL STUD STOPS WITH
                  CONCRETE NAILS TO SLAB.

STEP TWO   CUT TOP CHORD AS SHOWN BELOW. DO NOT
                  CUT TOP FLANGE. BEND TO FIT TEMPLATE.

VARIES

VARIES   MEASURE TO
                          CENTER OF STUD

├── 2" ──┤

CUT FACE AND BOTTOM
FLANGE AT ANGLE
(DO NOT CUT TOP FLANGE)

6" STUD UNPUNCHED
20 GA. PER TRUSS PLAN

STEP THREE   ATTACH WEB MEMBERS FLANG UP
                          WITH #8 SCREWS PER PLAN

FIGURE 11.30 Making your own truss with a jig.

Use the details in Figure 11.31 to make and install a gable end truss.

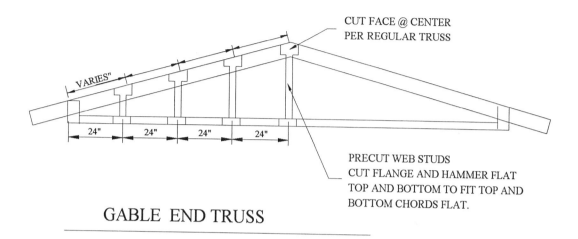

CUT FACE @ CENTER
PER REGULAR TRUSS

VARIES"

24"   24"   24"   24"

PRECUT WEB STUDS
CUT FLANGE AND HAMMER FLAT
TOP AND BOTTOM TO FIT TOP AND
BOTTOM CHORDS FLAT.

GABLE END TRUSS

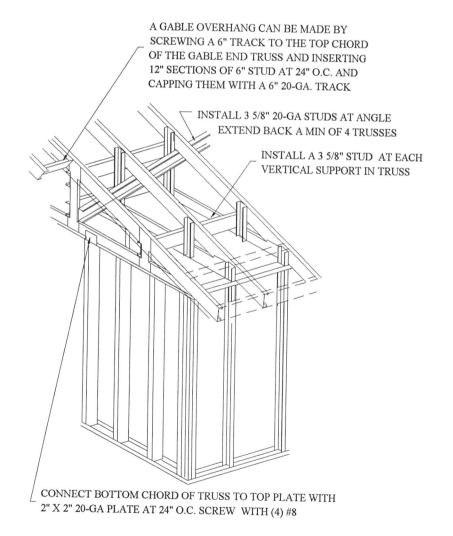

A GABLE OVERHANG CAN BE MADE BY
SCREWING A 6" TRACK TO THE TOP CHORD
OF THE GABLE END TRUSS AND INSERTING
12" SECTIONS OF 6" STUD AT 24" O.C. AND
CAPPING THEM WITH A 6" 20-GA. TRACK

INSTALL 3 5/8" 20-GA STUDS AT ANGLE
EXTEND BACK A MIN OF 4 TRUSSES

INSTALL A 3 5/8" STUD AT EACH
VERTICAL SUPPORT IN TRUSS

CONNECT BOTTOM CHORD OF TRUSS TO TOP PLATE WITH
2" X 2" 20-GA PLATE AT 24" O.C. SCREW WITH (4) #8

FIGURE 11.31 Gable end construction.

Providing a connection between the truss and wall section

To prevent uplift, provide a tie between the truss and the wall studs by using a commercial connector or installing a strip of strapping with four #8 screws to the 10-in web stiffener and stud. To transfer the roof diaphragm shear to the wall and foundation, the trusses must be blocked. Figure 11.32 below details one method of truss blocking. If gutters are required, the detail for boxing eaves provides a vertical face for installing the fascia. If gutters are not required, use 9-ft sheets of OSB to cover blocking. The required strength of truss tie-down connections to resist wind uplift forces can be obtained from Table 802.12 of the CABO code. The sample house, with a 20# design wind load and a 28-ft roof width, including 2-foot overhangs, requires a connection that resists 224 lb.

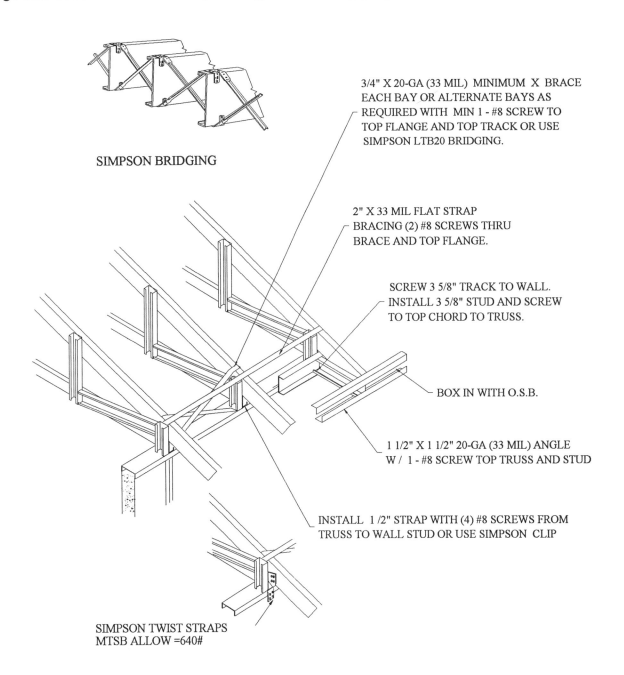

SIMPSON BRIDGING

3/4" X 20-GA (33 MIL) MINIMUM X BRACE EACH BAY OR ALTERNATE BAYS AS REQUIRED WITH MIN 1 - #8 SCREW TO TOP FLANGE AND TOP TRACK OR USE SIMPSON LTB20 BRIDGING.

2" X 33 MIL FLAT STRAP BRACING (2) #8 SCREWS THRU BRACE AND TOP FLANGE.

SCREW 3 5/8" TRACK TO WALL. INSTALL 3 5/8" STUD AND SCREW TO TOP CHORD TO TRUSS.

BOX IN WITH O.S.B.

1 1/2" X 1 1/2" 20-GA (33 MIL) ANGLE W / 1 - #8 SCREW TOP TRUSS AND STUD

INSTALL 1 /2" STRAP WITH (4) #8 SCREWS FROM TRUSS TO WALL STUD OR USE SIMPSON CLIP

SIMPSON TWIST STRAPS MTSB ALLOW =640#

FIGURE 11.32 Truss blocking and tie-downs.

Finishing the roof

Once the steel frame for the roof is complete, install the roof sheeting. Possible materials include metal decking and common boards, plywood, and other nonwood materials. Plywood is ideal for sheathing. The large-size panels add rigidity, are easy to install, and provide a sound base for the roof covering. Cover the roof with asphalt shingles, tile, wooden shakes or shingles, or metal.

Installing windows and doors

You're now ready to install windows and doors. When installing exterior doors, cut base track in the center of the opening. Cut flanges at the edge of the doorframe and bend the bottom track up over the lower part of the jack stud in the doorframe.

Keep in mind, the main objective when installing windows and doors is to create an airtight seal between the window or doorframe and the exterior wall. Use caulking and weather-stripping, as instructed in Chapter 4.

Installing gypsum board

Steel framing works well with a variety of sizes and types of gypsum board. The gypsum board can be applied horizontally or vertically.

To make your home more energy efficient, install gypsum board on the exterior walls and the ceiling before framing interior walls, with the exception of a portion of the ceiling above a plumbing wall. This method allows installation of plumbing, electrical connections, and other wiring into wall sections before installing the rest of the ceiling. As mentioned in Chapter 4, minimize electrical runs in the insulated wall section by installing within interior walls whenever possible, with minimum penetration into the attic. Not only will you save on electric wiring, but you'll eliminate many "home runs to the electric meter" with penetrations from the walls into the attic. Whenever possible, plumbing should be installed within the interior wall and sconces should be used to minimize ceiling fixtures.

When penetrations must be made in the attic for wiring, drill a hole through the exterior top plate where the OSB strip was installed. The hole should not be larger than necessary to allow passage of the romex or other wiring. Seal these connections tightly, since punched out holes in the studs permit circulation of air within the wall. Plastic grommets can be used to protect the romex going through the punch-outs.

As discussed in Chapter 2, the use of fastening systems can save money and reduce heat transfer and thermal bridging. Use Prest-on corner back fasteners for corner construction instead of back-up studs (Figure 13.33). The clips reduce the number of studs and eliminate cut wall insulation at the corners. Prest-on clips can also be used to put on the entire lid of the house (Figure 11.34). If Prest-on clips are not available, standard steel construction details should be modified by installing a stud flat to provide backing to attach gypsum board, as shown in Figure 11.33.

The Eliminator Track, a patented system used in commercial work, is another useful tool, especially when utilizing a suspended roof. The track is screwed to the gypsum board ceiling or to the T-bar grid in a suspended ceiling, allowing you to snap the studs in place without screws. For more information, contact your drywall supplier.

Some building components, such as Prest-on clips, are not described in the CABO code. However, the item can be used if it has a listing with the model code. For example, Prest-on clips are approved for use with steel studs, per BOCA Research Report No. 88-13, ICBO Research Report No. 2451, SBCCI Report No.8877, CMHC (Canada) 5710/11761, and CABO code (one-and two-family dwelling code).

Soffits, as discussed in Chapter 4, can be built with gypsum board, providing an economical way to keep ducts within the conditioned space. (See the following chapter for more information on soffits.)

STANDARD STEEL CONSTRUCTION DETAILS

ENERGY EFFICIENT STEEL CONSTRUCTION DETAILS

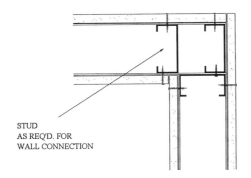

STUD
AS REQ'D. FOR
WALL CONNECTION

TYPICAL CORNER FRAMING

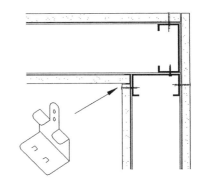

CORNER FRAMING PREST-ON CORNER BACK

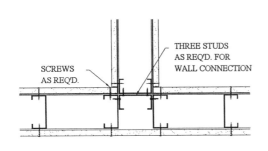

SCREWS
AS REQ'D.

THREE STUDS
AS REQ'D. FOR
WALL CONNECTION

TYPICAL WALL INTERSECTION FRAMING

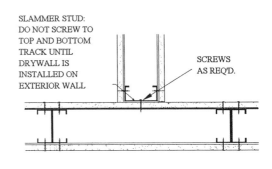

SLAMMER STUD:
DO NOT SCREW TO
TOP AND BOTTOM
TRACK UNTIL
DRYWALL IS
INSTALLED ON
EXTERIOR WALL

SCREWS
AS REQ'D.

WALL INTERSECTION FRAMING SLAMMER STUD

Attaching gypsum board to the walls.

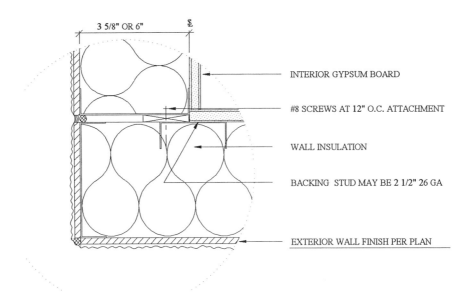

3 5/8" OR 6"

INTERIOR GYPSUM BOARD

#8 SCREWS AT 12" O.C. ATTACHMENT

WALL INSULATION

BACKING STUD MAY BE 2 1/2" 26 GA

EXTERIOR WALL FINISH PER PLAN

FIGURE 11.33 Attaching gypsum board .

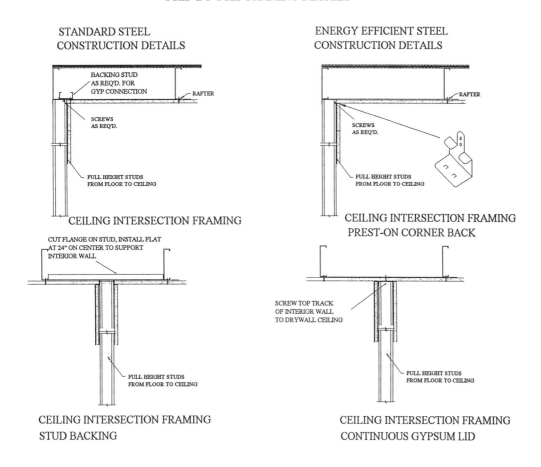

FIGURE 11.34 Attaching gypsum board to the ceiling.

Tips For Installing Interior Walls

The following are useful tips for installing interior walls in steel-framed houses:

- Shot pins or concrete nails can be used to attach the bottom plate to the floor. Screw the top plate directly to the gypsum board. While this method of attachment may not seem sturdy, it has been used with satisfactory results in commercial framing for years.
- The code permits the use of 3-5/8-in 25-ga or 26-ga studs to frame interior nonbearing walls, with the exception of both sides of door openings. Although some framers use wood studs, 20-ga studs are more than adequate. Install the hinge side of the frame first, making sure it's plumb. Leave the opposite king stud as a floating or slammer stud. Install the door, push the floating stud flush with the door, and screw into place. This method eliminates the need for wood shims to plumb the door.
- Many experienced steel framers put a single screw on one side of the interior wall studs where they attach to the top and bottom plate. Once again, this method has been used successfully for years in commercial framing. When you attach the gypsum board the drywall screws will anchor the stud in place.
- In Chapter 10 under the tool section, we indicate a brad nailer could be used to attach trim to walls. The brad generally penetrates the drywall and bends like a staple when it hits the stud. As a result, the brad normally doesn't penetrate the steel stud. While this makes a good connection, you must take care to properly align the trim and hold it firmly in place. Removing the trim tears the drywall and you generally cannot tighten the brad with a nail punch if the trim is loose.
- Ask your steel supplier for a box of the hole remnants left when punching out the studs. Use a minimum of 20-ga. The 18 and 16 gauge make excellent clips to tie the trusses to the top plate.

12

Energy-Efficient Details

In the first section of this book, we discussed various ways to make a steel home energy efficient. Since you probably won't be able to use all the suggestions we've outlined, this chapter lists the specific energy-saving components we feel are the most beneficial.

The effectiveness of these energy saving components is shown in the two energy analyses (using the Right-J software program) based on the sample house (Figure 11.17 in Chapter 11) found at the end of this chapter. The first calculation uses slab edge insulation, ducts in conditioned space, and "best construction methods." The result of omitting these items is shown in the second calculation, which uses standard construction methods. The effect on the energy use is outlined below (Figures 12.1 and 12.2).

Heating (Using Energy-Efficient Techniques)

Component	Btuh/Ft²	Btuh	% of Btuh	
Walls	5.6	4476	27.5	R-19 steel framed wall
Windows	35.8	3351	20.6	Dual vinyl
Doors	22.4	896	5.5	Steel foam core door
Ceilings	3.1	2507	15.4	R-22 steel truss
Floors	17.3	2005	12.3	Slab edge insulation R-7
Infiltration.	23.0	3070	18.8	Best construction methods
Ducts		0	0.0	In conditioned space
Total heat loss		**16305**		

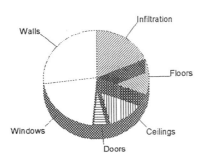

FIGURE 12.1 Distribution of heat losses in the sample house.

Heating (without energy efficient techniques)

Component	Btuh/Ft²	Btuh	% of Btuh	
Walls	5.6	4476	15.4	R-19 steel framed wall
Windows	35.8	3551	11.5	Dual vinyl
Doors	22.4	896	3.1	Steel foam core door
Ceilings	3.1	2507	8.6	R-22 steel truss
Floors	51.8	6015	20.7	No slab edge insulation
Infiltration	65.8	9210	31.6	Standard construction methods
Ducts		2645	9.1	In unconditioned space
Total heat loss		**29300**		

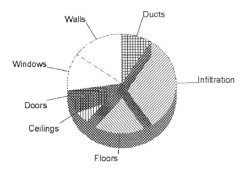

FIGURE 12.2 The distribution of heat losses in the sample house without slab edge insulation.

As you can see from the above comparison, the uninsulated slab and infiltration increased energy losses by **12,995 btuh.** Contrary to what people think, the major energy losses are not in the steel walls and trusses. As we stated at the beginning of the book, a well-designed home can more than compensate for energy lost through steel construction. The following are techniques I've used in my award-winning homes to make steel-framed homes *more* energy efficient than wood framed homes.

We'll start with construction techniques.

Structural Design

As discussed in Chapter 1, to reduce heat transfer and thermal bridging, steel must be kept to a minimum. Space studs as far apart as possible without affecting the integrity of the structure. Minimize the thickness of steel. Construct your own trusses. The use of Prest-on corner back clips eliminates the use of a stud backing at gable ends, interior walls, and building corners. In my homes, interior walls are constructed with 26-ga,2½-in studs, which saves steel while increasing living area. The top track is screwed directly to the gypsum board on the ceiling and wall, eliminating stud backing.

Insulation and foamed cement construction

- As seen from Figures 12.1 and 12.2, an insulated slab makes a tremendous difference. See Chapter 11 for construction details.
- OTW concrete panels, composed of Portland cement, fly ash, and polyester, were used in the award-winning homes. The foamed concrete not only provides insulation but adds thermal mass, increases indoor comfort levels, and reduces noise. In addition, the panels are fire-resistant (2-hour rated) and environmentally friendly. The foamed concrete blocks have been tested and proved to be a nontoxic and safe product. (See the following chapter for construction details.)

Infiltration and ventilation

- Install ducts within the conditioned space. Soffits can be used to keep the ducts out of the attic or crawl spaces and to add architectural interest. Figure 12.3 shows how air conditioning ducts are installed in a lowered ceiling in the hallway. If you must install ducts in unconditioned spaces, take great care to seal all penetrations of the ceiling, and provide connections for the ducts as outlined in Chapter 7. In the sample house supply ducts were installed in the conditioned space below the gypsum board ceiling. A dropped ceiling was used in the hallway, seven feet in height. Ducts should be insulated with R-4 insulation.

- Minimize mechanical and electrical runs in insulated wall sections. Eliminate holes in the top track by installing a conduit to a subpanel located in an interior wall. Install wiring from the subpanel, running utility lines in interior walls and electrical wiring in raceway conduits. If possible, install plumbing within interior walls. Use sconces whenever possible to minimize ceiling fixtures.

- Give careful attention to caulking and weather-striping. Remember, the overall thermal performance of the completed structure depends on the quality of the construction job.

- In the sample house, the water heater was located outside the conditioned area. As previously stated, if at all possible use a gas-fired water heater, even if you need to install a propane tank. Electric resistant water heaters are not energy efficient, and gas-fired water heaters require outside air for combustion. If the water heater compartment is not sealed from the conditioned space, it can be a major source of heat loss or gain and allow infiltration.

- Last but not least, provide adequate ventilation to the attic as required by code.

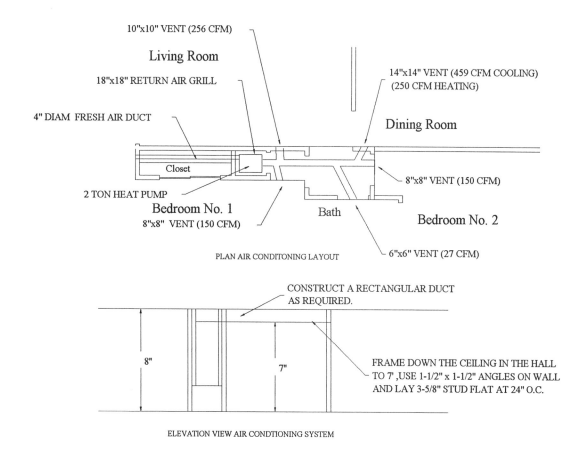

FIGURE 12.3 Creating a soffit for the air conditioning ducts

Windows and doors

In the award-winning homes, the following doors and windows were utilized:

- Vinyl windows filled with argon gas
- A hollow metal insulated door, with an R-value of 3 to 6
- An insulated foamcore R-5.2 garage door

Solar Design

- Provide thermal mass by using foamed concrete walls and an insulated R-7 concrete slab. Heavyweight materials can smooth out indoor temperature swings and make use of solar heat gain without causing overheating. An insulated slab is a good source of thermal mass; however an insulated slab loses effectiveness when covered by carpet. In the sample house, the kitchen, hallway, and bath are covered with vinyl flooring, for a total of about 250 ft^2 or 30 percent of the building area. Cover at least 20 percent of slab floors with hard surfaces, typically tile or linoleum. A larger percentage provides additional benefits in energy savings.
- To take advantage of solar heating, move as many windows to the south as possible. Reduce windows facing the east and west to avoid overheating, and the north to reduce winter heat loss. In an energy analysis, calculated to demonstrate the importance of proper orientation, we installed more west facing windows than necessary in the kitchen. As a result, the room-by-room calculation indicated 459 CFM of air conditioning was needed to cool the kitchen, while only 250 CFM was needed to heat the room. As President Harry Truman used to say, "If you can't take the heat, stay out of the kitchen." The appliances and people in the kitchen room generate heat. In order to provide an energy-efficient and comfortable house, carefully consider the orientation of the house and size the windows to balance the heat load. For example, in the sample house we could rotate the front to the north, and relocate one of the windows from the kitchen to the living room. (To cut the airflow in the winter, occupants will need to adjust the dampers in the supply vent for the kitchen.)
- Light colored vertical or white mini blinds should be installed in all the windows to block the sun, especially in warmer climates. These window coverings are economical and reduce the air conditioning load considerably.

Heating and Cooling Systems

- In the sample home, a heat pump with heat strips as back up heating was utilized, taking into consideration that resistance heat strips are not energy efficient with a COP of 1, considerably less than a heat pump. The insulated slab stores heat during the day and helps bridge cold nights when the temperature drops. A lower temperature at night is usually acceptable with a setback thermostat. Generally, a heat pump provides the majority of the heating required, and the resistance heat strips are used only in low temperatures. The minimum size normally available for a gas unit is about 40,000 btu, which oversizes the required heating unit for this size of house and has a tendency to remove moisture. Additionally, a gas unit requires combustion air from outside, which means penetrations in the ceiling for makeup air and venting the combustion. To provide the best construction, seal and insulate the area where the unit is installed. In larger homes, with heating demands of 40,000 or more, a gas-fired unit is probably the most economical. In this case, the unit should be installed in the garage or another area outside the conditioned space, to eliminate the need to seal the area. If heating is not the primary concern because you live in a warmer climate, we strongly recommend heat pumps as the most efficient method of providing heat.
- In hot climates, we recommend installing an evaporative cooler and separate thermostat with up-ducts directing the washed and conditioned air through the attic, which reduces the heat load on the ceiling dramatically.

- Do not oversize the air conditioner. As previously stated in Chapter 7, the old rule of thumb of installing one ton of air conditioning for each 400 ft² in a house is not valid with today's energy-efficient, "tighter" homes. Another rule of thumb, "bigger is better," not only raises utility bills unnecessarily but also can result in an uncomfortable house. These rules of thumb are like the thumb used for getting a free ride; they don't work today like in the past. Design an efficient and effective air conditioning system by using one of the many energy programs available or use professional services.

Lighting

We recommend the following:

- Interior fluorescent fixtures featuring electronic ballasts
- Hard-wired down lights and sconces designed for fluorescent lamps in bedrooms
- Photocell controls for outdoor safety lighting

Energy-Efficient Appliances

- A high-efficiency refrigerator
- A built-in microwave
- Natural gas water heating and laundry dryer

Landscaping

- Plant exterior trees to provide shading of south-facing windows in the summer. In cooler climates, deciduous trees should be used on the west, east, and south sides of a house to take advantage of the heat gain from the sun during the winter.
- Plant evergreen trees for windbreaks on the north side of the house.
- Minimize heat-absorbent material like asphalt, especially near the house.

Shower Heads and Water Closets

- If a low flush water closet does not empty the bowl with one flush, try a different brand. We had good results with a reasonably priced low-flow water closet manufactured by Western Pottery Company, Incorporated.
- Low-flow showerheads are required by most building codes, and flow reducers are generally built into the new showerheads.

Marketing Energy-Efficient Homes

By taking all the steps we've outlined in this chapter, you can build an energy efficient steel frame home. Not only will we conserve forests and save important energy resources, but also an energy-efficient home has significant potential as a marketing tool.

When marketing, present energy-efficient features along with other benefits of homes, such as the quality of materials, the floor plan, and low operating costs. Keep in mind, the home buyer must like the house for all its features, not simply because it's energy efficient. Builders have found the best approach is to first interest a person in buying the house, using standard sales approach and procedures, then try to close the sale by promoting energy features, without overdoing it. Let the open floor plan, natural lighting, and sense of comfort sell the house, with energy efficiency as an added benefit.

Energy analysis

Figure 12.4 shows the two energy calculations based on the sample house.

ENERGY CALCULATIONS USING ENERGY-EFFICIENT TECHNIQUES

RIGHT-J LOAD AND EQUIPMENT SUMMARY
File name: SAMPLE.bld   JUNE 97
For: RESIDENTIAL STEEL DESIGN AND CONSTRUCTION
       ENERGY EFFICIENCY, COST SAVINGS, CODE COMPLY

By: JOHN H HACKER PE
       2127 THORNHILL RD
       PUYALLUP   WA 98374

energy efficient steel frame house   Job #
                                       Wthr St._Louis_AP   MO
                                       Zone Entire House

WINTER DESIGN CONDITIONS		SUMMER DESIGN CONDITIONS	
Outside db:	6 Deg F	Outside db:	94 Deg F
Inside db:	70 Deg F	Inside db:	75 Deg F
Design TD:	64 Deg F	Design TD:	19 Deg F
		Daily Range	M
		Rel. Hum. :	50 %
		Grains Water	37 gr

HEATING SUMMARY		SENSIBLE COOLING EQUIP LOAD SIZING	
Bldg. Heat Loss	16305 Btuh	Structure	15513 Btuh
Ventilation Air	0 CFM	Ventilation	1045 Btuh
Vent Air Loss	0 Btuh	Design Temp. Swing	3.0 Deg F
Design Heat Load	16305 Btuh	Use Mfg. Data	n
		Rate/Swing Mult.	1.00
		Total Sens Equip Load	16558 Btuh

INFILTRATION		LATENT COOLING EQUIP LOAD SIZING	
Method	Simplified		
Construction Quality	Best	Internal Gains	920 Btuh
Fireplaces	0	Ventilation	1258 Btuh
		Infiltration	549 Btuh
	HEATING COOLING	Tot Latent Equip Load	2727 Btuh
Area (sq.ft.)	816 816		
Volume (cu.ft.)	6528 6528	Total Equip Load	19285 Btuh
Air Changes/Hour	0.4 0.2		
Equivalent CFM	44 22		

HEATING EQUIPMENT SUMMARY		COOLING EQUIPMENT SUMMARY	
Make DAY&NITE		Make DAY&NITE	
Model 4.83 kW		Model 693DN024-A	
Type elec		Type ashp	
Efficiency / HSPF	7.30	COP/EER/SEER	10.50
Heating Input	16499 Btuh	Sensible Cooling	16100 Btuh
Heating Output	16499 Btuh	Latent Cooling	7700 Btuh
Heating Temp Rise	18 Deg F	Total Cooling	23800 Btuh
Actual Heating Fan	830 CFM	Actual Cooling Fan	830 CFM
Htg Air Flow Factor	0.051 CFM/Btuh	Clg Air Flow Factor	0.053 CFM/Btuh
Space Thermostat		Load Sens Heat Ratio	91

MANUAL J: 7th Ed.   RIGHT-J: V1 3.0.00S/N 10238

Printout certified by ACCA to meet all requirements of Manual Form J

FIGURE 12.4a Energy calculations.

S/N 10238 RIGHT-J BUILDING ANALYSIS REPORT JUNE 97
File name: SAMPLE.bld Zone: Entire House

				Htg	Clg
Job #:					
For: RESIDENTIAL STEEL DESIGN AND CONSTRUCTION			Outside db	6	94
ENERGY EFFICIENCY, COST SAVINGS, CODE COMPLY			Inside db	70	75
			Design TD	64	19
			Daily Range	-	M
			Inside Humid.	-	50
By: JOHN H HACKER PE			Grains Water	-	37
2127 THORNHILL RD			INFILTRATION		
PUYALLUP	WA	98374	Method	Simplified	
			Const. qlty		Best
			Fireplaces		0

HEATING

Component	Btuh/SqFt	Btuh	% of Btuh
Walls	5.6	4476	27.5
Windows	35.8	3351	20.6
Doors	22.4	896	5.5
Ceilings	3.1	2507	15.4
Floors	17.3	2005	12.3
Infilt.	23.0	3070	18.8
Ducts		0	0.0

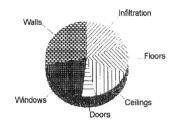

COOLING

Component	Btuh/SqFt	Btuh	% of Btuh
Walls	2.0	1581	10.2
Windows	46.2	4318	27.8
Doors	7.9	316	2.0
Ceilings	2.1	1684	10.9
Floors	0.0	0	0.0
Infilt.	3.4	456	2.9
Ducts		0	0.0
Int.Gains		4800	30.9

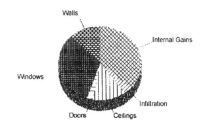

Clg Tons at	91 %	SHR =	1.6	Clg CFM/Ton	=	516
Clg Tons at	70 %	SHR =	2.0	Clg Tons at 400 CFM/Ton =		2.1
		Overall U-Value =	0.111			

Data entries checked.

MANUAL J: 7th Ed.   RIGHT-J: V3.0.00

FIGURE 12.4b Energy calculations.

```
S/N 10238          RIGHT-J COMPONENT CONSTRUCTION REPORT              JUNE 97
               File name:      SAMPLE.bld        Zone:  Entire House
Job #                                                                Htg    Clg
For:  RESIDENTIAL STEEL DESIGN AND CONSTRUCTION        Outside db     6     94
      ENERGY EFFICIENCY, COST SAVINGS, CODE COMPLY     Inside db     70     75
                                                       Design TD     64     19
                                                       Daily Range    -      M
                                                       Inside Humid.  -     50
By:   JOHN H HACKER PE                                 Grains Water   -     37
      2127 THORNHILL RD                                    INFILTRATION
      PUYALLUP                    WA   98374           Method          Simplified
                                                       Const. qlty          Best
                                                       Fireplaces              0
```

WALLS

CST DESCRIPTION	AREA	R-VAL	A/R	LOSS	GAIN
51A R-19 STEEL FRAMED WALL	795	11.4	69.9	4476	1581

WINDOWS

CST DESCRIPTION	AREA	R-VAL	A/R	LOSS	GAIN
61A DUAL VINYL	94	1.8	52.4	3351	4785

DOORS

CST DESCRIPTION	AREA	R-VAL	A/R	LOSS	GAIN
71A DEFAULT STEEL FOAM CORE DOOR	40	2.9	14.0	896	316

CEILINGS

CST DESCRIPTION	AREA	R-VAL	A/R	LOSS	GAIN
16E Under Unconditioned Room, R-22 Insul.	816	20.8	39.2	2507	1684

FLOORS

CST DESCRIPTION	AREA	R-VAL	A/R	LOSS	GAIN
22C Slab Floor on Grade, R-8,1 1/2"Edge Ins	116	3.7	31.3	2005	0

MANUAL J: 7th Ed.   RIGHT-J: V3.0.00

FIGURE 12.4c Energy calculations.

File name: SAMPLE.bld

		Htg	Clg
Job #:			
For:	RESIDENTIAL STEEL DESIGN AND CONSTRUCTION / ENERGY EFFICIENCY, COST SAVINGS, CODE COMPLY		

	Htg	Clg
Outside db	6	94
Inside db	70	75
Design TD	64	19
Daily Range	–	M
Inside Humid.	–	50
Grains Water	–	37
Method	Simplified	
Const. qlty		Best
Fireplaces		0

By: JOHN H HACKER PE
      2127 THORNHILL RD
      PUYALLUP   WA   98374

HEATING EQUIPMENT

Make	DAY&NITE	
Model	4.83 kW	
Type	elec	
Efficiency / HSPF	7.30	
Heating Input	16499	Btuh
Heating Output	16499	Btuh
Heating Temp Rise	18	Deg F
Actual Heating Fan	830	CFM
Htg Air Flow Factor	0.051	CFM/Btuh

Space Thermostat

COOLING EQUIPMENT

Make	DAY&NITE	
Model	693DN024-A	
Type	ashp	
COP/EER/SEER	10.50	
Sensible Cooling	16100	Btuh
Latent Cooling	7700	Btuh
Total Cooling	23800	Btuh
Actual Cooling Fan	830	CFM
Clg Air Flow Factor	0.053	CFM/Btuh

Load Sensible Heat Ratio   91

ROOM NAME	AREA SQ.FT.	HTG BTUH	CLG BTUH	HTG CFM	CLG CFM
Living Room	210	5032	3217	256	172
Dining Room	198	4912	8575	250	459
Bedroom 1	176	2950	1766	150	94
Bath and hall	64	529	217	27	12
Bedroom 2	168	2882	1739	147	93
Entire House p	816	16305	15513	830	830
Ventilation Air		0	1045		
Equip. @ 1.00 RSM			16558		
Latent Cooling			2727		
TOTALS	816	16305	19285	830	830

MANUAL J: 7th Ed.   RIGHT-J: V3.0.00

FIGURE 12.4d Energy calculations.

ENERGY CALCULATIONS WITHOUT ENERGY-EFFICIENT TECHNIQUES

S/N 10238   RIGHT-J BUILDING ANALYSIS REPORT   JUNE 97
            File name: sampleA.bld   Zone: Entire House

Job #:

For: RESIDENTIAL STEEL DESIGN AND CONSTRUCTION
      ENERGY EFFICIENCY, COST SAVINGS, CODE COMPLY

By: JOHN H HACKER PE
      2127 THORNHILL RD
      PUYALLUP   WA 98374

	Htg	Clg
Outside db	6	94
Inside db	70	75
Design TD	64	19
Daily Range	-	M
Inside Humid.	-	50
Grains Water	-	37
INFILTRATION		
Method	Simplified	
Const. qlty	Average	
Fireplaces		0

HEATING

Component	Btuh/SqFt	Btuh	% of Btuh
Walls	5.6	4476	15.4
Windows	35.8	3351	11.5
Doors	22.4	896	3.1
Ceilings	3.1	2507	8.6
Floors	51.8	6015	20.7
Infilt.	69.0	9210	31.6
Ducts		2645	9.1

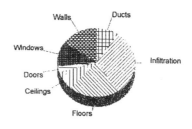

COOLING

Component	Btuh/SqFt	Btuh	% of Btuh
Walls	2.0	1581	10.4
Windows	46.2	4318	28.4
Doors	7.9	316	2.1
Ceilings	2.1	1684	11.1
Floors	0.0	0	0.0
Infilt.	8.5	1139	7.5
Ducts		1384	9.1
Int. Gains		4800	31.5

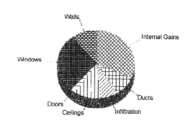

Clg Tons at 87 % SHR =   1.5   Clg CFM/Ton   =   558
Clg Tons at 70 % SHR =   1.8   Clg Tons at 400 CFM/Ton =   2.0
                      Overall U-Value =   0.145

Data entries checked.

WALLS
CST DESCRIPTION	AREA	R-VAL	A/R	LOSS	GAIN
51A R-19 STEEL FRAMED WALL	795	11.4	69.9	4476	1581

WINDOWS
CST DESCRIPTION	AREA	R-VAL	A/R	LOSS	GAIN
61A DUAL VINYL	94	1.8	52.4	3351	4785

DOORS
CST DESCRIPTION	AREA	R-VAL	A/R	LOSS	GAIN
71A DEFAULT STEEL FOAM CORE DOOR	40	2.9	14.0	896	316

CEILINGS
CST DESCRIPTION	AREA	R-VAL	A/R	LOSS	GAIN
16E Under Unconditioned Room, R-22 Insul.	816	20.8	39.2	2507	1684

FLOORS
CST DESCRIPTION	AREA	R-VAL	A/R	LOSS	GAIN
22A Slab Floor on Grade, No Edge Insulation	116	1.2	94.0	6015	0

FIGURE 12.4e Energy calculations.

13

Special Steel Frame Applications

Affordable Housing for Areas of High Wind, Seismic Activity, Floods, and Fire Hazards

As pointed out in Chapter 1, steel housing has some very important international implications, particularly in third world countries. Countries such as Mexico and Thailand have shown great interest in steel framing. We must also think about the importance of steel framing in providing low-cost housing in this country.

Several items must be considered when providing housing at a reasonable cost. Many areas that require affordable housing are in hot areas and may be subject to floods, high winds, seismic activity, and termites. In addition, water may not be available in developing nations for fire control. A properly designed steel house can mitigate these items at a reasonable cost, and at the same time provide comfort and energy efficiency. The following sample house (Figure 13.1) was designed for use in these areas.

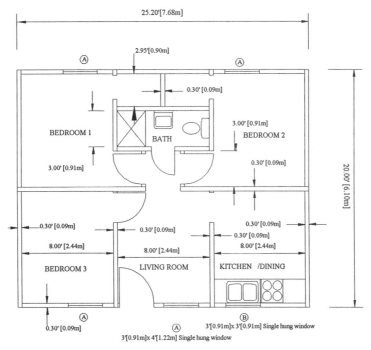

FLOOR PLAN - (CAN BE CONNECTED FOR UP TO FOUR UNITS)

FIGURE 13.1 Floor plan for low cost housing.

The homes don't utilize combustible materials, resulting in fire- and termite-resistant houses. The structural design is based on 125-mph winds, exposure C in seismic zone 4. The simple design can be built with local unskilled labor.

Structural framing details for low-cost housing

The framing details below (Figure 13.2) demonstrate how to build a low-cost steel frame house. Half-inch gypsum boards provide shear support for wind and seismic forces. See Figure 11.25 in Chapter 11 for specifications for required anchor bolt spacing, strapping, and required screw spacing.

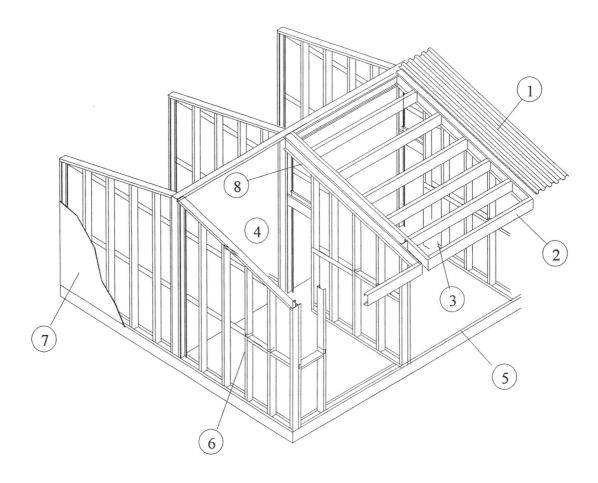

1. Corrugated metal roofing, galvanized 26 ga
2. 6" x 1 5/8" x 33 mil (20-ga) roof rafters at 24" O.C.
3. 1/2" gypsum board with R-19 insulation between rafters
4. 1/2" gypsum board typical wall - R-11 insulation
5. Concrete slab with 6" 20-ga steel form.
6. 3 1/2"or 3 5/8" x 1 5/8" 20-ga stud at 24" O.C.
7. 7/8" stucco over wire and paper per UBC specs
8. 3 5/8" x 1 1/4" 20-ga top track, with 6" x 1 1/4" track on side

FIGURE 13.2 Framing for low cost housing.

The following calculations (Figure 13.3 and 13.4) are for 125-mph winds, using various interior walls sheeted with ½-in gypsum board as shear walls to provide lateral support.

Date: 07/30/97   Page:

MULTI—STORY WIND FORCE DISTRIBUTION

—————————————— 1994 UBC WIND FORCE CALCULATION ——————————————

Exposure for Calc of Ce; Enter 2–>'B', 3–>'C', 4–>'D'	=	3			
	Cq: Press Coeff.	=	1.40	Basic Wind Speed	=	125.0 mph
Qs: Wind Stagnation Press 43.30 psf	Importance Factor	=	1.00	Parapet Height	=	2.00 ft

————————————————— TABLE OF WIND FORCES @ BUILDING LEVEL —————————————————

LEVEL #	Level Height (ft)	Exposed Width (ft)	Ce	Cq	Design Pressure (psf)	Lateral Force (k)	Story Shear (k)	Story Moment (k)
1	8.00	50.00	1.060	1.40	64.26	19.28		

Total Base Shear = 32.13 k
Base Moment = 154.22 k

FIGURE 13.3 Wind force distribution.

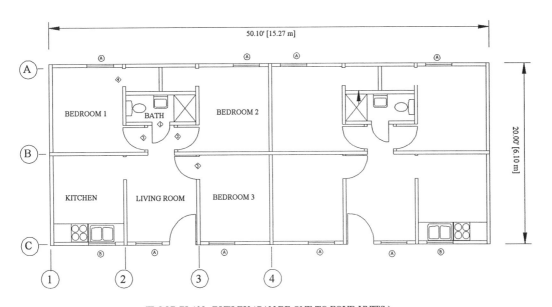

FLOOR PLAN - DUPLEX (CAN BE ONE TO FOUR UNITS)

REQUIRED SHEAR PANELS FOR 125 MPH WIND EXPOSURE "C"

(1)   4.38' X 386# = 1690 / 20 = 84# per L.F.   Use 1/2" gyp #6 screws at 7" o.c.

(2)(3) 8.38' X 386# = 3235 / 12 = 269# per L.F.   Use 1/2" gyp (2 sides) #6 screws at 4" o.c.

(4)   8.38' X 386# = 3235 / 20 = 161# per L.F.   Use 1/2" gyp (2 sides) #6 screws at 7" o.c.

(A)   23 sq.ft. X 65# = 1495 / 38 = 39# per L.F.   Use 1/2" gyp #6 screws at 7" o.c.

(B)   55 sq ft X 65# = 3575 / 44 = 82# perp L.F.   Use 1/2" gyp #6 screws at 7" o.c.

(C)   23 sq ft X 65# = 1495 / 26 = 57# per L.F.   Use 1/2" gyp #6 screws at 7" o.c.

FIGURE 13.4 Shear panels required for a low-cost duplex.

Flood-proofing the building

In the United States, areas that are subject to flooding are shown on FIRM (flood insurance rate maps) provided by HUD under the National Flood Insurance Program. The first floor of houses built in these areas must be elevated to above the 100-year flood elevation. For information, call the National Flood Insurance Program at (800) 638-6620 or (800) 424-8872.

Figure 13.5 details a method used on hundreds of homes in the southwestern United States to provide protection. (See Chapter 11 for details on constructing and insulating the foundation.) Using a concrete floor on imported fill material can alleviate several problems. Houses on raised foundations may be subject to flotation, and the crawl space may fill with ground water. The sloped earth berm around the foundation stem walls provides a measure of protection against scouring to the footing from floodwaters.

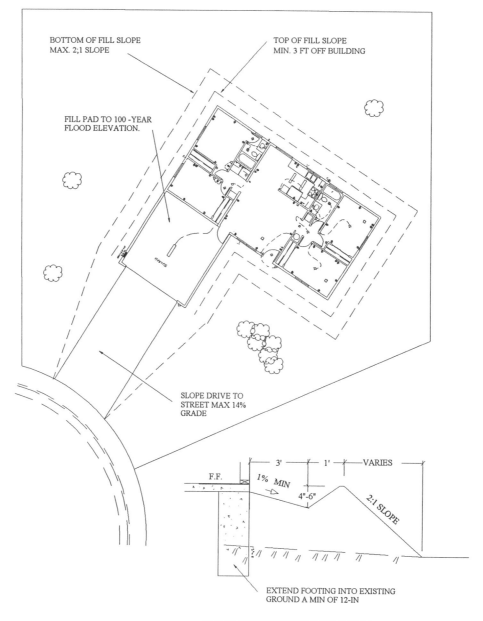

FIGURE 13.5 Foundation design for areas subject to flood hazard.

Steel framing in foreign countries

In exporting steel-framing technology to other countries, I discovered local people found hollow walls of steel frame buildings objectionable. When Noble Steel Tech in Thailand built several model steel frame homes, they noticed people knocking on the walls. People in Thailand are accustomed to the "solid" feel of concrete or block; however, concrete/block homes are not energy efficient and require a long time to build. If resonant walls can be avoided, there is a large market for steel frame homes,

Noble Steel Tech built two additional models using the OTW wall panel system (see Chapters 3 and 13 for more details) and found this type of construction acceptable. The product looks like concrete and produces a solid wall section. Noble was so impressed with the results, they obtained a license, built a factory, and are now manufacturing panels in Thailand.

Using the OTW (Other Than Wood) building system (U.S. Pat No. 5,596,860)

The next few pages provide construction details for the OTW Building System. The OTW panels are manufactured in 8-, 9-, and 10-ft heights with an electric conduit built in 2 ft from the top and bottom of each panel. The panels slip into the bottom track and are aligned like dominoes. The panels can be tilted in place by two workers with a third person guiding the panel and securing it with screws. Figure 13.6 below is a section for a typical two-story apartment building. The use of the OTW panels (with a single wall) between the units provide a 50-decibel reduction of sound and a fire rating of 2 hours using one layer of 5/8-in gypsum.

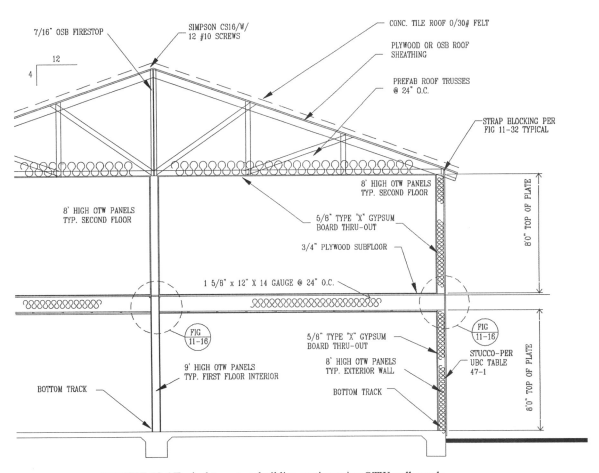

FIGURE 13.6 Typical two-story building section using OTW wall panels.

Headers are easy to install in the wall panels using the following details (Figure 13.7). Size the headers per Table 16 through 22 of Appendix B, or Tables 603.5a to 603.5d of the CABO code.

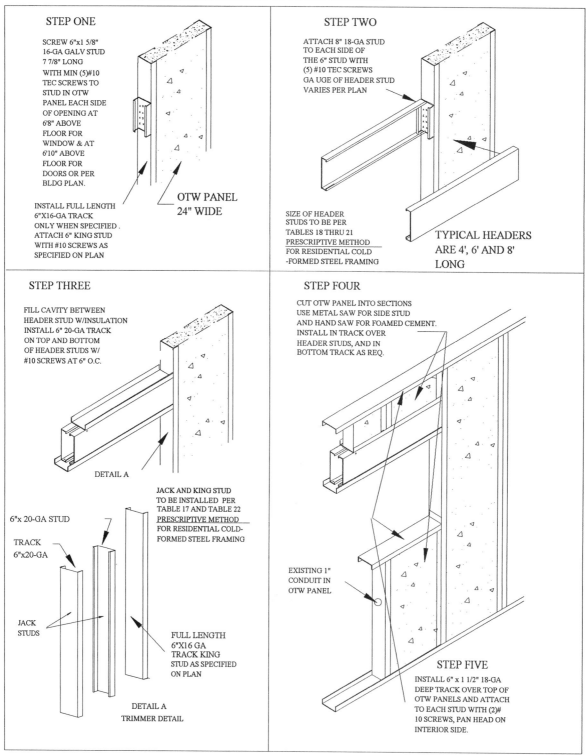

FIGURE 13.7 Installing headers with the OTW wall panels.

To cut panels for headers over windows and doors (Figure 13.8), cut the metal stud on both sides of the panel using a saw with a metal cutting blade. Mark the cut with a scribe in the foamed concrete. Use a wood guide on the backside and cut the foamed concrete with a drywall saw, producing a straight and clean cut. Cut the panel placed under the window framing from each end of the OTW panel, so the conduit is 24-in off the floor and aligned with the conduit in the adjacent panel. Cut sections for over the door and window framing from the center of the panel.

FIGURE 13.8 A Habitat for Humanity volunteer cutting an OTW panel.

The OTW building system was developed without grants from any government agency. However, one organization, Habitat for Humanity, the Coachella Valley chapter, provided volunteers who worked hard and long in perfecting the various details used in the building system (Figure 13.9). The work was accomplished without any promise of financial gain on their part from the future commercial sale of the building system. As indicated in a letter from the cofounder of Habitat for Humanity (Figure 13.10), the organization's motive is "to provide decent affordable housing for those who otherwise would not have a good place in which to live," striving to do so in an energy-efficient and environmentally sound way. It is, in part, because of their encouragement and support that this book is written.

FIGURE 13.9 Four-bedroom Habitat for Humanity house in Cathedral City, California.

Habitat for Humanity International

Building houses in partnership with God's people in need

August 18, 1997

I have seen several OTW houses under construction as well as completed in the Coachella Valley near Palm Springs, California. An engineer/building, John Hacker, has worked closely on a volunteer basis with that Habitat affiliate in developing the OTW fly ash-formed concrete panels for Habitat homes.

Last November, I organized an all-women build with more than 40 women participating from all over the country. Only the foundation existed when we arrived on a Sunday afternoon. Six days later, we had finished a 1250 sq. foot OTW home, including landscaping. My thought was if a group of a few skilled, but mostly unskilled, women can construction an OTW house, then that should prove the building system volunteer-friendly for Habitat's use. It took four women to lift the 150-pound panels in place. Also, the steel trusses were easier since they weigh 60% less than wood trusses.

The main reasons I am sold on the OTW product (and believe me, I don't own stock in the company or any company for that matter) is because it produces an extremely energy-efficient, fire-proof, earthquake-resistant, sound-proof home by making use of waste or recycled materials. The waste product is fly ash which is the residue from the coal industry and the recycled steel comes from things like old junk cars.

The cost of an OTW house is very compatible with traditional wood frame construction. Any type of exterior can be applied...from stucco to brick to aluminum siding. When the house is completed, it <u>looks</u> no different from a wood frame house. The big thing the family living in the house notices is untility bills are less than half all year around.

On the all women build, our biggest challenge was learning to use electric screw guns in place of hammers. We tried many different ones and found the Milwaukee 4000 model worked the best for us.

Thank you so much for your interest in providing solid, decent and affordable housing for those who otherwise would not have a good place in which to live.

In exciting Christian partnership,

Linda Fuller

FIGURE 13.10 A letter from the cofounder of Habitat for Humanity.

The OTW building system has been used in Mexico, Thailand, as well as around the United States in Arizona, California, Washington, Oregon, and Pennsylvania (Figure 13.11). Additional projects are under construction in Nevada and Georgia.

FIGURE 13.11 OTW house in Southern California.

If you'd like more information on the OTW Building System contact: Omega Transworld, Ltd., 2400 Leechburg Road, New Kensington PA 15068, 1-800-541-1575 or 412-339-2511, fax 412-339-9880.

Conclusion

As we've stressed throughout this book, building with steel can be both energy efficient and cost-effective. Despite popular belief, a well-designed home can more than compensate for energy lost through steel construction. Builders do not have to invest an extraordinary amount of money to achieve efficiency. To keep expenses down, always keep cost versus benefit in mind. Generally a modest investment in energy-efficient features, when used in combination with low-cost energy-saving design and siting features, reduces net monthly electrical and fuel outlays considerably.

Included in Appendix A through D are public domain publications downloaded from the Internet. These documents are invaluable when designing energy-efficient steel homes. Rather than try to duplicate the excellent work professionals produced in these documents, it is our desire to share their fine efforts by making them available in this book. We commend their work and have directed our efforts to illustrate and explain various ways you can use the information that they have provided.

A

MECcheck Manual

US DEPARTMENT OF ENERGY
BUILDING STANDARDS AND GUIDELINES PROGRAM

1995
MEC*check*™ Manual

1995 Model Energy Code Compliance Guide
Version 2.0

December 1995

MEC*check*™ was developed by the Building Standards and Guidelines Program at
Pacific Northwest National Laboratory (PNNL) for use by the U.S. Department of
Housing and Urban Development (HUD) and the Rural Economic and Community
Development (RECD, formerly Farmer's Home Administration) under contract with
the U.S. Department of Energy's Office of Codes and Standards. Pacific Northwest
National Laboratory is operated by Battelle Memorial Institute for the U.S. Department
of Energy under Contract DE-AC06-76RLO 1830.

"A Better Climate for Jobs"

Acknowledgements Our thanks to the following individuals who contributed to the development of the MEC *check*™ materials: Stephen Turchen, U.S. Department of Energy, for his direction and guidance; William Freeborne, U.S. Department of Housing and Urban Development, Richard Davis and Sam Hodges, Rural Economic and Community Development, Ronald Burton, National Association of Home Builders, Ronald Nickson, National Multi Housing Council, Dick Tracey, D&R International, Ken Eklund and Pat Gleason, Idaho Department of Water Resources, and Dave Conover, Pacific Northwest National Laboratory, for their review and thoughtful suggestions; Linda Connell, Craig Conner, and Eric Makela, Pacific Northwest National Laboratory, for development of the materials. Thank-you also to the many persons who contributed their comments and support.

Contents

Figures

Tables

Chapter 1
Overview

What is the Model Energy Code?

The 1995 Model Energy Code (MEC)[a] contains energy-related building requirements applying to many new U.S. residences. The U.S. Department of Housing and Urban Development (HUD) loan guarantee program requires compliance with the MEC. The Rural Economic and Community Development (RECD, formerly the Farmer's Home Administration) loan guarantee program requires that single-family buildings comply with the MEC. Several states have also adopted the MEC as their residential energy code.

A major focus of the MEC provisions is on the building envelope insulation and window requirements,[b] which are more stringent in colder climates. Other requirements focus on the heating and cooling system (including ducts), water heating system, and air leakage.

What Buildings Must Comply?

The MEC applies to new residential buildings, three stories or less in height, and additions to such buildings. Residential buildings are defined as detached one- and two-family buildings (referred to as single-family buildings) and multifamily buildings (such as apartments, condominiums, townhouses, and rowhouses). Multifamily buildings have three or more attached dwelling units (see Appendix A for definitions of single-family and multifamily buildings and dwelling unit). Throughout this manual, generic references to "building(s)" signify residential buildings three stories or less in height.

Exemptions: The following building categories are exempted from the provisions of the MEC:

- existing buildings

- very low-energy buildings (<3.4 Btu/h·ft^2 or 1 W/ft^2)

- buildings (or portions of buildings) that are neither heated nor cooled

- buildings designated as historical.

---

[a] The MEC is maintained by the Council of American Building Officials (CABO). CABO is an "umbrella" organization consisting of three U.S. model code groups: the Building Officials and Code Administrators International, Inc. (BOCA), the International Conference of Building Officials (ICBO), and the Southern Building Code Congress International, Inc. (SBCCI). This manual is applicable only to the 1995 MEC and refers to the 1995 MEC simply as the MEC.

[b] The building envelope consists of all building components that enclose heated and/or cooled spaces. The phrase "insulation and window requirements" is used throughout this manual to represent the building envelope thermal requirements, as opposed to requirements relating to the heating and cooling system, water heating system, and air leakage.

About This Manual

This manual is applicable only to the 1995 MEC and refers to the 1995 MEC simply as the MEC. It provides guidance on how to meet the MEC requirements using the MEC*check*™ materials. Making the MEC simple and understandable was the major motivation for developing the MEC*check*™ materials. The desire for simplicity and clarity led to changes in format, deletion of redundant material, and deletion of materials that had no impact. If you are familiar with the MEC, you will note that this manual differs significantly in format from the MEC.

The MEC*check*™ materials were created for HUD and RECD. Check with your building department or other state or local building code enforcement authority to verify that the MEC*check*™ materials are accepted in your jurisdiction, because some of the requirements may be superseded by state laws or local ordinances.

It is not necessary to have a copy of the MEC to use this manual. Although Chapter 2 lists MEC section numbers for cross reference, it is not necessary to refer to the referenced sections. All references to figures and tables refer to figures and tables located in this manual unless specifically stated otherwise.

What's in This Manual?

Chapters 1 and 2 of this manual apply to all residential buildings and should be read by all users of MEC*check*™ materials. Home builders and designers can then use one of the three MEC*check*™ approaches discussed in Chapters 3, 4, and 5 to show compliance with the insulation and window requirements. Chapters 6 and 7 will be of particular interest to code officials.

Chapter 2, *Basic Requirements*, discusses all of the basic requirements except for the insulation and window requirements (which are covered in Chapters 3, 4, and 5). The basic requirements represent minimum criteria that must be met regardless of which insulation compliance approach you choose. These criteria include provisions that limit air leakage through the building envelope and regulate heating and cooling systems and duct insulation levels.

Chapter 3, *Prescriptive Package Approach*, describes the simplest of the three compliance approaches. With this approach, you select a package of insulation and window requirements from a list of packages developed for a specific climate zone. Each package specifies insulation levels, glazing areas, glazing U-values,[a] and sometimes heating and cooling equipment efficiency. Once selected, simply meet or exceed all requirements listed in the package to achieve compliance. Few calculations are required. Chapter 3 briefly summarizes the prescriptive package approach. See the separate document on the prescriptive packages for a detailed description.

---

[a]   U-values are a measure of how well a material or series of materials conduct heat (higher U-values indicate more heat loss). For door and window assemblies, the U-value is the reciprocal of the R-value (U-value=1/R-value).

Chapter 4, *Trade-Off Approach*, describes a "pencil-and-paper" compliance approach. The trade-off approach enables you to trade off insulation and window efficiency levels in different parts of the building. You can trade off ceiling, wall, floor, basement wall, slab-edge, and crawl space wall insulation; glazing and door areas; and glazing and door U-values. The trade-off approach calculates whether your home as a whole meets the overall MEC insulation and window requirements.

Chapter 5, *Software Approach*, describes the most flexible of the three compliance approaches. The MEC*check*™ software is designed to run on most DOS-based computers and allows trade-offs between all building envelope components and heating and cooling equipment efficiencies. With minimal input, you can quickly compare different insulation levels to select a package that works best for your proposed building. Unlike the prescriptive package and trade-off approaches, the software approach enables you to trade off basement wall, slab-edge, and crawl space wall insulation depth as well as insulation R-value. The software automatically generates a report that can be submitted to enforcement personnel to document compliance. Chapter 5 briefly summarizes the software approach. See the *Software User's Guide* for a detailed description.

Chapter 6, *Plan Check*, provides guidance on verifying building plans and specifications to ensure that the proposed building complies with the MEC*check*™ requirements and, therefore, with the MEC.

Chapter 7, *Field Inspection*, provides guidance for ensuring that required energy conservation measures are properly installed in the building and are in accordance with the building plans and specifications.

Appendix A, *Definitions*, provides definitions for some of the terms used in this manual.

Appendix B, *The Building Envelope*, discusses which building components make up the building envelope and contains a table that is useful in determining which elements of a building are considered to be ceiling, wall, and foundation components.

Appendix C, *Counties By Climate Zone*, lists counties by state and their corresponding climate zone number. Some requirements vary depending on the climate zone in which your building is located.

Who Should Use This Manual?

This manual is designed as a stand-alone document to guide you through the MEC*check*™ compliance process. All necessary compliance forms, reference materials, and explanations are included (although the prescriptive packages and software are bundled separately from this manual). The manual is designed to provide guidance to builders and designers, plan check personnel, and field inspectors.

Builders and Designers can follow each step of the compliance process presented in this manual. Chapter 2 describes MEC requirements that must be satisfied by all residences. Chapters 3, 4, and 5 offer a choice of approaches, any of which can be used to show compliance with the insulation and window requirements of the MEC.

Plan Check Personnel can use Chapter 6, *Plan Check*, as a guide to ensure that building plans and specifications comply with the MEC. If questions arise, the plan reviewer can trace the compliance steps used by the applicant and reference the steps in this manual.

Field Inspectors and Site Superintendents can use Chapter 7, *Field Inspection*, to ensure that all of the applicable MEC requirements have been installed in a building. The features that meet these requirements should be included on the building plans or specifications and on compliance forms. Chapter 2 will also be of interest to field inspectors and site superintendents. Chapter 2 describes the features that must be installed in the building regardless of the compliance approach chosen.

MEC*check*™ Forms and Checklists

Several forms, worksheets, and lists are included with this manual to help determine and document compliance. You may want to remove them for reference while using the manual. You may make multiple copies of the forms and distribute them freely. Alternative forms that provide the same information may also be used if they are approved by your jurisdiction.

The following is a brief description of the forms and lists provided with this manual:

- *Summary of Basic Requirements* - Summarizes the MEC mandatory requirements applicable to all buildings as outlined in Chapter 2. The features that meet these requirements should be documented on the building plans or specifications.

- *Prescriptive Package Worksheet* - Documents that MEC insulation and window requirements are met using the prescriptive package approach described in Chapter 3. The *R-Value/U-Value Weighted Average Worksheet*, on the back side of the *Prescriptive Package Worksheet*, is used to average R-values when more than one insulation level is used in a component and to average U-values when more than one window or door U-value is used. (Although the prescriptive packages are distributed separately from this manual, the *Prescriptive Package Worksheet* is included here for easy removal and reference.)

- *Trade-Off Worksheet* - Documents that MEC insulation and window requirements are met using the Trade-Off approach described in Chapter 4.

- *Field Inspection Checklist* - Guides field inspectors in verifying that required energy efficiency features are installed in a building and are in accordance with the building plans and specifications.

• *Energy Label* - Describes the energy efficiency features installed in the residence. This label is optional. It may be posted at the building site or provided to the home buyer.

MEC*check*™ Compliance Process

Figure 1-1 illustrates the steps you should follow to determine compliance with the MEC.

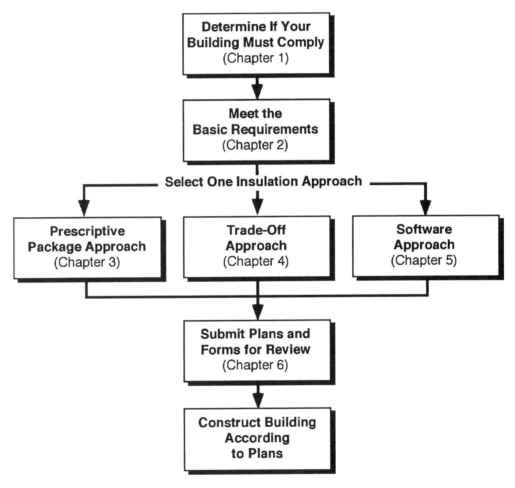

Figure 1-1. MECcheck™ Compliance Path

Step 1:   **Determine If Your Building Must Comply With the MEC.** (See *What Buildings Must Comply?* in this chapter.)

Step 2:   **Meet the Basic Requirements.** The basic requirements discussed in Chapter 2 must be incorporated into the design.

Step 3:   **Use One of Three Compliance Approaches for Insulation and Windows.** Select one of the three compliance approaches described in Chapters 3, 4, and 5. Examining the prescriptive packages for the building location will give an idea of the insulation requirements. Use the selected approach to determine the insulation and window requirements. Document compliance on the form(s) provided for the selected approach.

Step 4:   **Submit Building Plans and Compliance Forms for Plan Review.** Submit MEC*check*™ forms or their equivalent, building plans, and specifications for plan review. The compliance forms must match the building plans and specifications.

Step 5:   **Construct the Building According to Approved Plans.** In most jurisdictions, construction may begin after a building permit is issued. It is important to have the approved set of plans and specifications at the job site for use by the field inspector. MEC*check*™ forms or their equivalent must be re-submitted if changes from the approved plans or specifications are made that increase the glazing area, decrease insulation R-values, or decrease equipment efficiencies of the building.

Chapter 2
Basic Requirements

The MEC specifies basic requirements that are mandatory for all buildings. Some of these requirements apply to the heating and cooling system (including ducts), hot water system, and electrical system. Other requirements apply to material and equipment identification and to sealing the building envelope. This chapter discusses the MEC basic requirements, except for the insulation and window requirements (which are covered in Chapters 3, 4, and 5). Each requirement in this chapter lists the corresponding MEC section number as a reference.

Figure 2-1 graphically illustrates several basic requirements. Refer to the *Summary of Basic Requirements* provided with this manual for a one-page listing of the requirements discussed in this chapter. The *Summary of Basic Requirements* and several other forms are bound separately from the manual for easy removal and copying.

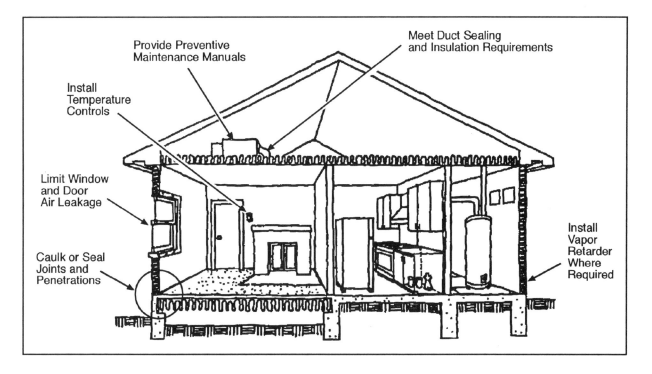

Figure 2-1. Some of the Basic Requirements

Building Envelope

Air Leakage

(Section 502.3.3 and 502.3.4) All joints and penetrations in the building envelope that are sources of air leakage must be caulked, gasketed, weatherstripped, or otherwise sealed in an approved manner.

The following areas should be sealed:

- exterior joints around window and door frames

- between wall sole plates, floors, and exterior wall panels

- openings for plumbing, electricity, refrigerant, and gas lines in exterior walls, floors, and roofs

- openings in the attic floor (such as where ceiling panels meet interior and exterior walls and masonry fireplaces)

- service and access doors or hatches

- all other similar openings in the building envelope

- recessed lighting fixtures.

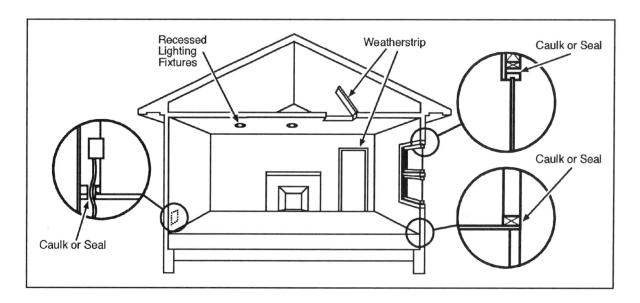

Figure 2-2. Typical Openings That Should Be Sealed

(Section 502.3.2) The maximum air leakage rates[a] for manufactured windows are given in Table 2-1. Windows and doors certified by an accredited lab (such as the

---

() When tested at 1.567 lb/ft^2 pressure difference.

National Wood Window and Door Association [NWWDA] or the Architectural
Aluminum Manufacturers Association [AAMA]) meet these requirements and will be
labeled. For non-certified doors and windows, check manufacturer's test reports to
verify compliance with these air leakage requirements.

Table 2-1. Maximum Leakage Rates for Manufactured Windows and Doors

Frame Type	Windows (cfm/ft of Operable Sash Crack)	Doors (cfm per ft² of Door Area)	
		Sliders	Swinging
Wood	0.34	0.35	0.5
Aluminum	0.37	0.37	0.5
PVC	0.37	0.37	0.5

Vapor Retarders

(Section 502.1.4) Vapor retarders must be installed in all non-vented framed ceilings,
walls, and floors. Non-vented areas are framed cavities without vents or other
openings allowing the free movement of air. The vapor retarder must have a perm
rating of 1.0 or less[a] and must be installed on the "warm-in-winter side" of the
insulation (between the insulation and the conditioned space).

Exemptions: The above requirements do not apply to the following locations (your
zone number can be found in Appendix C or on the state map included with the
prescriptive packages):

- Texas . . . . Zones 2-5
- Alabama, Georgia, N. Carolina, Oklahoma, S. Carolina . . . . Zones 4-6
- Arkansas, Tennessee . . . . Zones 6-7
- Florida, Hawaii, Louisiana, Mississippi . . . . All Zones

Vapor retarders are not required where moisture or its freezing will not damage the
materials.

Materials and Equipment Information

(Section 104.2) Insulation R-values and glazing and door U-values must be clearly
marked on the building plans or specifications. If two or more different insulation
levels exist for the same component, record each level separately on the plans or
specifications. For example, if the walls adjacent to the garage have less insulation
than the other walls, both insulation levels must be noted. If credit is taken for high-

---

[a]   Tested in accordance with the American Society for Testing and Materials (ASTM) Standard E96-80.

efficiency heating or cooling equipment, the equipment efficiency, make and model number must also be marked on the plans or specifications.

(Section 102.1) Materials and equipment must be identified so that compliance with the MEC can be determined. There are several ways to label materials and equipment to satisfy this requirement.

- Provide labels on all pertinent materials and equipment. For example, the R-value of the insulation is often pre-printed directly on the insulation or can be determined from a striping code. Window U-values are often included on the manufacturer label posted directly on the window.

- Provide contractor statements certifying the products they have installed. For example, the insulation contractor should certify the R-value of the installed insulation.

- An optional *Energy Label* is included with this manual. Materials and equipment can be identified on this label which should then be posted in the residence (e.g., on the main fuse box, on a garage wall, in the utility room) to document the energy efficiency features of the building.

For blown or sprayed insulation, the initial installed thickness, the settled thickness, the coverage area, and the number of bags must be clearly posted at the job site. In attics, thickness markers must be placed at least once every 300 ft².

Check with your local building official to determine what is required in your jurisdiction.

(Section 102.2) Manufacturer manuals for all installed heating and cooling equipment and service water heating equipment must be provided.

Heating and Cooling

Heating and Cooling Equipment Efficiencies

The MEC defines heating and cooling equipment efficiency requirements. However, federal regulations have restricted manufactured equipment efficiency minimums to levels above these MEC requirements.[a] Because new equipment with efficiencies below the MEC requirements can no longer be manufactured, these requirements have been omitted from the MEC*check*™ manual.

Duct Insulation

---

0   The National Appliance Energy Conservation Act of 1987 (NAECA) sets federally mandated equipment efficiency minimums for most residential equipment. NAECA does not, however, specify minimum efficiencies for groundwater-source heat pumps, duct furnaces, and unit heaters. For groundwater-source heatpumps, the MEC specifies a minimum COP of 3.4 (heating mode) and a minimum EER of 11.0 (cooling mode) for high temperature rating conditions (70°F). For low temperature rating conditions (50°F), the MEC specifies a minimum COP of 3.0 (heating mode) and a minimum EER of 11.5 (cooling mode). See the MEC for requirements for duct furnaces and unit heaters.

(Section 503.7.1) Supply and return ductwork for heating and cooling systems located in unconditioned spaces (spaces neither heated nor cooled) must be insulated to the minimum R-value specified in Table 2-2. Select the zone number for your building location and find the R-value requirement from Table 2-2 based on where the ducts are located. Your zone number can be found in Appendix C or on the state map included with the prescriptive packages.

When ducts are located in exterior building cavities, either

- the full insulation R-value requirement for that building component must be installed between the duct and the building exterior, in which case the ducts do not require insulation, or

- the ducts must be insulated to the duct R-value requirement given in Table 2-2 and the duct area must be treated as a separate component. For example, if ducts insulated to R-6 are located in an exterior wall insulated to R-19, the area of the wall minus the duct area is a wall component with R-19 insulation, and the area of the ducts is a wall component with R-6 insulation.

Table 2-2. Duct Insulation R-Value Requirements

Zone Number	Ducts in Unconditioned Spaces (i.e. Attics, Crawl Spaces, Unheated Basements and Garages, and Exterior Cavities)	Ducts Outside the Building
Zones 1-4	R-5	R-8
Zones 5-14	R-5	R-6.5
Zone 15-19	R-5	R-8

Exceptions: Duct insulation is not required in the following cases:

- within heating, ventilating, and air conditioning (HVAC) equipment

- for exhaust air ducts

- when the design temperature difference between the air in the duct and the surrounding air is 15°F or less.

Additional insulation with vapor barrier must be provided if condensation will create a problem.

Duct Construction

(Section 503.8.2) Ducts must be sealed using mastic with fibrous backing tape. For fibrous ducts, pressure-sensitive tape may be used. Other sealants may be approved by the building official. Duct tape is not permitted.

(Section 503.5) The HVAC system must provide a means for balancing air and water systems. For air systems, this requirement can be met by installing manual dampers at each branch of the ductwork or by installing adjustable registers that can constrict the airflow into a room. For water systems, balancing valves can be installed to control the water flow to rooms or zones.

Temperature Controls

(Section 503.6.3.1) For one- and two-family buildings, at least one thermostat must be provided for each separate system (heating, cooling, or combination heating and cooling). Electric baseboard heaters can be individually controlled by separate thermostats or several baseboard heaters can be controlled by a single thermostat.

(Section 503.6.3.2) For multifamily buildings, each dwelling unit must have a separate thermostat and a readily accessible, manual or automatic means to restrict or shut off the heating and/or cooling input to each room must be provided. Operable diffusers or registers that can restrict or shut off the airflow into a room meet this requirement. At least one thermostat must be provided for each system or each zone in the non-dwelling portions of multifamily buildings. For example, separate systems serving interior corridors or attached laundry rooms must have their own thermostat.

(Section 503.6.1) Each heating and cooling system must have a thermostat with at least the following range:

- heating only   55°F to 75°F

- cooling only   70°F to 85°F

- heating and cooling   55°F to 85°F
                             The thermostat must be capable of operating the
                             system heating and cooling in sequence (i.e.,
                             simultaneous operation is not permitted).

Heat Pump Thermostats

(Section 503.3.2.2) Heat pump installations must include a thermostat that can prevent the back-up heat from turning on when the heating requirements can be met by the heat pump alone. A two-stage thermostat that controls the back-up heat on its second stage meets this requirement.

HVAC Piping Insulation

(Sections 503.9) All HVAC piping (such as in hydronic heating systems) installed in unconditioned spaces and conveying fluids at temperatures greater than 120°F or chilled fluids at less than 55°F must be insulated to the thicknesses specified in Table 2-3. Pipe insulation is not required for piping installed within HVAC equipment.

Table 2-3. Minimum HVAC Piping Insulation Thickness[a]

Piping System Types	Fluid Temp Range (°F)	Insulation Thickness in Inches by Pipe Sizes[b]			
		Runouts 2 in.[c]	1 in. and Less	1.25 in. to 2 in.	2.5 in. to 4 in.
Heating Systems					
Low Pressure/Temperature	201-250	1.0	1.5	1.5	2.0
Low Temperature	120-200	0.5	1.0	1.0	1.5
Steam Condensate (for feed water)	Any	1.0	1.0	1.5	2.0
Cooling Systems					
Chilled Water	40-55	0.5	0.5	0.75	1.0

(a) The pipe insulation thicknesses specified in this table are based on insulation R-values ranging from R-4 to R-4.6 per inch of thickness. For materials with an R-value greater than R-4.6, the insulation thickness specified in this table may be reduced as follows:

$$\text{New Minimum Thickness} = \frac{4.6 \times \text{Table 2-3 Thickness}}{\text{Actual R-Value}}$$

For materials with an R-value less than R-4, the minimum insulation thickness must be increased as follows:

$$\text{New Minimum Thickness} = \frac{4.0 \times \text{Table 2-3 Thickness}}{\text{Actual R-Value}}$$

(b) For piping exposed to outdoor air, increase thickness by 0.5 in.
(c) Applies to runouts not exceeding 12 ft in length to individual terminal units.

Service (Potable) Water Heating

Swimming Pools

(Section 504.5) All heated swimming pools must be equipped with an on/off pool heater switch mounted for easy access. Heated pools require a pool cover unless over 20% of the heating energy is from non-depletable sources (such as solar heat).

(Section 504.5.3) All swimming pool pumps must be equipped with a time clock.

Circulating Service Hot Water Systems

(Section 504.6) Circulating hot water systems must have automatic or manual controls that allow the pumps to be conveniently turned off when the hot water system is not in operation.

(Section 504.7) Piping in circulating hot water systems must be insulated to the levels specified in Table 2-4 unless an engineering calculation is provided that demonstrates that insulation will not reduce the annual energy requirements of the building.

Table 2-4. Minimum Insulation Thickness for Recirculation Piping

Heated Water Temperature (°F)	Insulation Thickness in Inches by Pipe Sizes[a]			
	Non-Circulating Runouts	Circulating Mains and Runouts		
	Up to 1 in.	Up to 1.25 in.	1.5 - 2.0 in.	Over 2 in.
170-180	0.5	1.0	1.5	2.0
140-160	0.5	0.5	1.0	1.5
100-130	0.5	0.5	0.5	1.0
(a) Nominal pipe size and insulation thickness.				

Electrical

(Section 505.1) All dwelling units in multifamily buildings must be equipped with separate electric meters.

Chapter 3
Prescriptive Package Approach

The prescriptive package approach requires minimal calculations and is the simplest method for demonstrating compliance with the MEC insulation and window requirements (refer to Chapter 2 for additional requirements that must also be satisfied). The prescriptive packages along with your state map and instructions are provided separately from this manual.

To use the prescriptive package approach, find the appropriate climate zone on the map of your state. Select one of several possible packages from the prescriptive packages table for that zone. Enter the package R-values and U-values on the *Prescriptive Package Worksheet* and document that your building complies with the requirements of the selected package.

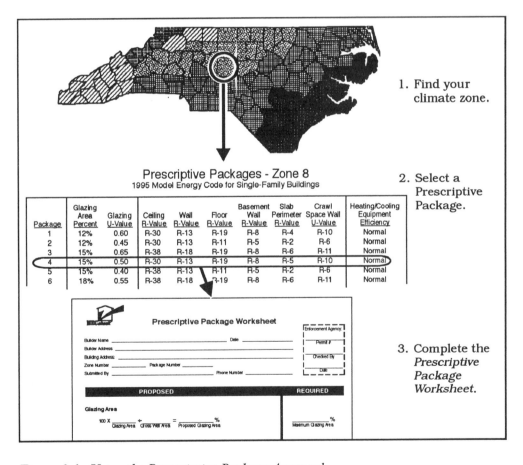

Figure 3-1. Using the Prescriptive Package Approach

Chapter 4

Trade-Off Approach

The trade-off approach is a pencil-and-paper method that can be used for one- and two-family (referred to as single-family) and multifamily residential buildings. This approach allows you to trade off insulation and window efficiency levels in different parts of the building envelope. You may trade off ceiling, wall, floor, basement wall, slab-edge, and crawl space wall insulation; window area; and window and door U-values. The trade-off approach determines whether your building as a whole meets the MEC insulation and window requirements (refer to Chapter 2 for additional requirements that must also be satisfied).

To determine compliance, you must complete the *Trade-Off Worksheet* provided with this manual. Refer to this worksheet while reading this chapter. To complete the *Trade-Off Worksheet*, enter the area, R-value, and U-value of each component in your proposed building and calculate the total proposed UA.[a]

Next, find the required U-values for your zone from Table 4-11 and calculate the

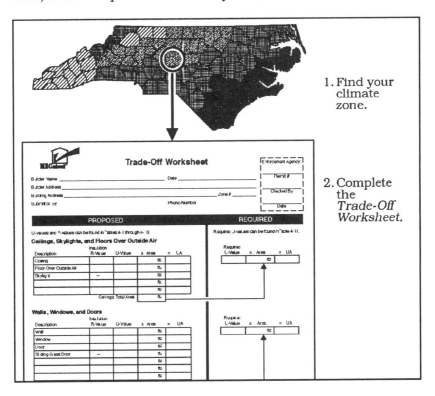

Figure 4-1. Using the Trade-Off Approach

---

[a]   UA is the U-value of a building assembly times the surface area of that assembly through which heat flows. The UA for the entire building is the sum of all UAs for each assembly in the building envelope, giving a total UA for the building envelope. A larger UA indicates more heat loss, either because of a larger surface area or higher U-values (or both).

total required UA. If the proposed UA is less than or equal to the required UA, your building complies with the MEC insulation and window requirements.

When applying the trade-off approach to multifamily buildings, the building can be considered as a whole or a separate worksheet can be completed for each individual dwelling unit. Where individual units are identical, one worksheet may be submitted as representative of the others. Contact the authority having jurisdiction to determine which approach to take.

Instructions for Using the Trade-Off Approach

The *Trade-Off Worksheet* documents compliance with the insulation and window requirements of the MEC. The following instructions explain how to complete this worksheet. Figure 4-2 shows an example *Trade-Off Worksheet*. The numbers in Figure 4-2 identify the various locations on this worksheet that correspond to the following steps.

Step 1: Find Your Climate Zone

Based on the county in which your building is to be located, determine your climate zone number. Your zone number can be found in Appendix C or on the state map included with the prescriptive packages.

Step 2: Complete the General Information Section

Fill in the information at the top of the worksheet. Be sure to record the climate zone number found in Step 1.

Step 3: Complete the PROPOSED Section

On the left side of the worksheet, provide the area, R-value, and U-value of each building component. U-values are a measure of how well a material conducts heat. Tables 4-1 through 4-10 can be used to determine the proposed U-values needed for completing the *Trade-Off Worksheet*. If your particular construction type is not included in these tables, use U-values derived through testing or calculation procedures accepted by your local jurisdiction.

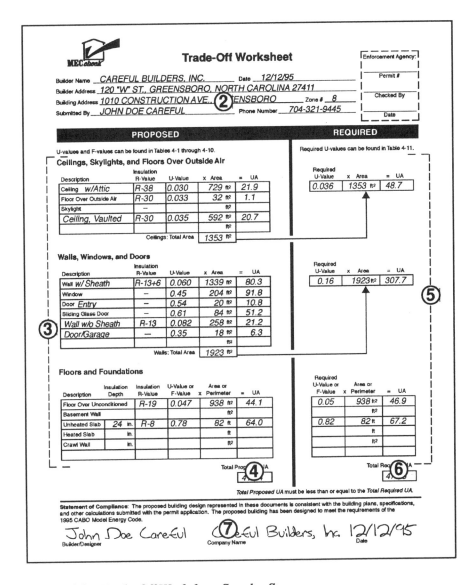

Figure 4-2. Trade-Off Worksheet Step-by-Step

Ceilings, Skylights, and Floors Over Outside Air

Ceilings Enter the R-value of the insulation to be installed in each ceiling component in the *Insulation R-Value* column. R-values for ceilings represent the sum of the cavity insulation plus insulating sheathing (if used). For ventilated ceilings, insulating sheathing must be placed between the conditioned space and the ventilated portion of the roof (typically applied to the trusses or rafters immediately behind the drywall or other ceiling finish material). Sheathing placed on the roof deck over a ventilated attic does not qualify.

Based on the insulation R-value of each ceiling component, find the corresponding U-value from Table 4-1 and enter it in the *Proposed U-Value* column. If the ceiling is to be constructed so the insulation achieves its full insulation thickness over the exterior walls, U-values from the *Raised Truss* column may be used and this should be noted on the worksheet. Otherwise, U-values from the *Standard Truss* column must be used.

Enter the net ceiling area (ft²) of each component in the *Area* column. The net ceiling area includes the following (see Appendix B for a more complete list):

- flat and cathedral ceilings (excluding skylights)
- dormer roofs
- bay window roofs.

Ceiling area should be measured on the slope of the finished interior surface. All ceiling components with the same U-value can be entered on the worksheet as a single component along with their combined area.

Floors Over Outside Air Enter the R-value of the cavity insulation to be installed in each floor over outside air component in the *Insulation R-Value* column. Based on the insulation R-value of the floor component, find the corresponding U-value from Table 4-5 and enter it in the *Proposed U-Value* column.

Floors over outside air include the following:

- floor cantilevers

- floors of an elevated building

- floors of overhangs (such as the floor above a recessed entryway or open car-port).

Skylights You are not required to enter skylight insulation R-values on the worksheet. Enter the proposed U-value of the skylight in the *U-Value* column. U-values for skylights should be tested and documented by the manufacturer in accordance with the NFRC[a] test procedure or taken from Table 4-9. Enter the area (ft²), as measured on the roof slope, of the skylight assembly (including frames and elevated curbs) in the *Area* column.

Additional Components Use the blank lines for additional entries if required. For example, if you have two different skylight types with different U-values, enter each type on a separate line. You can modify or write in descriptions to better fit your proposed building.

Sum the areas of all ceiling, floor over outside air, and skylight components and enter this sum in the space labeled *Ceilings: Total Area.*

(a)   National Fenestration Rating Council. *NFRC 100-91: Procedure for Determining Fenestration Product Thermal Properties.* Silver Spring, Maryland.

Walls, Windows, and Doors

Walls Enter the R-value of the insulation to be installed in each wall component in the *Insulation R-Value* column. R-values for walls represent the sum of the wall cavity insulation plus insulating sheathing (if used). For example, R-13 cavity insulation plus R-6 sheathing is considered R-19 wall insulation. However, the use of insulating sheathing should be specifically indicated on the worksheet. For example, if R-13 batt insulation is to be used with R-6 insulating sheathing, enter "R-13 + R-6" in the *Insulation R-Value* column.

Based on the insulation R-value of the wall component and the type of construction, find the corresponding proposed U-value for each component in Tables 4-2, 4-3 or 4-4 and enter it in the *U-Value* column. For above-grade concrete, masonry, and log walls, use the 16 in. O.C. column of Table 4-2.

Enter the net area (ft²) of each wall component in the *Area* column. The net wall area includes the following:

- the opaque area of all above-grade walls enclosing conditioned spaces (excluding doors and windows)

- the area of the band joist and subfloor between floors

- the opaque wall area of conditioned basements with an average depth less than 50% below grade (excluding basement doors and windows but including the below-grade portion of the wall). For further clarification, refer to the basement wall examples given later in this chapter.

The areas of all wall components with the same U-value may be combined and entered as a single component on the worksheet.

Glazing Glazing R-values are not required. Enter the proposed U-values of glazing assemblies (such as windows and sliding glass doors) in the *U-Value* column. U-values for glazing must be tested and documented by the manufacturer in accordance with the NFRC test procedure or taken from Table 4-9. Center-of-glass U-values cannot be used.

In the *Area* column, enter the total area (ft²) of all glazing assemblies located in the building envelope. The area of a glazing assembly is the interior surface area of the entire assembly, including glazing, sash, curbing, and other framing elements. The nominal area or rough opening is also acceptable for flat windows. The area of windows in the exterior walls of conditioned basements should be included (windows in unconditioned basements are *NOT* included). Do not include the area of skylights; skylights are entered in the *Ceilings, Skylights, and Floors Over Outside Air* section of the worksheet.

Doors In the *U-Value* column, enter the proposed U-values for all doors in the building envelope. U-values for doors must be tested and documented by the manufacturer in accordance with the NFRC test procedure or taken from Table 4-10.

If an opaque door with glazing is rated with an aggregate R-value (an R-value that includes both the glass and opaque area), the following equation applies:

$$U\text{-Value} = \frac{1}{R\text{-Value}}$$

If a door contains glass and an aggregate R-value or U-value rating for that door is not available, include the glass area of the door with your glazing and use the opaque door R-value or U-value to determine compliance of the door. The U-values listed in Table 4-10 are only for doors without glass.

Enter the nominal area (ft²) or rough opening area of all doors in the *Area* column. Include doors located in the walls of conditioned basements.

Additional Components Use the blank lines for additional entries if required. For example, if you have two different window types with different U-values, enter each type on a separate line. You can modify or write in descriptions to better fit your proposed building.

Sum the areas of all wall components and enter this sum in the space labeled *Walls: Total Area.*

Floors and Foundations

Floors Floors over unconditioned spaces include floors over unconditioned crawl spaces, basements, and garages. Floors over outside air should be entered in the *Ceilings, Skylights, and Floors Over Outside Air* section. Enter the R-value of the insulation to be installed in each floor over unconditioned space component in the *Insulation R-Value* column. Enter the corresponding U-value from Table 4-5 in the *U-Value or F-Value* column. Enter the floor area (ft²) in the *Area or Perimeter* column.

Basement Walls Enter the R-value of the insulation to be installed in each basement wall component in the *Insulation R-Value* column. If you intend to install insulation on both the exterior and interior of the wall, enter the sum of both R-values. Enter the corresponding proposed U-value from Table 4-6 in the *U-Value or F-Value* column. Basement walls must be insulated from the top of the basement wall to 10 ft below ground level or to the basement floor, whichever is less. The MEC*check*™ software enables you to trade off the basement wall insulation depth as well as the insulation R-value.

Enter the opaque wall area (ft²) of the basement walls in the *Area or Perimeter* column. Include the entire opaque area of any individual wall with an average depth 50% or more below grade that encloses a conditioned space. The entire area of any basement wall less than 50% below grade is included with above-grade walls and is subject to the above-grade wall requirement. The following examples help to clarify the treatment of wood kneewalls, walk-out basements, and basement walls constructed from specialty foundation systems.

Example 1: Wood Kneewalls

Assume a basement is to be constructed in Zone 8 with 3-ft-high wood kneewalls built on a 5-ft-high concrete foundation. R-13 insulation will be installed in the wood kneewall cavities and R-5 rigid insulation will be installed on the concrete foundation walls. The wood kneewalls are completely above grade and fully insulated. The concrete foundation walls are 4 ft below grade and fully insulated.

Each basement wall (as measured from the top of the kneewalls to the basement floor) is at least 50% below grade. Therefore, both the masonry foundation walls and the wood kneewalls must be entered in the *Trade-Off Worksheet* as basement wall components under the *Floors and Foundations* section. Since the kneewalls will be insulated to a different level than the masonry foundation walls, you will need to enter the basement on two lines in the *Trade-Off Worksheet* (use the line labeled *Basement Wall* and use the blank line for the second entry). Refer to Table 4-6 for U-values that corresponds to both the masonry foundation wall insulated to R-5 (0.115) and the wood kneewalls insulated to R-13 (0.059). All proposed basement wall U-values should be taken from Table 4-6, including wood-frame basement walls. Table 4-11 lists the basement U-value requirement for Zone 8 as 0.09. This requirement applies to both the masonry and wood portions of the wall, and should be entered on the right side of the worksheet across from both basement wall entries.

Example 2: Walk-Out Basement

Assume an 8-ft basement is to be built in Zone 8 on a slope so that the front wall is 7 ft below grade and the rear wall is totally above grade. The ground level along both side walls is sloped so that approximately 50% of each wall is below grade. The rear basement wall will be wood-frame construction with R-19 insulation. The other three walls will be concrete walls with R-10 insulation. All four walls will be fully insulated.

Because the front and side walls are at least 50% below grade, they must be entered on the *Trade-Off Worksheet* as a basement wall component under the *Floors and Foundations* section. Refer to Table 4-6 for the basement U-value that corresponds to R-10 insulation (0.072). The rear wall is not 50% below grade, however, and should be entered as an above-grade wall under the *Walls, Windows, and Doors* section. Refer to Table 4-2 for the U-value that corresponds to an above-grade wood-frame wall insulated to R-19 (0.060).

Note that the basement floor along the rear wall should be considered a slab-on-grade component. Slab insulation should be installed along the basement floor for the length of the rear wall. Refer to Table 4-7 for the U-value corresponding to the R-value and depth of insulation that will be installed.

Example 3: Specialty Foundation Systems

Manufacturers of insulating foam concrete form systems and premanufactured concrete panels with integrated insulation generally supply R-value ratings for the entire wall, not just the insulation. Where the R-value of the insulation alone is not know, the manufacturer overall wall R-value rating may be used in place of the insulation R-value. Refer to Table 4-6 for the U-value corresponding to this R-value. For example, if the manufacturer reports an overall wall R-value of R-19, the corresponding U-value from Table 4-6 is 0.043.

Slabs Enter the R-value of the slab insulation to be installed around the perimeter of each slab-on-grade component in the *Insulation R-Value* column. Enter the corresponding F-value (slab-edge U-values are sometimes referred to as F-values) from Table 4-7 in the *U-Value or F-Value* column. Table 4-7 offers U-values for slab insulation depths of 24 in. or 48 in., installed using any of the following configurations.

- The slab insulation extends from the top of the slab downward to the required depth.

- The slab insulation extends from the top of the slab downward to the bottom of the slab and then horizontally underneath the slab for a minimum total distance equal to or greater than the required depth.

- The slab insulation extends from the top of the slab downward to the bottom of the slab and then horizontally away from the slab for a minimum total distance equal to or greater than the required depth. The horizontal insulation must be covered by pavement or at least 10 in. of soil.

The top edge of insulation installed between the exterior wall and the interior slab can be cut at a 45° angle away from the exterior wall.

The slab description on the worksheet should indicate the proposed insulation depth (24 in. or 48 in.). The software approach enables you to select additional slab perimeter insulation depths. The slab description should also indicate whether or not

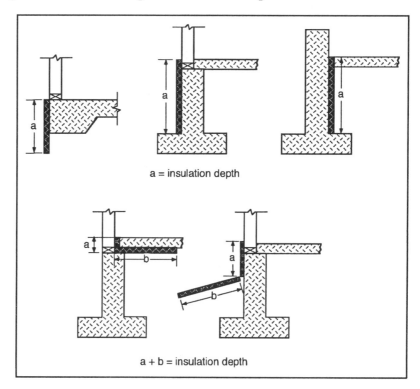

Figure 4-3. Slab Insulation Depth Requirements

the slab is heated. A heated slab is a slab with ducts or hydronic heating elements in or under the slab.

Enter the slab perimeter (ft) in the *Area or Perimeter* column of the worksheet.

Crawl Space Walls The crawl space wall R-value requirements are for walls of unventilated crawl spaces (i.e. not directly vented to the outside). Enter the R-value to be installed in each crawl space wall component in the *Insulation R-Value* column. Enter the corresponding proposed U-value from Table 4-8 in the *U-Value or F-Value* column. The crawl space wall insulation must extend from the top of the wall to at least 12 in. below the outside finished grade. If the distance from the outside finished grade to the top of the footing is less than 12 in., the insulation must extend a total vertical plus horizontal distance of 24 in. from the outside finished grade. Enter the total vertical plus horizontal distance of the insulation to be installed in the *Insulation Depth* column.

Enter the opaque wall area (ft²) of the crawl space walls in the *Area or Perimeter* column.

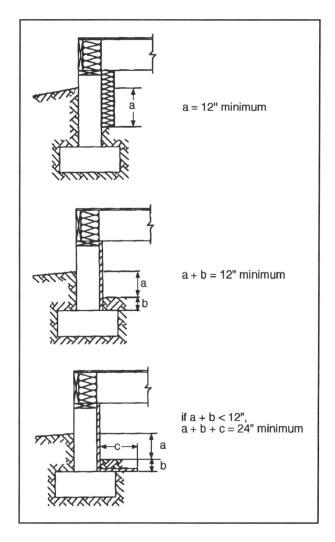

Figure 4-4. Crawl Space Wall Insulation Depth Requirements

Step 4: Compute the Total Proposed UA

Multiply all proposed U-values by their corresponding area; multiply all proposed slab F-values by their corresponding perimeter. Enter the results in the *UA* column. Sum the proposed UAs on the left side of the worksheet and enter this sum in the *Total Proposed UA* box. This sum is the total UA of your proposed building.

Step 5: Complete the *REQUIRED* Section

Table 4-11 lists ceiling, wall, floor, basement wall, slab-edge, and crawl space wall U-value and F-value requirements for each climate zone (your climate zone number can be found in Appendix C or on the state map included with the prescriptive packages). Enter the required U-values and F-values for your zone in the appropriate *Required U-Value* column.

Copy the total ceiling and wall areas to the *Area* column on the right side of the worksheet. Copy the area or perimeter of all floor and foundation components to the corresponding box in the *Area or Perimeter* column on the right side of the worksheet.

Step 6: Compute the Total Required UA

Multiply U-values in the *Required U-Value* column by their corresponding area; multiply F-values by their corresponding perimeter. Enter the results in the *UA* column. Sum the UAs on the right side of the worksheet. Record this sum in the *Total Required UA* box. This value is the total UA of the code building (a building with the same dimensions as your building but insulated to the minimum MEC requirements).

Step 7: Check for Compliance

If the *Total Proposed UA* (Step 4) is less than or equal to the *Total Required UA* (Step 6), then your building complies with the MEC insulation and window requirements. If not, you must adjust the insulation R-values, window or door U-values, or areas in your proposed building. For example, increasing insulation R-values or reducing the glass area may bring the building design into compliance.

When you are satisfied that your building complies, sign and date the worksheet in the blanks provided. Transfer the proposed R-value and U-value information from your worksheet to your building plans or specifications.

Quick Compliance

If your proposed ceiling, floor, basement wall, and crawl space wall
U-values and slab-edge F-values are all less than or equal to the required U-values and
F-values, no further calculations are required for ceilings and foundations. You do
not need to calculate proposed or required building UAs or enter areas in the *Ceilings,
Skylights, and Floors Over Outside Air* or *Floor and Foundations* sections. However,
you will still need to demonstrate that your wall, window, and door components meet
the MEC requirements by completing the *Walls, Windows, and Doors* section.

If your building meets the following criteria, you can use the Quick Compliance m-
ethod to demonstrate that your building complies.

- There are no skylight components.

- All ceiling U-values are less than or equal to the ceiling U-value requirement.

- All floor, basement wall, and crawl space wall U-values are less than or equal
  to their corresponding U-value requirements.

- All slab F-values are less than or equal to the slab F-value requirement.

- The total proposed UA for the *Walls, Windows, and Doors* section is less than
  or equal to the required UA for the *Walls, Windows, and Doors* section.

Figure 4-5 shows a *Trade-Off Worksheet* filled out using the Quick Compliance
method. All proposed ceiling, floor, and foundation components have U-values and
F-values that are less than their corresponding required U-values and
F-values. Therefore, areas and UA calculations are not required for these components. Note that the total proposed UA for the *Walls, Windows, and Doors* section
(219.3) is less than the total required UA (309.7). Hence, the building complies.

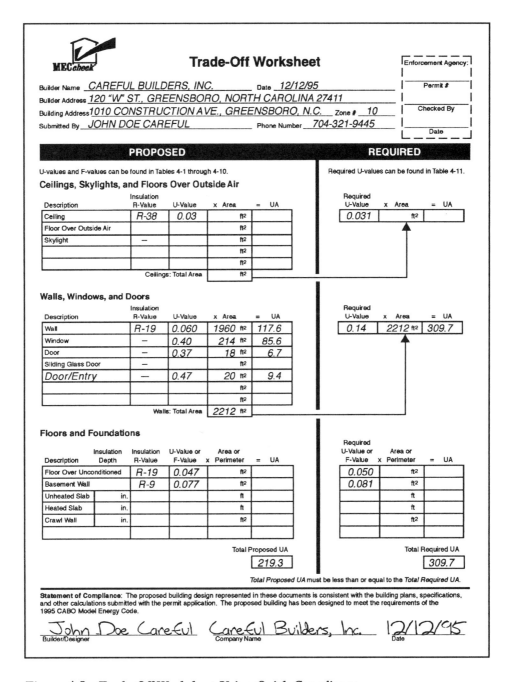

Figure 4-5. Trade-Off Worksheet Using Quick Compliance

Table 4-1. Ceiling U-Values[a]

Insulation R-Value	Standard Truss U-Value	Raised Truss[b] U-Value	Insulation R-Value	Standard Truss U-Value	Raised Truss[b] U-Value
R-0	0.568	0.568	R-33	0.033	0.029
R-7	0.119	0.119	R-34	0.032	0.028
R-8	0.108	0.108	R-35	0.032	0.028
R-9	0.098	0.098	R-36	0.031	0.027
R-10	0.089	0.089	R-37	0.031	0.026
R-11	0.082	0.082	R-38	0.030	0.025
R-12	0.076	0.076	R-39	0.030	0.025
R-13	0.070	0.070	R-40	0.029	0.024
R-14	0.066	0.066	R-41	0.029	0.024
R-15	0.062	0.061	R-42	0.028	0.023
R-16	0.059	0.058	R-43	0.028	0.023
R-17	0.056	0.055	R-44	0.027	0.022
R-18	0.053	0.052	R-45	0.027	0.022
R-19	0.051	0.049	R-46	0.027	0.021
R-20	0.048	0.047	R-47	0.026	0.021
R-21	0.047	0.045	R-48	0.026	0.020
R-22	0.045	0.043	R-49	0.026	0.020
R-23	0.043	0.041	R-50	0.026	0.020
R-24	0.042	0.040	R-51	0.025	0.019
R-25	0.040	0.038	R-52	0.025	0.019
R-26	0.039	0.037	R-53	0.025	0.019
R-27	0.038	0.035	R-54	0.025	0.018
R-28	0.037	0.034	R-55	0.024	0.018
R-29	0.036	0.033	R-56	0.024	0.018
R-30	0.035	0.032	R-57	0.024	0.018
R-31	0.034	0.031	R-58	0.024	0.017
R-32	0.034	0.030	R-59	0.024	0.017

(a)   R-values represent the sum of the ceiling cavity insulation plus the R-value of insulating sheathing (if used). For example, R-19 cavity insulation plus R-2 sheathing is reported as R-21 ceiling insulation. For ventilated ceilings, insulating sheathing must be placed between the conditioned space and the ventilated portion of the roof (typically applied to the trusses or rafters immediately behind the drywall or other ceiling finish material).

(b)   To receive credit for a raised truss, the insulation must achieve its full insulation thickness over the exterior walls.

Table 4-2. Wood-Frame Wall U-Values[a,b]

Insulation R-Value[c]	16-in. O.C. Wall U-Value	24-in. O.C. Wall U-Value
R-0	0.238	0.241
R-7	0.105	0.104
R-8	0.099	0.097
R-9	0.094	0.092
R-10	0.090	0.088
R-11	0.089	0.087
R-12	0.085	0.083
R-13	0.082	0.080
R-14	0.079	0.077
R-15	0.077	0.074
R-16	0.066	0.064
R-17	0.064	0.062
R-18	0.062	0.060
R-19	0.060	0.059
R-20	0.059	0.057
R-21	0.057	0.056
R-22	0.056	0.054
R-23	0.055	0.053
R-24	0.054	0.052
R-25	0.053	0.051
R-26	0.052	0.050
R-27	0.051	0.049
R-28	0.050	0.048

(a)   U-values are for uncompressed insulation.

(b)   U-values in this table were developed for wood-frame walls, but the *16-in. O.C. Wall U-Value* column can also be used for above-grade concrete, masonry, and log walls. Mass wall R-value to U-value conversion tables are planned for future versions of the MEC*check*™ Manual.

(c)   Wall R-values are the sum of the cavity insulation plus insulating sheathing (if used).

Table 4-3. 16-in. O.C. Metal-Frame Wall U-Values

Cavity R-Value	Insulating Sheathing R-Value										
	R-0	**R-1**	**R-2**	**R-3**	**R-4**	**R-5**	**R-6**	**R-7**	**R-8**	**R-9**	**R-10**
R-0	0.270	0.258	0.205	0.170	0.146	0.127	0.113	0.101	0.092	0.084	0.078
R-11	0.120	0.118	0.106	0.096	0.087	0.080	0.074	0.069	0.065	0.061	0.057
R-13	0.114	0.111	0.100	0.091	0.084	0.077	0.072	0.067	0.063	0.059	0.056
R-15	0.109	0.107	0.096	0.088	0.081	0.075	0.070	0.065	0.061	0.058	0.054
R-19	0.101	0.099	0.090	0.083	0.077	0.071	0.066	0.062	0.059	0.055	0.052
R-21	0.098	0.096	0.088	0.081	0.075	0.070	0.065	0.061	0.058	0.054	0.052
R-25	0.094	0.093	0.085	0.078	0.073	0.068	0.063	0.060	0.056	0.053	0.051

Table 4-4. 24-in. O.C. Metal-Frame Wall U-Values

Cavity R-Value	Insulating Sheathing R-Value										
	R-0	**R-1**	**R-2**	**R-3**	**R-4**	**R-5**	**R-6**	**R-7**	**R-8**	**R-9**	**R-10**
R-0	0.270	0.258	0.205	0.170	0.146	0.127	0.113	0.101	0.092	0.084	0.078
R-11	0.106	0.104	0.095	0.086	0.080	0.074	0.069	0.064	0.060	0.057	0.054
R-13	0.100	0.098	0.090	0.082	0.076	0.071	0.066	0.062	0.058	0.055	0.052
R-15	0.094	0.093	0.085	0.078	0.073	0.068	0.063	0.060	0.056	0.053	0.051
R-19	0.088	0.086	0.080	0.074	0.069	0.064	0.060	0.057	0.054	0.051	0.049
R-21	0.085	0.084	0.077	0.072	0.067	0.063	0.059	0.056	0.053	0.050	0.048
R-25	0.081	0.080	0.074	0.069	0.064	0.060	0.057	0.054	0.051	0.049	0.046

Table 4-5. Floor U-Values

Insulation R-Value	Floor U-Value
R-0	0.249
R-7	0.096
R-11	0.072
R-13	0.064
R-15	0.057
R-19	0.047
R-21	0.044
R-26	0.037
R-30	0.033

Table 4-6. Basement U-Values[a]

Insulation R-Value	Basement Wall U-Value	Insulation R-Value	Basement Wall U-Value
R-0	0.360	R-10	0.072
R-1	0.244	R-11	0.067
R-2	0.188	R-12	0.062
R-3	0.155	R-13	0.059
R-4	0.132	R-14	0.055
R-5	0.115	R-15	0.052
R-6	0.102	R-16	0.050
R-7	0.092	R-17	0.047
R-8	0.084	R-18	0.045
R-9	0.077	R-19	0.043
		R-20	0.041

(a)   Insulation R-values represent the sum of exterior and/or interior insulation. Basement walls must be insulated from the top of the basement wall to 10 ft below ground level or to the floor of the basement, whichever is less.

Table 4-7. Slab F-Values

Perimeter Insulation R-Value	Slab F-Value	
	24-in. Insulation Depth	48-in. Insulation Depth
R-0	1.04	1.04
R-1	0.91	0.89
R-2	0.86	0.83
R-3	0.83	0.79
R-4	0.82	0.76
R-5	0.80	0.74
R-6	0.79	0.73
R-7	0.79	0.71
R-8	0.78	0.70
R-9	0.77	0.69
R-10	0.77	0.68
R-11		0.68
R-12		0.67
R-13		0.66
R-14		0.66
R-15		0.65
R-16		0.65
R-17		0.65
R-18		0.64
R-19		0.64
R-20		0.64

Table 4-8. Crawl Space Wall U-Values

Insulation R-Value	Crawl Space Wall U-Value
R-0	0.477
R-1	0.313
R-2	0.235
R-3	0.189
R-4	0.158
R-5	0.136
R-6	0.120
R-7	0.107
R-8	0.096
R-9	0.088
R-10	0.081
R-11	0.075
R-12	0.069
R-13	0.065
R-14	0.061
R-15	0.057
R-16	0.054
R-17	0.051
R-18	0.049
R-19	0.047
R-20	0.045

Table 4-9. U-Values for Windows, Glazed Doors, and Skylights[a]

Frame/Glazing Features	Single Pane	Double Pane
Metal Without Thermal Break		
Operable	1.30	0.87
Fixed	1.17	0.69
Door	1.26	0.80
Skylight	2.02	1.30
Metal With Thermal Break		
Operable	1.07	0.67
Fixed	1.11	0.63
Door	1.10	0.66
Skylight	1.93	1.13
Metal-Clad Wood		
Operable	0.98	0.60
Fixed	1.05	0.58
Door	0.99	0.57
Skylight	1.50	0.88
Wood/Vinyl		
Operable	0.94	0.56
Fixed	1.04	0.57
Door	0.98	0.56
Skylight	1.47	0.85
Glass Block Assemblies	0.60	

Table 4-10. U-Value Table for Non-Glazed Doors[a]

Steel Doors		
Without Foam Core	0.60	
With Foam Core	0.35	
Wood Doors	**Without Storm**	**With Storm**
Panel With 7/16-in. Panels	0.54	0.36
Hollow Core Flush	0.46	0.32
Panel With 1-1/8-in. Panels	0.39	0.28
Solid Core Flush	0.40	0.26

(a)   The U-values in these tables can be used in the absence of test U-values. The product cannot receive credit for a feature that cannot be clearly detected. Where a composite of materials from two different product types is used, the product must be assigned the higher U-value.

Table 4-11. U-Value and F-Value Requirements by Climate Zone

Climate Zone	Ceiling U-Value	Single-Family Wall U-Value	Multi-Family Wall U-Value	Floor U-Value	Basement Wall U-Value	Unheated Slab F-Value	Heated Slab F-Value	Crawl Space Wall U-Value
1	0.047	0.25	0.38	0.08	0.360	1.04	1.04	0.477
2	0.044	0.23	0.35	0.08	0.360	1.04	0.79	0.137
3	0.042	0.21	0.31	0.07	0.360	1.04	0.79	0.137
4	0.039	0.20	0.28	0.07	0.121	1.04	0.79	0.137
5	0.036	0.18	0.25	0.07	0.113	1.04	0.79	0.124
6	0.036	0.17	0.22	0.05	0.106	0.82	0.79	0.111
7	0.036	0.16	0.22	0.05	0.098	0.82	0.79	0.098
8	0.036	0.16	0.22	0.05	0.090	0.82	0.79	0.085
9	0.033	0.15	0.22	0.05	0.082	0.82	0.79	0.071
10	0.031	0.14	0.22	0.05	0.081	0.81	0.79	0.058
11	0.028	0.13	0.22	0.05	0.080	0.81	0.79	0.058
12	0.026	0.13	0.22	0.05	0.079	0.80	0.79	0.058
13	0.026	0.12	0.20	0.05	0.078	0.74	0.71	0.058
14	0.026	0.11	0.18	0.05	0.077	0.73	0.70	0.058
15	0.026	0.11	0.15	0.05	0.075	0.72	0.69	0.058
16	0.026	0.11	0.15	0.05	0.052	0.71	0.69	0.058
17	0.026	0.11	0.12	0.05	0.052	0.69	0.67	0.058
18	0.026	0.10	0.12	0.05	0.052	0.68	0.66	0.058
19	0.025	0.10	0.12	0.04	0.052	0.66	0.65	0.058

Example of How to Use the Trade-Off Approach

The trade-off approach is illustrated in this section. Assume that you plan to build the house shown in Figure 4-6 on a lot located in Greensboro, North Carolina. Greensboro is in Guilford County and is designated as Zone 8 on the North Carolina state map.

Table 4-12 lists the components that make up the building envelope, the dimensions of each of these components, and the proposed insulation R-values and window and door U-values. Figure 4-7 shows how to determine proposed and required U-values using Tables 4-1 through 4-11. The completed *Trade-Off Worksheet* is shown in Figure 4-8.

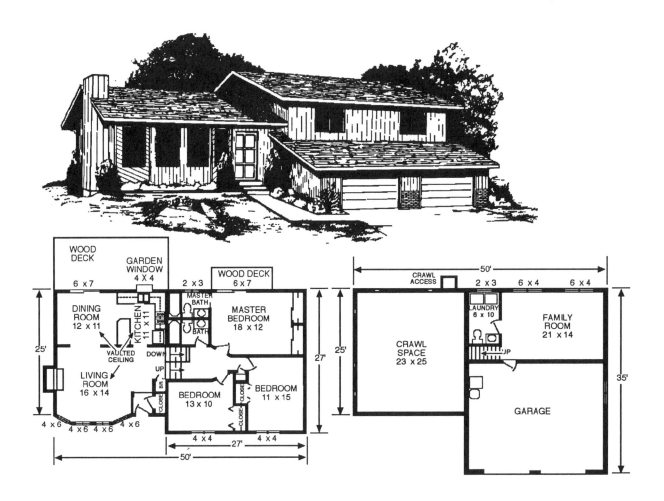

Figure 4-6. Example House

Table 4-12. Example House Specifications

Building Component	Area	Insulation Level
Ceilings		
With Attic (Std. Truss)	729 ft²	R-38
Vaulted	592 ft²	R-30
Walls (2x4 @ 16-in. O.C.)		
Without Sheathing[a]	276 ft² (gross)	R-13
With Sheathing	1647 ft² (gross)	R-19 (R-13 cavity + R-6 sheathing)
Windows	204 ft²	U-0.45
Sliding Glass Doors	84 ft²	U-0.61
Doors		
Entrance	20 ft²	U-0.54
Garage to Family Room	18 ft²	U-0.35
Floors		
Over Garage	363 ft²	R-19
Over Crawl Space	575 ft²	R-19
Slab (Unheated)	82 ft (perimeter)	R-8 (24-in. depth)
Bay Window Floor	32 ft²	R-30
(a) Walls without sheathing are located between the family room and the garage, the laundry room and the crawl space, and the garage and the living room.		

Determine Which Components Are Part of the Building Envelope

Only the building components that are part of the building envelope are entered on the *Trade-Off Worksheet*. Building envelope components are those that separate conditioned spaces (heated or cooled rooms) from unconditioned spaces (rooms that are not heated or cooled) or from outside air. Walls, floors, and other building components separating two conditioned spaces are *NOT* part of the building envelope.

Walls In this example, the garage is unconditioned, so the exterior garage walls are not part of the building envelope. The wall between the conditioned family room and the unconditioned garage is part of the building envelope. Likewise, the wall between the garage and the living room is part of the building envelope.

Part of the laundry room wall separates the laundry room from the crawl space and the other part separates the laundry room from the kitchen. The portion adjacent to the crawl space is part of the building envelope because it separates the conditioned laundry from the unconditioned crawl space. The portion adjacent to the kitchen can be ignored. The wall between the portion adjacent to the kitchen can also be ignored. Likewise, the wall between the upstairs bathrooms and the kitchen and the wall between the center bedroom and the living room are not part of the building envelope. Portions of both of these walls are also adjacent to outside air, and those portions are part of the building envelope. Table 4-13 lists the walls that are part of the building envelope and indicates whether sheathing is installed on them (which is relevant when determining the R-value of the wall).

Table 4-13. Walls Comprising the Building Envelope

Wall	Sheathing?	Gross Area	Net Area[a]
All walls between interior conditioned space and outside air	Yes	1647	1339
The wall between the family room and the garage	No	192	174
The wall between the garage and the living room	No	44	44
The wall between the laundry and the crawl space	No	40	40
(a) Net area does not include doors or windows.			

Floors The example house has a conditioned floor area of 1714 ft², but 378 ft² of the floor area is located over the family room and is not part of the building envelope (both the family room and the rooms above it are conditioned). The living room, dining room, and kitchen are over an unheated crawl space. The family room and garage both have slab-on-grade floors. The floor of the bay window is a floor over outside air which is subject to the ceiling requirement rather than the requirement for floors over conditioned space.

Glazing and Doors There are two sliding glass doors in the building envelope -- one leading from the dining room to the larger deck and one leading from the master bedroom to the smaller deck. There are two opaque doors in the building envelope -- the front entry door and the door leading from the garage into the family room.

Complete the PROPOSED Section

Areas and proposed R-values for the example house are given in Table 4-12. Refer to the example *Trade-Off Worksheet* illustrated in Figure 4-8 when reading the following sections.

Ceilings, Skylights, and Floors Over Outside Air The ceiling over the dining room, living room, and kitchen is vaulted -- the rest is a flat ceiling under an attic space. Enter R-38 as the proposed insulation R-value for the portion of the ceiling with an attic. Use the blank space to create a second ceiling component for the vaulted ceiling and enter R-30 as the proposed insulation R-value for the vaulted ceiling. The floor of the bay window is a floor over outside air component. Enter R-30 as the proposed insulation R-value for the *Floor Over Outside Air* component.

Refer to Table 4-1 for the U-values corresponding to your proposed ceiling R-values. Table 4-1 lists R-value to U-value conversions for ceilings with standard and raised trusses. You are building a standard truss with R-38 insulation in the ceiling with an attic space, and R-30 insulation in the vaulted ceiling. R-38 insulation with a standard truss corresponds to a U-value of 0.030. R-30 insulation corresponds to a U-value of 0.035. Enter 0.030 in the *U-Value* column for the *Ceiling w/Attic* component and enter 0.035 in the *U-Value* column for the *Ceiling, Vaulted* component.

Table 4-5 lists the U-value for floors having R-30 insulation as 0.033. Enter 0.033 in the *U-Value* column for the *Floor Over Outside Air* component.

In the *Area* column, enter 729 ft² for the ceiling with attic, 592 ft² for the vaulted ceiling, and 32 ft² for the bay window floor. Sum the areas of all ceiling components and enter this sum (1353 ft²) in the space labeled *Ceilings: Total Area.*

There are no skylights in the example house, so the space for skylight components may be left blank.

Walls, Windows, and Doors All walls that make up the building envelope will have R-13 cavity insulation, but not all of these walls will have R-6 exterior sheathing (see Table 4-13). Therefore, you must enter two wall components on the *Trade-Off Worksheet*. Enter R-13 + R-6 (R-13 cavity insulation plus R-6 sheathing) as the proposed insulation R-value for the walls with sheathing and R-13 as the proposed insulation R-value for the walls without sheathing. Use a blank line to record the second component. The example worksheet in Figure 4-8 illustrates how these two wall components are labeled.

Refer to Table 4-2 and the *16-in. O.C. Wall U-Value* column for the U-values that correspond to the proposed R-values. The walls with sheathing have an effective R-value of R-19, which corresponds to a U-value of 0.060. The walls without sheathing have an R-value of R-13, which corresponds to a U-value of 0.082. Enter 0.060 in the *U-Value* column for the *Wall w/Sheathing* component and enter 0.082 in the *U-Value* column for the *Wall w/o Sheathing* component.

Enter the net opaque wall area for each wall component in the *Area* column. The net wall area for each component will be equal to the gross wall area minus openings for windows, doors, sliding glass doors, etc. In this example, the only opening in the walls without sheathing is the door from the garage to the living room. The entry door, the windows, and the sliding glass doors are all subtracted from the gross area of the walls with sheathing.

```
Net Wall Area (with sheathing) = 1647 - 204 - 84 - 20 = 1339

Net Wall Area (without sheathing) = 276 - 18 = 258
```

Enter 1339 ft² for the wall with sheathing and 258 ft² for the wall without sheathing.

You are not required to enter an R-value for windows and doors. The windows and sliding glass doors have been rated and labeled by the manufacturer in accordance with the NFRC test procedure. The windows have a U-value of 0.45 and the sliding glass doors have a U-value of 0.61. Enter the window and sliding glass door areas and their corresponding U-values.

Neither of the doors (entrance and garage) have tested U-values, so use Table 4-10 to determine their default U-values. The entry door is an opaque wood panel door with 7/16-in. panels and no storm door, corresponding to a U-value of 0.54. The door from the garage to the family room is a metal door with foam core, corresponding to a U-value of 0.35. Because the doors have different U-values, each must be entered on a separate line on the *Trade-Off Worksheet*. Enter the door areas and their corresponding U-values.

Sum the areas of all wall components (including the windows and doors) and enter this sum (1923 ft²) in the space labeled *Walls: Total Area.*

Floors and Foundations Two floor and foundation components must be considered. The floor over the garage and the floor over the crawl space are floors over unconditioned spaces. R-19 insulation will be installed in both. Therefore, these floor components can be combined. The slab floor, however, must be considered separately. Enter R-19 as the proposed insulation R-value for the floors over conditioned spaces and R-8 as the proposed insulation R-value for the slab perimeter.

Refer to Table 4-5 for the floor U-value corresponding to your proposed floor R-value. A floor R-value of R-19 corresponds to a U-value of 0.047. Refer to Table 4-7 for the slab F-value corresponding to your proposed slab R-value. Note that Table 4-7 lists R-value to F-value conversions for insulation installed to either 24 in. or 48 in. The slab perimeter insulation will be installed to 24 in. A slab R-value of R-8 corresponds to an F-value of 0.78 under the *24-in. Insulation Depth* column.

In the *Area* column, enter 938 ft² as the combined area of the two floor components over unconditioned space and 82 ft for the slab perimeter. Note that slab components require a perimeter -- not an area.

Compute the Total Proposed UA

For each building component, multiply the proposed U-value or F-value by its corresponding area or perimeter and enter the result in the proposed *UA* column. For example, the *Ceiling w/Attic* component has a U-value of 0.030 and an area of 729 ft², resulting in a UA of 21.9.

```
Ceiling w/Attic UA = 0.030 × 729 = 21.9
```

Sum the UAs on the left side of the worksheet and enter this sum (413.4) in the space labeled *Total Proposed UA*.

Complete the *REQUIRED* Section

Table 4-11 lists the required U-values and F-values for each climate zone. Refer to the row for Zone 8 that lists the ceiling U-value requirement as 0.036. Enter this U-value in the *Required U-Value* column in the *Ceilings, Skylights, and Floors Over Outside Air* section. This is the U-value requirement for all ceiling components (including skylights and floors over outside air). Next, enter 0.16 as the required U-value for walls. This is the U-value requirement for all wall components (including opaque walls, windows, and doors). Finally, enter 0.05 as the U-value requirement for floors over unconditioned spaces and 0.82 as the F-value requirement for unheated slabs.

Enter the area of all ceiling components in the *Area* column. This sum (1353 ft²) was previously computed and entered on the left side of the worksheet in the box labeled *Ceilings: Total Area*. Follow the arrow that indicates this same sum should be entered on the right side of the worksheet. Likewise, the sum of all wall components (1923 ft²) was previously computed and entered on the left side of the worksheet in the box labeled *Walls: Total Area*. Follow the arrow that indicates this same sum should be entered on the right side of the worksheet. The areas of each floor and foundation component should be entered individually on the right side of the worksheet. Enter 938 ft² in the *Area or Perimeter* column for the floor component and 82 ft for the slab component.

Compute the Total Required UA

For each entry on the right side of the worksheet, multiply the required U-value or F-value by its corresponding area or perimeter and enter the result in the required *UA* column. For example, the *Ceilings, Skylights, and Floors Over Outside Air* U-value requirement is 0.036 and the total area of all ceiling components is 1353 ft², resulting in a UA of 48.7.

```
Ceilings Required UA = 0.036 × 1353 = 48.7
```

Sum the UAs on the right side of the worksheet, and enter this sum (470.5) in the box labeled *Total Required UA*.

Check for Compliance

The sum in the *Total Proposed UA* box (413.4) is less than the sum in the *Total Required UA* box (470.5), indicating that your building complies with the MEC insulation and window requirements (congratulations!). Sign and date the worksheet.

Table 4-1. Ceiling U-Values

Insulation R-Value	Standard Truss U-Value	Raised Truss U-Value	Insulation R-Value	Standard Truss U-Value	Raised Truss U-Value
R-0	0.568	068	R-33	0.033	0.029
R-7	0.119	0.119	R-34	0.032	0.028
R-8	0.108	0.108	R-35	0.032	0.028
R-9	0.098	0.098	R-36	0.031	0.027
R-10	0.089	0.089	R-37	0.031	0.026
R-11	0.082	0.082	R-38	0.030	0.025
R-12	0.076	0.076	R-39		0.025
R-13	0.070	0.070	R-40	029	0.024
R-14	0.066	0.066	R-41	029	0.024
R-15	0.062	0.061	R-42	028	0.023
R-16	0.059	0.058	R-43	0.28	0.023
R-17	0.056	0.055	R-44	0.0	0.022
R-18	0.053	0.052	R-45	0.02	0.022
R-19	0.051	0.049	R-46	0.02	0.021
R-20	0.048	0.047	R-47	0.026	0.021
R-21	0.047	0.045	R-48	0.026	0.020
R-22	0.045	0.043	R-49	0.026	0.020
R-23	0.043	0.041	R-50	0.026	0.020
R-24	0.042	0.040	R-51	0.025	0.019
R-25	0.040	0.038	R-52	0.025	0.019
R-26	0.039	0.037	R-53	0.025	0.019
R-27	0.038	0.035	R-54	0.025	0.018
R-28	0.037	0.034	R-55	0.024	0.018
R-29	0.036	0.033	R-56	0.024	0.018
R-30	0.035	0.032	R-57	0.024	0.018

Table 4-11. U-Value/F-Value Requirements by Climate Zone

Climate Zone	Ceiling U-Value	Single-Family Wall U-Value	Multi-Family Wall U-Value	Floor U-Value	Basement Wall U-Value	Unheated Slab F-Value	Heated Slab F-Value	Crawl Space Wall U-Value
1	0.047	0.25	0.38	0.08	0.360	1.04	1.04	0.477
2	0.044	0.23	0.35	0.08	0.360	1.04	0.79	0.137
3	0.042	0.21	0.31	0.07	0.360	1.04	0.79	0.137
4	0.039	0.20	0.28	0.07	0.121	1.04	0.79	0.137
5	0.036	0.18	0.25	0.07	0.113	1.04	0.79	0.124
6	0.036	0.17	0.22	0.05	0.106	0.82	0.79	0.111
7	0.036	0.16	0.22	0.05	0.098	0.82	0.79	0.098
8	0.036	0.16	0.22	0.05	0.090	0.82	0.79	0.085
9	0.033	0.15	0.22	0.05	0.082	0.82	0.79	0.071
10	0.031	0.14	0.22	0.05	0.081	0.81	0.79	0.058
11	0.028	0.13	0.22	0.05	0.080	0.81	0.79	0.058
12	0.026	0.13	0.22	0.05	0.079	0.80	0.79	0.058
13	0.026	0.12	0.20	0.05	0.078	0.74	0.71	0.058
14	0.026	0.11	0.18	0.05	0.077	0.73	0.70	0.058
15	0.026	0.11	0.15	0.05	0.075	0.72	0.69	0.058
16	0.026	0.11	0.15	0.05	0.052	0.71	0.69	0.058
17	0.026	0.11	0.12	0.05	0.052	0.69	0.67	0.058

Table 4-2. Wood-Frame Wall U-Values

Insulation R-Value	16-in. O.C. Wall U-Value
R-0	0.238
R-7	0.105
R-8	0.099
R-9	0.094
R-10	0.090
R-11	0.089
R-12	0.085
R-13	0.082
R-14	0.079
R-15	0.077
R-16	0.066
R-17	0.064
R-18	0.062
R-19	0.060
R-20	0.058
R-21	0.057

Table 4-5. Floor U-Values

Insulation R-Value	Floor U-Value
R-0	0.249
R-7	0.096
R-11	0.072
R-13	0.064
R-15	0.057
R-19	0.047
R-21	0.044
R-26	0.037
R-30	0.033

Table 4-7. Slab F-Values

Perimeter Insulation R-Value	24-in. Insulation Depth
R-0	1.04
R-1	0.91
R-2	0.86
R-3	0.83
R-4	0.82
R-5	0.80
R-6	0.79
R-7	0.78
R-8	0.78
R-9	0.77
R-10	0.77

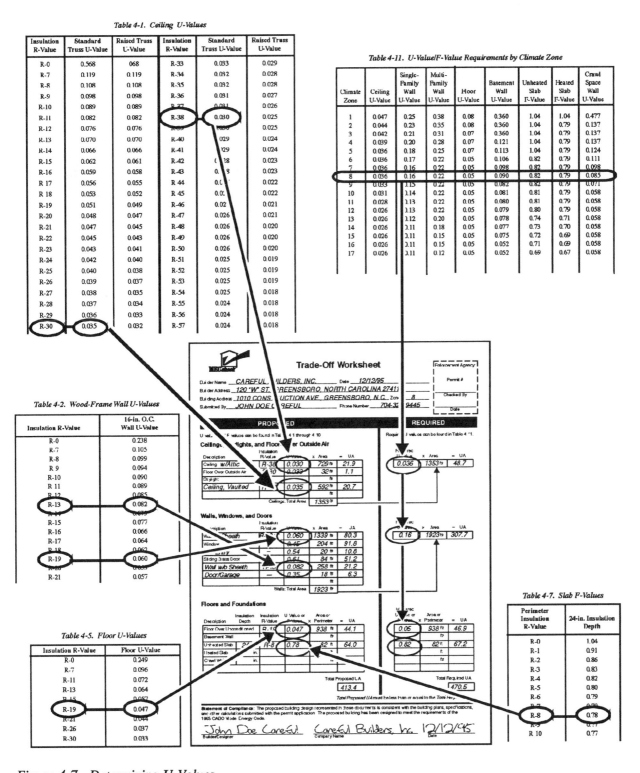

Figure 4-7. Determining U-Values

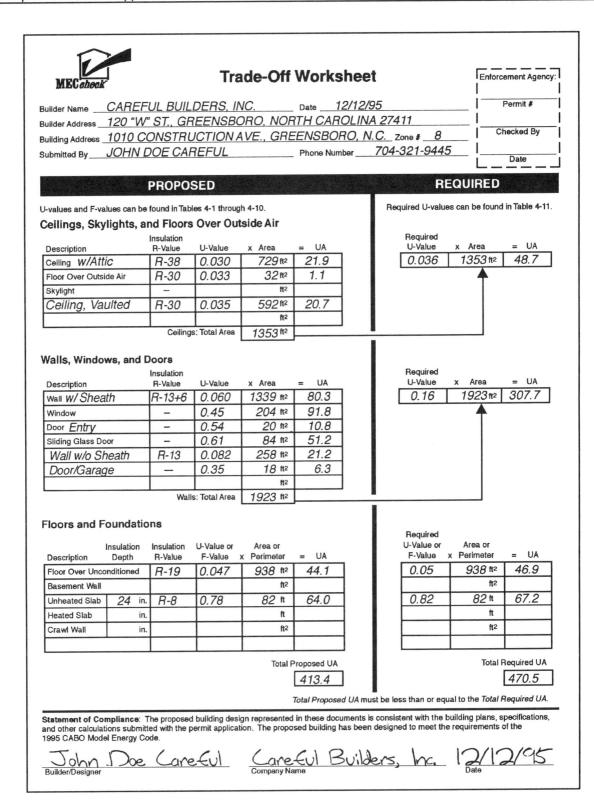

Figure 4-8. Completed Trade-Off Worksheet

Chapter 5
Software Approach

The software approach uses the MEC*check*™ software to automate the trade-off approach and to incorporate additional features (such as high-efficiency heating and cooling equipment trade-offs). It is the most flexible approach for meeting the MEC insulation and window requirements (refer to Chapter 2 for additional requirements that must also be satisfied). This approach can be used for one- and two-family (referred to as single-family) and multifamily residential buildings. The MEC*check*™ software is designed to run on most DOS-based computers and requires minimal input.

The software enables you to quickly compare different insulation levels in different parts of your building to arrive at a package that works best for you. A report that can be submitted with your building plans for plan review is automatically generated. The MEC*check*™ software and user's guide are provided separately from this manual. See the software user's guide for complete details on using this compliance approach.

Figure 5-1 illustrates the main software screen. You specify component types (e.g., 16"-O.C. wood-frame walls) and their area, insulation R-values, and glazing and door U-values. You must also enter the location of your building. The software calculates the rest, including your proposed home UA and the MEC-required UA, and determines if the building complies.

```
MECcheck 2.0 / 1995 MEC     Building Description:  example.mec
 File  Trade-Offs  Project  Help

 Ceilings │ Walls │ Glazing │ Doors │ Floors │ Basement │ Slab │ Crawl │

 Compliance │Passes       │          Required UA │467│   Your Home │416│

                              Area or   Insulation Sheathing Glazing/Door
                              Perimeter R-Value    R-Value   U-Value      UA

 CEILINGS                       │729│     │38│       │0│                  22

 CEILINGS                       │592│     │30│       │0│                  21

 WALLS: Wood Frame, 16" O.C.    │1339│    │13│       │6│                  82

 WALLS: Wood Frame, 16" O.C.    │258│     │13│       │0│                  21

 GLAZING: Windows or Doors      │204│                          │0.45│    92

 GLAZING: Windows or Doors      │84│                           │0.61│    51

 DOORS                          │20│                           │0.54│    11

 DOORS                          │10│                           │0.35│     G

 FLOORS: Over Unconditioned Space │938│  │19│                           45

 FLOORS: Over Outside Air       │32│     │30│                            1

 SLAB FLOORS: Unheated, 24.0" insul. │02│ │0│                           G4


 Use the buttons at the top of the screen to create a building description.
```

Figure 5-1. Software Screen

Chapter 6
Plan Check

Plan checkers can use this chapter as a guide in verifying that building plans and specifications comply with the basic requirements given in Chapter 2 and with one of the three compliance approaches given in Chapters 3, 4, and 5. This chapter recommends which code-related information be documented and submitted for plan review. The MEC requires this information be provided on the plans and specifications. For simplicity, this chapter refers to the "plans." However, this reference can be loosely interpreted to allow the same information to be provided on specifications, schedules, and/or other documents accepted by your jurisdiction.

Prior to performing a plan check, verify that the MEC requirements apply to the building under consideration (see *What Buildings Must Comply?* in Chapter 1). You should also be familiar with the basic requirements outlined in Chapter 2.

Building Envelope

An applicant may use any of the three compliance approaches to demonstrate compliance with the MEC insulation and window requirements -- the prescriptive package approach, the trade-off approach, or the software approach. The requirements listed under the *General* section (directly below) apply to all three of these approaches. Requirements specific to a given approach are listed next, each under their own section.

General

- Verify that the general information section of the compliance forms have been filled out and signed by the applicant. Note the name and phone number of the compliance author if the builder did not complete the forms. The author will be a valuable source of information if questions arise concerning compliance.

- Verify that the location on the compliance form agrees with the actual building location.

- Verify that all proposed R-values and U-values on the compliance form correspond to the values listed on the plans. These values are usually found in the building sections and construction details. The actual proposed R-values and U-values are required on the plans, not the weighted average R-values and U-values.

- Verify that the insulation levels specified on the compliance forms, plans, and/or specifications will fit into the framing cavities without compressing the insulation. For example, R-19 fiberglass batt insulation (6-in. thickness) will not fit in a 2x4 stud cavity (3 1/2 in. depth) without compressing the insulation. Insulation ratings are based on uncompressed insulation -- compressed insulation has a lower R-value. Thus, credit cannot be taken for manufacturer rated

R-values in the calculations if the insulation is compressed. Where loose fill insulation is proposed, the plans should specify the material to be used and the minimum depth required to meet the proposed R-value.

- Verify that the plans indicate the type of insulation to be used. For example, if proposed R-values for walls or ceilings are based on cavity insulation plus insulating sheathing, verify that the R-values of both materials are identified.

- Verify that the U-values for windows are less than or equal to the U-values indicated on the compliance forms. U-values for all windows should be written on the building plans or a window schedule. This notation will help the field inspector verify that windows with the correct U-value are installed in the building. If default window U-values from Table 4-9 are used, verify that the features (frame type, number of panes, and thermal break) are written on the plans.

- Refer to Table 2-1 for maximum air leakage rates for manufactured doors and windows. Windows and doors certified by an accredited lab (such as NWWDA or AAMA) will be labeled and can be accepted as meeting these requirements. For non-certified doors and windows, the applicant should submit manufacturer's test reports showing that the product meets these requirements. The field inspector should verify that installed windows and doors are either labeled by an accredited lab or that the make and model of the installed windows and doors correspond to the test report(s) submitted.

Prescriptive Package Approach

Check the following additional items if the Prescriptive Package approach is used.
- Verify that the *Prescriptive Package Worksheet* has been submitted.

- Verify that the values in the *Minimum R-Value* column and *Maximum U-Value* column on the right side of the *Prescriptive Package Worksheet* correspond to the values specified in the prescriptive package selected by the applicant (the applicant must record the package number at the top of the worksheet). Only one value for each component type may be listed in the *Minimum R-Value* and *Maximum U-Value* columns. If the required values on the right side of the worksheet differ from the values specified in the selected prescriptive package, have the applicant explain and/or correct the form.

- All insulation R-values in the *Proposed R-Value* column on the left side of the worksheet must be greater than or equal to corresponding values in the *Minimum R-Value* column on the right. All U-values in the *Proposed*

Prescriptive Package Worksheet

Builder Name _CAREFUL BUILDERS, INC._ Date _12/12/95_

Builder Address _120 "W" ST., GREENSBORO, NORTH CAROLINA 27411_

Building Address _1010 CONSTRUCTION AVE. GREENSBORO, NORTH CAROLINA_

Zone Number _8_ Package Number _4_

Submitted By _JOHN DOE CAREFUL_ Phone Number _704-321-9445_

Enforcement Agency:

Permit #

Checked By

Date

PROPOSED			REQUIRED

Glazing Area

$100 \times \dfrac{288}{\text{Glazing Area}} \div \dfrac{1923}{\text{Gross Wall Area}} = \dfrac{15.0}{\text{Proposed Glazing Area}}$ %

$\dfrac{15}{\text{Maximum Glazing Area}}$ %

R-Value

Description	Comments	Proposed R-Value	Minimum R-Value
Ceiling		R-30	R-30
Wall		R-13	R-13
Floor Over Unconditioned Space		R-19	R-19
Floor Over Outside Air		R-30	R-30
Basement Wall		R-N/A	R-8
Slab Floor	Unheated, 24" Depth	R-8	R-5
Crawl Space Wall		R-N/A	R-10

U-Value

Description	Comments	Proposed U-Value	Maximum U-Value
Glazing	See back	U-0.50	U-0.50
Opaque Door	Front door exempt	U-0.35	U-0.35

Equipment Efficiency (This section may be left blank if *Normal* is selected on the right.)

Heating _____ AFUE/HSPF _____

Cooling _____ SEER _____
    Efficiency   Make & Model Number

Check One
☒ Normal
☐ High Heating
☐ High Cooling
☐ High Heating & Cooling

Statement of Compliance: The proposed building design represented in these documents is consistent with the building plans, specifications, and other calculations submitted with the permit application. The proposed building has been designed to meet the requirements of the 1995 CABO Model Energy Code.

John Doe Careful _Careful Builders, Inc._ _12/12/95_
Builder/Designer   Company Name   Date

Figure 6-1. Example Prescriptive Package Worksheet

U-Value column of the worksheet must be less than or equal to the corresponding values in the *Maximum U-Value* column. In some cases, more than one insulation level will be installed in the same component (e.g., flat ceilings and vaulted ceilings). Unless all insulation levels exceed the requirement, the applicant must verify that the area-weighted average R-value for the component meets or exceeds the required R-value (or U-value for windows and doors) by completing the *R-Value/U-Value Weighted Average Worksheet* located on the back side of the *Prescriptive Package Worksheet*. Verify that the average R-value or U-value was copied to the front of the worksheet. If the proposed values on the left side of the worksheet differ from the values specified on the plans, have the applicant explain and/or correct the form.

- If a raised-truss is installed as part of the roof assembly, R-30 batt insulation may be used to meet an R-38 requirement and R-38 batt insulation may be used to meet an R-49 requirement. To receive this credit, the insulation must extend to the full thickness over the exterior walls, and the plans must be clearly marked to show the truss type.

- Verify that the area of the glazing assemblies shown on the plans matches the *Glazing Area* specified on the worksheet. Verify that the gross wall area shown on the plans matches the *Gross Wall Area* specified on the worksheet. Verify that the *Proposed Glazing Area* percentage specified on the worksheet is equal to or less than the *Maximum Glazing Area* percentage.

- If the chosen package requires high-efficiency equipment, the efficiency of the proposed unit must be specified on the worksheet in the space(s) labeled *Efficiency* and the make and model number(s) must be specified in the space(s) labeled *Make & Model Number*. The building plans should include a note designating the make, model number, and efficiency of any credited equipment. High-efficiency heating equipment must have an AFUE of at least 90% or an HSPF of at least 7.8, and high-efficiency cooling equipment must have a SEER of at least 12. AFUE, HSPF, and SEER ratings can be obtained from manufacturer data sheets or certified product directories (such as directories published by the Air Conditioning and Refrigeration Institute [ARI] or the Gas Appliance Manufacturers Association [GAMA]).

Trade-Off Approach

Check the following additional items if the trade-off approach is used.

- Verify that the *Trade-Off Worksheet* has been submitted.

- The proposed U-values listed on the left side of the *Trade-Off Worksheet* must come from Tables 4-1 through 4-8 unless derived through an alternative test procedure or calculation procedure accepted by your local jurisdiction. If alternative U-values are used, ask the applicant to submit supporting documentation. When U-values from Tables 4-1 through 4-8 are used, verify that the U-values listed on the worksheet match the U-values in these tables, given the proposed R-values and framing indicated on the plans. Window and door U-values must be tested and documented by the manufacturer in accordance with the NFRC test procedure or taken from Table 4-9 and 4-10.

- Verify that all building envelope components are included on the worksheet and that the proposed building areas on the plans agree with those listed on the worksheet.

- Verify that the U-values in the *Required U-Value (/F-Value)* column of the worksheet are correct for the building location (zone number). These values come from Table 4-11. The climate zone numbers referred to in Table 4-11 can be found in Appendix C or on the state map included with the prescriptive packages.

- Values in the *Area* column on the right side of the worksheet should be equal to the sum of the proposed component areas on the left side for a given assembly type (i.e., ceilings, walls, or floors/foundations). For example, the wall area on the right side of the worksheet is the sum of the opaque wall, window, and door components listed on the left side.

- The *Total Proposed UA* for the proposed building must be less than or equal to the *Total Required UA*. Verify that both totals are calculated correctly by spot-checking the UA calculations.

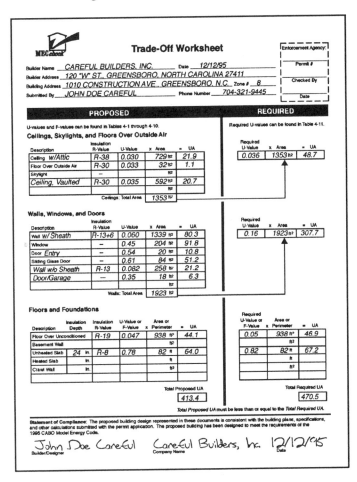

Figure 6-2. Example Trade-Off Worksheet

Software Approach

Check the following additional items if the software approach is used.

- Verify that the software report has been submitted.

- Verify that the proposed building component areas on the plans agree with the areas listed on the software report. Verify that the insulation and sheathing

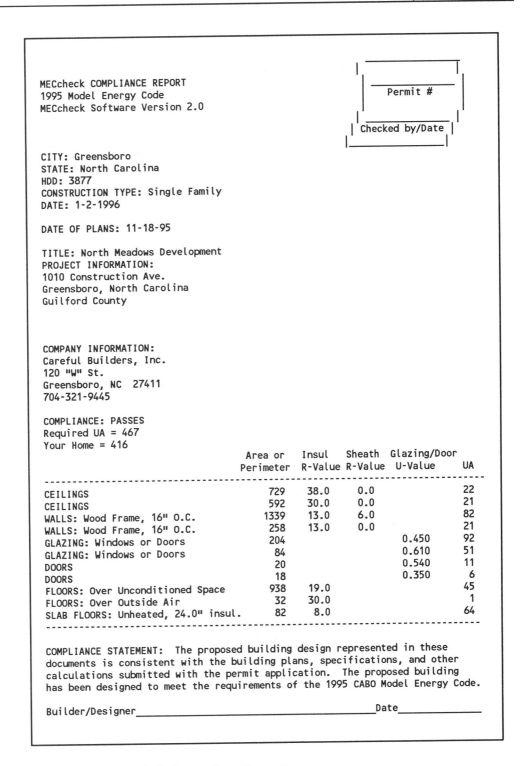

```
MECcheck COMPLIANCE REPORT                           _____
1995 Model Energy Code                              |_____|
MECcheck Software Version 2.0                       |   Permit #     |
                                                    |_____|
                                                    |                |
                                                    | Checked by/Date|
                                                    |_____|
CITY: Greensboro
STATE: North Carolina
HDD: 3877
CONSTRUCTION TYPE: Single Family
DATE: 1-2-1996

DATE OF PLANS: 11-18-95

TITLE: North Meadows Development
PROJECT INFORMATION:
1010 Construction Ave.
Greensboro, North Carolina
Guilford County


COMPANY INFORMATION:
Careful Builders, Inc.
120 "W" St.
Greensboro, NC  27411
704-321-9445

COMPLIANCE: PASSES
Required UA = 467
Your Home = 416
                            Area or    Insul  Sheath  Glazing/Door
                            Perimeter  R-Value R-Value U-Value    UA
                         ---------------------------------------------------
CEILINGS                      729       38.0    0.0                22
CEILINGS                      592       30.0    0.0                21
WALLS: Wood Frame, 16" O.C.  1339       13.0    6.0                82
WALLS: Wood Frame, 16" O.C.   258       13.0    0.0                21
GLAZING: Windows or Doors     204                       0.450      92
GLAZING: Windows or Doors      84                       0.610      51
DOORS                          20                       0.540      11
DOORS                          18                       0.350       6
FLOORS: Over Unconditioned Space  938   19.0                       45
FLOORS: Over Outside Air       32       30.0                        1
SLAB FLOORS: Unheated, 24.0" insul.  82  8.0                       64
                         ---------------------------------------------------

COMPLIANCE STATEMENT:  The proposed building design represented in these
documents is consistent with the building plans, specifications, and other
calculations submitted with the permit application.  The proposed building
has been designed to meet the requirements of the 1995 CABO Model Energy Code.

Builder/Designer_____Date_____
```

Figure 6-3. Example Software Compliance Report

R-values and window and door U-values in the software report are shown on the plans. The applicant must correct inaccurate information and resubmit the software report.

- If high-efficiency heating or cooling equipment is indicated in the software report, verify that the plans designate equipment that meets or exceeds what is claimed in the software report. AFUE, HSPF, and SEER ratings can be obtained from manufacturer data sheets or certified product directories (such as directories published by the Air Conditioning and Refrigeration Institute [ARI] or the Gas Appliance Manufacturers Association [GAMA]).

- Verify that the software report shows that the building passed.

Heating and Cooling

Heating and cooling system requirements must be called out on the plans. These requirements include thermostats, duct and piping insulation R-values, and the make, model, and efficiency of the heating and/or cooling equipment if credit is taken for high-efficiency equipment.

- Verify that at least one thermostat is planned for each system in one- and two-family buildings and that a separate thermostat is planned for each dwelling unit in multifamily buildings.

- Verify that the duct insulation levels documented on the plans match the insulation levels specified in Table 2-2. If any insulation levels differ, note the correct level on the plans.

- All piping associated with HVAC systems installed in unconditioned spaces and conveying fluid at temperatures greater than 120°F or chilled fluids at less than 55°F must be insulated to the levels specified in Table 2-3. Verify that pipe sizes, insulation thicknesses, and R-values are documented on the plans.

Service (Potable) Water Heating

If a swimming pool is to be installed, verify a time clock for the pump has been indicated on the plans. If heated, verify that the plans call for an on/off pool heater switch. A pool cover is required for heat pools unless over 20% of the heating energy is from non-depletable sources.

Circulating hot water systems require pump controls and pipe insulation to the levels specified in Table 2-4. Verify that the controls and insulation levels are specified on the plans.

Electrical

Verify that each dwelling unit in a multifamily building is equipped with a separate electric meter. These should be individual service meters or, where allowed by the utility, one master meter to the building and individual submeters to each unit. Check the electric plans for the location and number of electric meters.

Field Notes

All new residences must comply with the basic requirements discussed in Chapter 2. Many of these requirements cannot be verified during plan check and must, instead, be verified during field inspection. Field notes should be included on the plans as a reminder to field inspectors. These might include the following:

- insulation type and R-value

- a window schedule listing window sizes, U-values, and make and model numbers

- heating and cooling equipment efficiency and make and model number

- duct locations and insulation levels

- HVAC piping locations and insulation levels

- circulating hot-water piping locations and insulation levels

The *Summary of Basic Requirements* included with this manual (or a similar list of applicable requirements) can be included on the building plans as a reminder to the field inspector.

You should also initiate the *Field Inspection Checklist* included with this manual. Transfer the proposed R-values from the plans to the appropriate location on the checklist. All of the basic requirements specified in Chapter 2 are summarized on this checklist. These requirements should be included on the plans and compliance forms. The field inspector should have the compliance forms during the inspection process.

Chapter 7
Field Inspection

Field inspections are necessary to ensure that required materials and equipment are properly installed at the site and are in accordance with the approved building plans and/or specifications. This chapter provides guidance to field inspectors performing site inspections on residential buildings that must comply with the MEC. Because the number and types of inspections vary throughout the country, you are encouraged to customize these guidelines for your jurisdiction.

This chapter at times refers to code-related information that was documented and submitted prior to plan review. The MEC requires this information be provided on the plans and specifications. For simplicity, this chapter refers to the "plans." However, this same information can instead be provided on specifications, schedules, and/or other documents accepted by your jurisdiction.

Four separate site inspections are commonly used to inspect energy features:

- foundation inspection

- framing inspection

- insulation inspection

- final inspection.

These should coincide with site visits typically required for general structural/mechanical/electrical inspections. The insulation inspection, however, is best done after installation of the electrical and plumbing systems and before the insulation is covered, and may require a separate trip.

You can use the *Field Inspection Checklist* provided with this manual to verify energy features. The checklist is divided into sections that reflect the four separate inspection visits and identifies the items to be inspected on each of these visits. For example, the *Foundation Inspection* section identifies slab-edge, basement wall, and crawl space wall insulation as items that should be inspected during the foundation inspection.

Pre-Inspection

Before beginning the field inspection, verify that the approved building plans and specifications are on site and that the MEC*check*™ forms (or alternate forms that provide the same information) are attached to the plans. The building plans, specifications, and compliance forms should provide you with all the information necessary to perform a field inspection properly.

Field Inspection Checklist

	Requirement	Installed (Y/N)	Comments

Pre-Inspection
- Approved Building Plans on Site (104.1)

Foundation Inspection   Inspection Date _____ Approved: Yes ___ No ___ Init._____
- Slab-Edge Insulation (502.2.1.4)   Depth: _____
- Basement Wall Exterior Insulation (502.2.1.6)   Depth: _____
- Crawl Space Wall Insulation (502.2.1.5)   Depth: _____

Framing Inspection   Inspection Date _____ Approved: Yes ___ No ___ Init._____
- Floor Insulation (502.2.1.3)
- Glazing and Door Area (502.2.1.1)
- Mass Walls (502.1.2)
- Caulking/Sealing Penetrations (502.4.3)
- Duct Insulation (503.9.1)
- Duct Construction (503.10.2)
- HVAC Piping Insulation (503.11)
- Circulating Hot-Water Piping Insulation (504.7)

Insulation Inspection   Inspection Date _____ Approved: Yes ___ No ___ Init._____
- Wall Insulation (502.2.1.1)
- Basement Wall Interior Insulation (502.2.1.6)   Depth: _____
- Ceiling Insulation (502.2.1.2)
- Glazing and Door U-Values (502.2.1.1)
- Vapor Retarder (502.1.4)

Final Inspection   Inspection Date _____ Approved: Yes ___ No ___ Init._____
- Heating Equipment (102.1)
    Make and Model Number
    Efficiency (AFUE or HSPF)
- Cooling Equipment (102.1)
    Make and Model Number
    Efficiency (SEER)
- Multifamily Units Separately Metered (505.2)
- Thermostats for Each System (503.8.3)
- Heat Pump Thermostat (503.4.2.3)
- Window and Door Air Leakage (502.4.2)
- Weatherstripping at Doors/Windows (502.4.3)
- Equipment Maintenance Information (102.2)

Figure 7-1. Field Inspection Checklist

Foundation Inspection

Slab-Edge Insulation

If slab-edge insulation is indicated on the compliance forms or the building plans, verify the installed R-value is greater than or equal to the specified R-value. The insulation R-value should be stamped on the slab insulation. If the R-value is not shown, the installation contractor should provide verification of the insulation R-value. In addition, slab-edge insulation must be protected from physical damage and ultraviolet light deterioration when installed on the exterior of the footing.

The required slab insulation depth will vary depending on the building location and the compliance approach. The insulation must extend from the top of the slab down to at least the bottom of the slab, and then optionally it may extend in under the slab or out away from the slab. The total vertical plus horizontal distance of the slab insulation must be at least as great as indicated on the plans or compliance forms. Insulation that extends horizontally away from the slab must be covered by pavement or at least 10 in. of soil.

Basement Wall Exterior Insulation

When the insulation is installed on the outside of the basement wall, verify that the R-value is greater than or equal to the R-value specified on the compliance forms or the building plans. (Interior basement insulation is usually checked during the insulation inspection.) The insulation should be securely fastened to the foundation wall and protected from weather and ultraviolet degradation. The prescriptive package and trade-off compliance approaches require basement walls to be insulated from the top of the basement wall to 10 ft below ground level or to the basement floor, whichever is less. The Software approach allows the insulation depth to vary -- the required depth will be indicated on the software report. If possible, observe backfilling to make sure the insulation is evenly applied and not damaged.

Crawl Space Wall Insulation

Crawl spaces below insulated floors do not require insulation nor a vapor retarder. If the crawl space is not directly ventilated to the outdoors, the builder may choose to insulate the crawl space walls instead of the floor above the crawl space. In this case, verify that the crawl space wall insulation R-value is greater than or equal to the R-value specified on the plans and that it extends the required distance. The prescriptive package and trade-off approaches require the crawl space wall insulation to extend from the top of the foundation wall to at least   12 in. below the outside finished grade. If the distance from the outside finished grade to the top of the footing is less than 12 in., the insulation must extend a total vertical plus horizontal distance of 24 in. from the outside finished grade.   The MEC*check*™ Software approach allows the depth of insulation to be traded off, and the required depth will be printed on the software report.

If the insulation is to be installed as part of the foundation form, the R-value should be verified prior to pouring concrete. Ask for manufacturer literature if the R-value is not printed on the insulation. The insulation should be securely attached to the foundation wall to ensure that it will remain in place. The insulation must be flush with the top of the foundation wall.

Framing Inspection

Floor Insulation

If the building has a raised floor and the floor insulation has been installed, verify that the installed insulation matches the plans. If the floor insulation has not yet been installed, it may be inspected during the insulation inspection. Insulation R-values are marked on the insulation either with an R-value designation (e.g., R-19) or striping. Table 7-1 lists stripe codes developed by the North American Insulation Manufacturers Association (NAIMA) as a voluntary guideline for identification of insulation R-values.

Properly installed floor insulation should be flush against the subfloor. The insulation should be adequately and uniformly supported with supports such as furring running perpendicular to the joists, piano wire stapled to the joists, or "tiger teeth". The insulation should not sag away from the floor.

Table 7-1. Stripe Codes on Unfaced Fiberglass Batts

Labeled R-Value	Number of Stripes
R-11	3
R-13	4
R-15 HP	2
R-19	5
R-21 HP	4
R-22	6
R-25	1
R-26	1
R-30	2
R-30 HP	3
R-38	4
R-38 HP	1
HP = High performance and/or high density insulation.	

Glazing and Door Area

Openings for windows, doors, and skylights will be roughed out during the framing inspection. Verify that the rough openings for these assemblies correspond to what is shown on the plans. Note any increase in area. If an increase is noted, the applicant must complete new compliance forms to show that the building is still in compliance.

Mass Walls

In some locations, the MEC specifies less-stringent requirements for above-grade heavy mass walls than for wood- or metal-frame walls.[a] Masonry, concrete, and log walls are examples of mass walls that are sometimes eligible to receive this credit. Check the framing details, floor plans, and structural plans to ensure that any mass walls shown on the plans are installed in the building and that insulation levels match those indicated.

Joints and Penetrations

Openings and joints in the building envelope must be caulked or sealed with appropriate sealants (such as caulking or foam) to ensure there is a continuous seal (no daylight is showing). Penetrations should be sealed on both the inside and outside when sidings or veneers are used. Potential air leakage points to check include:

- exterior joints around window and door frames

- between wall sole plates, subfloors, and exterior wall panels

- openings for plumbing, electricity, refrigerant, and gas lines in exterior walls, floors, and roofs

- openings in the attic floor (such as where ceiling panels meet interior and exterior walls and masonry fireplaces)

- recessed lighting fixtures

- between tops of foundations and sills

- all other similar openings in the building envelope.

Duct Insulation

Ducts installed outside the building and in unconditioned spaces are subject to the minimum duct insulation R-value requirements given in Table 2-2. Check the mechanical plans (if available) or the compliance forms to ensure that the minimum insulation level has been installed.

Ducts located in unconditioned spaces include ducts located in attics, ventilated crawl spaces, unheated basements and garages, and inside building envelope assemblies

---

[a] In Version 2.0, only the software approach implements the MEC mass wall credit. It is not necessary to inspect for mass walls if the prescriptive package or trade-off approach is used.

(e.g., exterior wall, ceiling, and floor cavities). When ducts are located in exterior building cavities, either

- the full insulation R-value requirement for that building component must be installed between the duct and the building exterior, in which case the ducts do not require insulation, or

- the ducts must be insulated to the duct R-value requirement given in Table 2-2 and the duct area must be treated as a separate component on the compliance documentation.

For flexible ductwork, verify that the R-value printed on the jacket meets or exceeds the R-value listed on the plans and the compliance forms. Fiberglass ductwork may have the R-value printed on the outside of the material.

Duct Construction

Verify that ducts located in unconditioned spaces are sealed using mastic with fibrous backing tape. Pressure-sensitive tape may be used for fibrous ducts. Duct tape is not permitted. Verify that all ducts are supported properly and duct runs have no sags. Also verify that the installed ducts are not crimped nor constricted.

HVAC Plumbing Insulation

All HVAC system piping installed in unconditioned spaces and conveying fluid temperatures greater than 120°F or chilled fluids less than 55°F must be insulated to the levels specified in Table 2-3. Pipes located in unconditioned spaces include those located inside building envelope assemblies and those located outside of the building. The pipe sizes, insulation thicknesses, and R-values must be identified on the plans. The installed insulation R-value must be greater than or equal to that shown on the plans.

Circulating Hot Water Systems

Circulating hot water systems must have manual or automatic controls that allow pumps to be conveniently turned off when the system is not in operation. Check the plans to determine the location of the pump control. If the pump is conveniently located (such as in a garage) the on/off switch may be located on or near the pump. If the pump is located in an inaccessible location (such as under the house or in an attic) the control must be located in a more convenient location -- not on the pump.

Check the plans for the required insulation R-value to be installed on circulation loop piping and verify that piping is insulated to the required level.

Insulation Inspection

Wall Insulation

Verify that the R-value of the installed wall insulation is greater than or equal to that specified on the building plans and compliance forms. The R-value is usually marked on the insulation by an R-value designation or striping (see Table 7-1). For the prescriptive package and trade-off compliance approaches, the cavity insulation R-value and the R-value of exterior rigid insulation may be added together to make a composite R-value which meets the wall requirement. In these cases, both insulation R-values should be marked on the plans or compliance forms. The software approach lists exterior rigid insulation (sheathing) separately on the software report. Some wall constructions may have rigid insulation on the interior of the framing.

Batt insulation must evenly fill the wall cavity. When kraft-faced batts are used, they should be properly stapled on the face of the studs to serve as a vapor retarder. Vapor retarders must be placed between the conditioned space and the insulation, except in exempted locations (see Chapter 2, *Vapor Retarders*).

Wall cavity insulation must extend into the perimeter floor joist (rim joist) cavities along the same plane as the wall (Figure 7-2). Verify that exterior walls behind bathtubs and showers, kitchen cabinets, and stairwells are fully insulated. Batts must be cut around switch and outlet boxes and split around plumbing and wiring to ensure that the insulation is not compressed and there are no gaps or holes in the insulation.

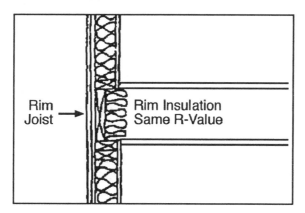

Figure 7-2. Wall Insulation Between Floors

Basement Wall Interior Insulation

When applicants use the prescriptive package approach or the trade-off approach, basement walls below uninsulated floors must be insulated from the top of the basement wall to 10 ft below ground level or to the basement floor, whichever is less. The software approach enables the user to select an alternate insulation depth, which is indicated on the software report. If basement wall insulation is installed on the wall interior, it can be inspected during the insulation inspection. Interior insulation typically consists of fiberglass batt or rigid board installed between furring strips or framing. Verify that the R-value of the installed basement wall insulation is at least that specified on the plans.

Ceiling Insulation

For fiberglass batt insulation, verify that the R-value printed on the batt is greater than or equal to that specified on the plans. Loose-fill insulation must be distributed evenly within the attic space. It is important that the insulation be installed to manufacturer specifications to minimize settling. Because of variations in material used for loose-fill insulation, the R-value may not be universally determined by direct measurement of installed depth. There should be clearly visible thickness markers installed about once every 300 ft^2. Verify that the depth and type of insulation matches what is indicated on the plans. Some jurisdictions may require that the installer provide a label at the attic access certifying the type of insulation used, the maximum coverage per bag at the installed R-value, the minimum thickness, and the number of bags installed.

Insulation should be kept at least 3 inches away from recessed lighting not approved for a zero-clearance lighting cover (i.e., not IC rated). Verify that insulation clearances from appliances meet manufacturer specifications and comply with any local code restrictions. Insulation should be kept away from vents, or baffling should be installed to direct the vent air up and over the ceiling insulation.

If the applicant has taken credit for a raised truss, verify that the insulation achieves its full insulation depth over the exterior walls.

Glazing and Door U-Values

If glazing and door assemblies are labeled, ensure that the label U-values are equal to or less than the U-values marked on the plans or window schedule. If not labeled, verify that the make and model numbers of the installed assemblies match those specified on the plans or window schedule. If the glazing assembly U-values were obtained from Table 4-9 of this manual the installed features (frame type, number of panes, and thermal break) must match the features described on the plans. Likewise, if door U-values were obtained from
Table 4-10, the door features must correspond to the features described on the plans. If the assemblies do not match the plans, the applicant must demonstrate that the replacements are equivalent (area and U-values are less than or equal to the originally specified products) or they must re-submit the compliance documentation based on the new assemblies.

The appropriate time to inspect glazing and door assemblies will vary depending on when they are installed and the types of inspections conducted in your jurisdiction.

Vapor Retarder

Vapor retarders having a perm rating of no more than 1.0 must be installed in all unvented frame walls, floors, and ceilings, except in exempted locations. For example, sealed floor cavities (such as floors of overhangs) require a vapor retarder, but floors over a ventilated crawl space do not. Floor vapor retarders will typically be inspected during the framing inspection. The vapor retarder must be placed between the conditioned space and the insulation.

Final Inspection

Heating and Cooling Equipment

If high-efficiency heating or cooling equipment is claimed, the equipment must be designated on the plans. To find the make and model number of an installed furnace, remove the front cover of the heating unit. The information is sometimes printed on the nameplate. If the numbers do not match, the installer must show that the installed unit meets the required efficiency. Where equipment efficiencies are not available, verify that the installed equipment make and model numbers match what is indicated on the plans. If high-efficiency equipment was not claimed, you do not need to verify equipment make, model number, or efficiency.

Electric Meters

Inspect the electrical service to each multifamily dwelling unit and make sure an electric meter is provided for each individual dwelling unit.

Thermostats

Each heating and cooling system must have a thermostat capable of manually or automatically reducing the temperature. Heat pumps must include a heat pump thermostat that can prevent the back-up heat from turning on when the heating requirements can be met by the heat pump alone. A two-stage thermostat that controls the back-up heat on its second stage meets this requirement.

Window and Door Air Leakage

Verify that windows and doors are either certified by an accredited lab (such as NWWDA or AAMA) or that the make and model numbers of installed manufactured windows and doors match the information on the plans. If proposed models of unlabeled products are different from installed models, the applicant must verify the air leakage requirements have been met. Check site-built windows for adequate sealing and caulking.

Weatherstripping

Verify that all doors between conditioned and unconditioned spaces have door boots and weatherstripping. Weatherstripping should also be installed around attic and crawlspace access panels if the panels are located in a conditioned room.

Equipment Maintenance and Operation Information

Manufacturer manuals must be provided for installed heating and cooling equipment and service water heating equipment. This information must be available at the building site during the final inspection.

Materials Information

If the *Energy Label* included with this manual is posted in the building, verify that the insulation levels and equipment efficiencies reported on the label correspond to what was actually installed.

Appendix A
Definitions

Addition The MEC applies to new residential buildings and additions to existing buildings. Additions can be shown to comply by themselves without reference to the rest of the building. Alternatively, the entire building (the existing building plus the new addition) can be shown to comply.

Basement Wall Basement walls that enclose conditioned spaces are part of the building envelope. Basement wall refers to the opaque portion of the wall (excluding windows and doors). To be considered a basement wall, the average gross wall area (including openings) must be at least 50% below grade. Treat walls on each side of the basement individually when determining if they are above grade or basement walls. For any individual wall less than 50% below grade, include the entire opaque wall area as part of the above-grade walls.

Building Envelope The building envelope includes all components of a building that enclose conditioned spaces (see the definition of conditioned space). Building envelope components separate conditioned spaces from unconditioned spaces or from outside air. For example, walls and doors between an unheated garage and a living area are part of the building envelope; walls separating an unheated garage from the outside are not. Although floors of conditioned basements and conditioned crawl spaces are technically part of the building envelope, the MEC does not specify insulation requirements for these components and they can be ignored.

Ceiling The ceiling requirements apply to portions of the roof and/or ceiling through which heat flows. Ceiling components include the interior surface of flat ceilings below attics, the interior surface of cathedral or vaulted ceilings, and skylights. The ceiling requirements also apply to floors over outside air, including floor cantilevers, floors of an elevated home, and floors of overhangs (such as the floor above a recessed entryway or open carport).

Conditioned Space A space is conditioned if heating and/or cooling is deliberately supplied to it or is indirectly supplied through uninsulated surfaces of water or heating equipment or through uninsulated ducts. For example, a basement with registers or heating devices designed to supply heat is conditioned.

Crawl Space The MECcheck™ crawl space wall insulation requirements are for the exterior walls of unventilated crawl spaces (i.e. not directly vented to the outside) below uninsulated floors. A crawl space wall component includes the opaque portion of a wall that encloses a crawl space and is partially or totally below grade.

Door Doors include all openable opaque assemblies located in exterior walls of the building envelope. Doors with glass can be treated as a single door assembly, in which case an aggregate U-value (a U-value that includes both the glass and opaque area) must be used; OR the glass area of the door can be included with the other glazing and an opaque door U-value can be used to determine compliance of the door.

Dwelling Unit   A single housekeeping unit of one or more rooms providing complete, independent living facilities, including permanent provisions for living, sleeping, eating, cooking and sanitation.

Envelope   See Building Envelope

Glazing   Glazing is any translucent or transparent material in exterior openings of buildings, including windows, skylights, sliding glass doors, the glass areas of opaque doors, and glass block.

Glazing Area   The area of a glazing assembly is the interior surface area of the entire assembly, including glazing, sash, curbing, and other framing elements. The nominal area or rough opening is also acceptable for flat windows and doors.

Floor Area   Not all floors in a building are considered when computing the floor area for compliance purposes:

- Floors over unconditioned spaces (such as floors over an unheated garage, basement, or crawl space) must be insulated.

- Floors over outside air (such as floors of overhangs and floors of an elevated home) must also be insulated but are subject to the ceiling requirements rather than the floor over unconditioned space requirements.

- In most locations, slab-on-grade floors of conditioned spaces must be insulated along the slab perimeter.

- Floors of basements and crawl spaces are not subject to an insulation requirement and do not have to be included as a building envelope component, even if the basement or crawl space is conditioned. In some cases, however, crawl space wall insulation is required to extend down from the top of the wall to the top of the footing and then horizontally a short distance along the floor.

- Floors separating two conditioned spaces are not subject to an insulation requirement and do not have to be included as a building envelope component.

Gross Wall Area   The gross wall area includes the opaque area of above-grade walls, the opaque area of any individual wall of a conditioned basement less than 50% below grade (including the below-grade portions), all windows and doors (including windows and doors of conditioned basements), and the peripheral edges of floors.

Multifamily   A multifamily building is a residential building three stories or less in height that contains three or more attached dwelling units. Multifamily buildings include apartments, condominiums, townhouses, and rowhouses. Hotels and motels are considered commercial rather than residential buildings.

Net Wall Area

The net wall area includes the opaque wall area of all above-grade walls enclosing conditioned spaces, the opaque area of conditioned basement walls less than 50% below grade (including the below-grade portions), and peripheral edges of floors. The net wall area does not include windows, doors, or other such openings, as they are treated separately.

Opaque Areas

Opaque areas as referenced in this guide include all areas of the building envelope except openings for windows, skylights, doors, and building service systems. For example, although solid wood and metal doors are opaque, they should not be included as part of the opaque wall area (also referred to as the net wall area).

Raised Truss

Raised truss refers to any roof/ceiling construction that allows the insulation to achieve its full thickness over the exterior walls. Several constructions allow for this, including elevating the heel (sometimes referred to as an energy truss, raised-heel truss, or Arkansas truss), use of cantilevered or oversized trusses, lowering the ceiling joists, or framing with a raised rafter plate.

R-Value

R-value ($h \cdot ft^2 \cdot {}^{\circ}F/Btu$) is a measure of thermal resistance, or how well a material or series of materials resists the flow of heat. R-value is the reciprocal of U-value:

$$R\text{-}Value = \frac{1}{U\text{-}Value}$$

Single Family

As defined by the MEC, a single-family building is a detached one- or two-family residential building.

Slab Edge

Slab edge refers to the perimeter of a slab-on-grade floor, where the top edge of the slab floor is above the finished grade or 12 in. or less below the finished grade.

U-Value

U-value ($Btu/h \cdot ft^2 \cdot {}^{\circ}F$) is a measure of how well a material or series of materials conducts heat. U-values for window and door assemblies are the reciprocal of the assembly R-value:

$$U\text{-}Value = \frac{1}{R\text{-}Value}$$

For other building assemblies (such as a wall), the R-value used in the above equation is the R-value of the entire assembly, not just the insulation.

Appendix B
The Building Envelope

The MEC requirements are intended to limit heat loss and air leakage through the building envelope. The building envelope includes all of the building components that separate conditioned spaces (conditioned space is defined in Appendix A) from unconditioned spaces or from outside air. For example, the walls and doors between an unheated garage and a living area are part of the building envelope; the walls separating an unheated garage from the outside are not. Walls, floors, and other building components separating two conditioned spaces (such as an interior partition wall) are *NOT* part of the building envelope, nor are common or party walls which separate dwelling units in multifamily buildings.

You can think of the building envelope as the boundary separating the inside from the outside and through which heat is transferred. Areas that have no heating or cooling sources are considered to be outside the building envelope. A space is conditioned if heating and/or cooling is deliberately supplied to it or is indirectly supplied through uninsulated surfaces of water or heating equipment or through uninsulated ducts.

To use the MEC*check*™ materials, you must specify proposed insulation levels for ceiling, wall, floor, basement wall, slab-edge, and crawl space wall components located in the building envelope. In some cases, it may be unclear how to classify a given building element. For example, are skylight shafts considered a wall component or a ceiling component? The following table can be used to help determine how a given building envelope assembly should be entered in the MEC*check*™ materials.

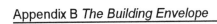

Table B-1. Building Envelope Components

Ceiling Components

Ceiling	Flat ceilings Cathedral or vaulted ceilings Dormer roofs Bay window roofs Overhead portions of an interior stairway to an attic Attic hatches
Floors Over Outside Air[a]	Floors of overhangs (such as the floor above a recessed entryway or carport) Floor cantilevers Floors of an elevated home
Skylights	Skylight assemblies

(a)   The insulation requirements for floors over outside air are the same as those for ceilings.

Wall Components

Wall	Opaque portions of above-grade walls Basement walls and kneewalls less than 50% below grade Peripheral edges of floors Gables walls bounding conditioned space Dormer walls Roof or attic kneewalls Through-wall chimneys Walls of an interior stairway to an unconditioned basement Skylight shafts
Glazing	Windows (including basement windows) Sliding glass doors Glass block Transparent portions of doors
Door	Opaque portions of all doors (including basement doors)

Floor and Foundation Components

Floor Over Unconditioned Space	Floors over an unconditioned crawl space, basement, garage, or similar unconditioned space
Basement Wall	Opaque portions of basement walls 50% or more below grade and basement kneewalls (if part of a basement wall 50% or more below grade)
Slab Floor	Perimeter edges of slab-on-grade floors
Crawl Space Wall	Walls of unventilated crawl spaces below uninsulated floors

Appendix C
Counties By Climate Zone

County	Zone	County	Zone	County	Zone	County	Zone	County	Zone
ALABAMA		Washington	5	Baxter	9	St Francis	7	Ventura	4
Autauga	6	Wilcox	5	Benton	9	Stone	9	Yolo	6
Baldwin	4	Winston	7	Boone	9	Union	6	Yuba	6
Barbour	5			Bradley	6	Van Buren	8		
Bibb	6	**ALASKA BUROUGHS**		Calhoun	6	Washington	9	**COLORADO**	
Blount	7	**AND REAA's**		Carroll	9	White	7	Adams	13
Bullock	5	Adak Region	16	Chicot	6	Woodruff	7	Alamosa	16
Butler	5	Alaska Gateway	19	Clark	6	Yell	7	Arapahoe	13
Calhoun	6	Aleutian Region	17	Clay	8			Archuleta	16
Chambers	6	Aleutians East	17	Cleburne	8	**CALIFORNIA**		Baca	11
Cherokee	7	Anchorage	17	Cleveland	6	Alameda	6	Bent	11
Chilton	6	Annette Island	15	Columbia	6	Alpine	15	Boulder	13
Choctaw	5	Bering Straits	19	Conway	7	Amador	8	Chaffee	16
Clarke	5	Bristol Bay	17	Craighead	8	Butte	6	Cheyenne	13
Clay	7	Chatham	16	Crawford	8	Calaveras	8	Clear Creek	17
Cleburne	7	Chugach	17	Crittenden	7	Colusa	6	Conejos	16
Coffee	4	Copper River	18	Cross	7	Contra Costa	6	Costilla	16
Colbert	8	Delta/Greely	18	Dallas	6	Del Norte	9	Crowley	11
Conecuh	5	Denali	18	Desha	6	El Dorado	8	Custer	16
Coosa	6	Fairbanks N. Star	18	Drew	6	Fresno	6	Delta	13
Covington	4	Haines	16	Faulkner	7	Glenn	6	Denver	13
Crenshaw	5	Iditarod Area	19	Franklin	8	Humboldt	9	Dolores	15
Cullman	7	Juneau	16	Fulton	8	Imperial	3	Douglas	13
Dale	4	Kashunamiut	18	Garland	7	Inyo	9	Eagle	15
Dallas	5	Kenai Peninsula	17	Grant	6	Kern	5	El Paso	13
De Kalb	8	Ketchikan		Greene	8	Kings	6	Elbert	13
Elmore	6	Gateway	15	Hempstead	7	Lake	8	Fremont	11
Escambia	4	Kodiak Island	16	Hot Spring	7	Lassen	13	Garfield	15
Etowah	7	Kuspuk	18	Howard	7	Los Angeles	4	Gilpin	13
Fayette	7	Lake & Peninsula	17	Independence	8	Madera	6	Grand	17
Franklin	8	Lower Kuskokwim	18	Izard	8	Marin	6	Gunnison	17
Geneva	4	Lower Yukon	18	Jackson	8	Mariposa	8	Hinsdale	17
Greene	5	Matanuska-		Jefferson	6	Mendocino	8	Huerfano	11
Hale	5	Susitna	17	Johnson	8	Merced	6	Jackson	17
Henry	4	North Slope	19	Lafayette	6	Modoc	15	Jefferson	13
Houston	4	Northwest Arctic	19	Lawrence	8	Mono	15	Kiowa	13
Jackson	8	Pribilof Islands	17	Lee	7	Monterey	6	Kit Carson	13
Jefferson	6	Sitka	15	Lincoln	6	Napa	6	La Plata	15
Lamar	7	Southeast Island	15	Little River	6	Nevada	11	Lake	17
Lauderdale	8	Southwest Region	17	Logan	7	Orange	4	Larimer	13
Lawrence	8	Yakutat	17	Lonoke	7	Placer	8	Las Animas	11
Lee	6	Yukon Flats	19	Madison	9	Plumas	13	Lincoln	13
Limestone	8	Yukon-Koyukuk	19	Marion	9	Riverside	4	Logan	13
Lowndes	5	Yupiit	18	Miller	6	Sacramento	6	Mesa	13
Macon	6			Mississippi	8	San Benito	6	Mineral	17
Madison	8	**ARIZONA**		Monroe	7	San Bernardino	4	Moffat	15
Marengo	5	Apache	13	Montgomery	8	San Diego	3	Montezuma	15
Marion	7	Cochise	6	Nevada	6	San Francisco	6	Montrose	13
Marshall	8	Coconino	14	Newton	9	San Joaquin	6	Morgan	13
Mobile	4	Gila	8	Ouachita	6	San Luis Obispo	6	Otero	11
Monroe	5	Graham	6	Perry	7	San Mateo	6	Ouray	15
Montgomery	6	Greenlee	6	Phillips	7	Santa Barbara	5	Park	17
Morgan	8	La Paz	3	Pike	7	Santa Clara	6	Phillips	13
Perry	5	Maricopa	3	Poinsett	8	Santa Cruz	6	Pitkin	17
Pickens	6	Mohave	7	Polk	8	Shasta	6	Prowers	11
Pike	5	Navajo	10	Pope	8	Sierra	11	Pueblo	11
Randolph	7	Pima	4	Prairie	7	Siskiyou	11	Rio Blanco	15
Russell	5	Pinal	4	Pulaski	7	Solano	6	Rio Grande	17
Shelby	6	Santa Cruz	6	Randolph	8	Sonoma	6	Routt	17
St Clair	6	Yavapai	10	Saline	7	Stanislaus	6	Saguache	16
Sumter	5	Yuma	3	Scott	7	Sutter	6	San Juan	17
Talladega	6			Searcy	9	Tehama	6	San Miguel	15
Tallapoosa	6	**ARKANSAS**		Sebastian	8	Trinity	9	Sedgwick	13
Tuscaloosa	6	Arkansas	6	Sevier	7	Tulare	6	Summit	17
Walker	6	Ashley	6	Sharp	8	Tuolumne	8	Teller	13

County	Zone	County	Zone	County	Zone	County	Zone	County	Zone
Washington	13	St Johns	3	Hall	7	Warren	6	Coles	12
Weld	13	St Lucie	2	Hancock	6	Washington	6	Cook	14
Yuma	13	Sumter	2	Haralson	7	Wayne	4	Crawford	11
		Suwannee	3	Harris	6	Webster	5	Cumberland	12
CONNECTICUT		Taylor	3	Hart	7	Wheeler	5	De Kalb	14
Fairfield	12	Union	3	Heard	6	White	8	De Witt	12
Hartford	13	Volusia	2	Henry	7	Whitfield	8	Douglas	12
Litchfield	14	Wakulla	4	Houston	5	Wilcox	5	Du Page	14
Middlesex	12	Walton	4	Irwin	5	Wilkes	7	Edgar	12
New Haven	12	Washington	4	Jackson	7	Wilkinson	5	Edwards	11
New London	12			Jasper	6	Worth	5	Effingham	11
Tolland	14	**GEORGIA**		Jeff Davis	4			Fayette	11
Windham	14	Appling	4	Jefferson	6	**HAWAII**		Ford	13
		Atkinson	4	Jenkins	5	Hawaii	1	Franklin	10
DELAWARE		Bacon	4	Johnson	5	Honolulu	1	Fulton	13
Kent	9	Baker	4	Jones	6	Kalawao	1	Gallatin	10
New Castle	10	Baldwin	6	Lamar	6	Kauai	1	Greene	11
Sussex	9	Banks	7	Lanier	4	Maui	1	Grundy	13
		Barrow	7	Laurens	5			Hamilton	10
DC		Bartow	7	Lee	5	**IDAHO**		Hancock	13
Washington	10	Ben Hill	5	Liberty	4	Ada	12	Hardin	10
		Berrien	4	Lincoln	6	Adams	15	Henderson	13
FLORIDA		Bibb	5	Long	4	Bannock	15	Henry	13
Alachua	3	Bleckley	5	Lowndes	4	Bear Lake	15	Iroquois	13
Baker	3	Brantley	4	Lumpkin	8	Benewah	14	Jackson	10
Bay	4	Brooks	4	Macon	5	Bingham	15	Jasper	11
Bradford	3	Bryan	4	Madison	7	Blaine	15	Jefferson	11
Brevard	2	Bulloch	5	Marion	5	Boise	15	Jersey	10
Broward	1	Burke	6	Mcduffie	6	Bonner	15	Jo Daviess	14
Calhoun	4	Butts	7	Mcintosh	4	Bonneville	15	Johnson	10
Charlotte	2	Calhoun	5	Meriwether	6	Boundary	15	Kane	14
Citrus	2	Camden	4	Miller	4	Butte	16	Kankakee	13
Clay	3	Candler	5	Mitchell	4	Camas	15	Kendall	13
Collier	1	Carroll	7	Monroe	6	Canyon	12	Knox	13
Columbia	3	Catoosa	8	Montgomery	5	Caribou	15	La Salle	13
Dade	1	Charlton	4	Morgan	6	Cassia	14	Lake	14
De Soto	2	Chatham	4	Murray	8	Clark	15	Lawrence	11
Dixie	3	Chattahoochee	5	Muscogee	5	Clearwater	12	Lee	14
Duval	3	Chattooga	8	Newton	7	Custer	16	Livingston	13
Escambia	4	Cherokee	8	Oconee	7	Elmore	13	Logan	12
Flagler	3	Clarke	7	Oglethorpe	7	Franklin	15	Macon	12
Franklin	4	Clay	5	Paulding	7	Fremont	16	Macoupin	11
Gadsden	4	Clayton	7	Peach	5	Gem	13	Madison	10
Gilchrist	3	Clinch	4	Pickens	8	Gooding	13	Marion	11
Glades	1	Cobb	7	Pierce	4	Idaho	15	Marshall	13
Gulf	4	Coffee	5	Pike	6	Jefferson	16	Mason	12
Hamilton	3	Colquitt	4	Polk	7	Jerome	14	Massac	10
Hardee	2	Columbia	6	Pulaski	5	Kootenai	14	Mcdonough	13
Hendry	1	Cook	4	Putnam	6	Latah	14	Mchenry	14
Hernando	2	Coweta	7	Quitman	5	Lemhi	15	Mclean	12
Highlands	2	Crawford	5	Rabun	8	Lewis	15	Menard	12
Hillsborough	2	Crisp	5	Randolph	5	Lincoln	15	Mercer	13
Holmes	4	Dade	8	Richmond	6	Madison	16	Monroe	10
Indian River	2	Dawson	8	Rockdale	7	Minidoka	15	Montgomery	11
Jackson	4	De Kalb	7	Schley	5	Nez Perce	12	Morgan	12
Jefferson	4	Decatur	4	Screven	5	Oneida	15	Moultrie	12
Lafayette	3	Dodge	5	Seminole	4	Owyhee	12	Ogle	14
Lake	2	Dooly	5	Spalding	7	Payette	12	Peoria	13
Lee	1	Dougherty	5	Stephens	7	Power	15	Perry	10
Leon	4	Douglas	7	Stewart	5	Shoshone	14	Piatt	12
Levy	2	Early	5	Sumter	5	Teton	16	Pike	12
Liberty	4	Echols	4	Talbot	5	Twin Falls	14	Pope	10
Madison	3	Effingham	4	Taliaferro	6	Valley	16	Pulaski	10
Manatee	2	Elbert	7	Tattnall	4	Washington	13	Putnam	13
Marion	2	Emanuel	5	Taylor	5			Randolph	10
Martin	1	Evans	4	Telfair	5	**ILLINOIS**		Richland	11
Monroe	1	Fannin	8	Terrell	5	Adams	12	Rock Island	13
Nassau	3	Fayette	7	Thomas	4	Alexander	10	Saline	10
Okaloosa	4	Floyd	7	Tift	5	Bond	11	Sangamon	12
Okeechobee	2	Forsyth	8	Toombs	4	Boone	14	Schuyler	12
Orange	2	Franklin	7	Towns	8	Brown	12	Scott	12
Osceola	2	Fulton	7	Treutlen	5	Bureau	13	Shelby	11
Palm Beach	1	Gilmer	8	Troup	6	Calhoun	11	St Clair	10
Pasco	2	Glascock	6	Turner	5	Carroll	14	Stark	13
Pinellas	2	Glynn	4	Twiggs	5	Cass	12	Stephenson	14
Polk	2	Gordon	8	Union	8	Champaign	12	Tazewell	12
Putnam	3	Grady	4	Upson	5	Christian	11	Union	10
Santa Rosa	4	Greene	6	Walker	8	Clark	12	Vermilion	12
Sarasota	2	Gwinnett	7	Walton	7	Clay	11	Wabash	11
Seminole	2	Habersham	8	Ware	4	Clinton	10	Warren	13

County	Zone	County	Zone	County	Zone	County	Zone	County	Zone
Washington	10	Ripley	11	Jones	14	Franklin	10	Adair	9
Wayne	11	Rush	12	Keokuk	13	Geary	11	Allen	9
White	10	Scott	11	Kossuth	15	Gove	12	Anderson	10
Whiteside	14	Shelby	12	Lee	13	Graham	12	Ballard	9
Will	13	Spencer	10	Linn	14	Grant	11	Barren	9
Williamson	10	St Joseph	13	Louisa	13	Gray	11	Bath	11
Winnebago	14	Starke	13	Lucas	13	Greeley	12	Bell	10
Woodford	13	Steuben	14	Lyon	15	Greenwood	10	Boone	11
		Sullivan	11	Madison	14	Hamilton	11	Bourbon	10
INDIANA		Switzerland	10	Mahaska	13	Harper	9	Boyd	11
Adams	13	Tippecanoe	13	Marion	13	Harvey	11	Boyle	10
Allen	13	Tipton	13	Marshall	14	Haskell	11	Bracken	11
Bartholomew	11	Union	12	Mills	13	Hodgeman	11	Breathitt	10
Benton	13	Vanderburgh	10	Mitchell	15	Jackson	11	Breckenridge	9
Blackford	13	Vermillion	12	Monona	14	Jefferson	11	Bullitt	10
Boone	12	Vigo	12	Monroe	13	Jewell	12	Butler	9
Brown	11	Wabash	14	Montgomery	13	Johnson	11	Caldwell	9
Carroll	13	Warren	12	Muscatine	13	Kearny	11	Calloway	9
Cass	13	Warrick	10	Obrien	15	Kingman	10	Campbell	11
Clark	10	Washington	11	Osceola	15	Kiowa	10	Carlisle	9
Clay	12	Wayne	12	Page	13	Labette	9	Carroll	10
Clinton	13	Wells	13	Palo Alto	15	Lane	12	Carter	11
Crawford	11	White	13	Plymouth	15	Leavenworth	11	Casey	10
Daviess	11	Whitley	14	Pocahontas	15	Lincoln	11	Christian	9
De Kalb	13			Polk	14	Linn	10	Clark	10
Dearborn	11	**IOWA**		Pottawattamie	14	Logan	12	Clay	10
Decatur	12	Adair	14	Poweshiek	14	Lyon	11	Clinton	10
Delaware	13	Adams	13	Ringgold	13	Marion	11	Crittenden	9
Dubois	11	Allamakee	15	Sac	15	Marshall	12	Cumberland	9
Elkhart	13	Appanoose	13	Scott	13	Mcpherson	11	Daviess	9
Fayette	12	Audubon	14	Shelby	14	Meade	10	Edmonson	9
Floyd	10	Benton	14	Sioux	15	Miami	10	Elliot	11
Fountain	12	Black Hawk	15	Story	14	Mitchell	12	Estill	10
Franklin	12	Boone	14	Tama	14	Montgomery	9	Fayette	10
Fulton	14	Bremer	15	Taylor	13	Morris	11	Fleming	11
Gibson	10	Buchanan	15	Union	13	Morton	10	Floyd	10
Grant	13	Buena Vista	15	Van Buren	13	Nemaha	11	Franklin	10
Greene	11	Butler	15	Wapello	13	Neosho	9	Fulton	9
Hamilton	12	Calhoun	15	Warren	14	Ness	12	Gallatin	11
Hancock	12	Carroll	14	Washington	13	Norton	13	Garrard	10
Harrison	10	Cass	14	Wayne	13	Osage	10	Grant	11
Hendricks	12	Cedar	14	Webster	15	Osborne	12	Graves	9
Henry	12	Cerro Gordo	15	Winnebago	15	Ottawa	11	Grayson	9
Howard	13	Cherokee	15	Winneshiek	15	Pawnee	11	Green	9
Huntington	14	Chickasaw	15	Woodbury	15	Phillips	12	Greenup	11
Jackson	11	Clarke	13	Worth	15	Pottawatomie	11	Hancock	9
Jasper	13	Clay	15	Wright	15	Pratt	10	Hardin	9
Jay	13	Clayton	15			Rawlins	13	Harlan	10
Jefferson	10	Clinton	13	**KANSAS**		Reno	11	Harrison	11
Jennings	11	Crawford	14	Allen	10	Republic	12	Hart	9
Johnson	12	Dallas	14	Anderson	10	Rice	11	Henderson	9
Knox	11	Davis	13	Atchison	11	Riley	11	Henry	10
Kosciusko	14	Decatur	13	Barber	9	Rooks	12	Hickman	9
La Porte	13	Delaware	15	Barton	11	Rush	11	Hopkins	9
Lagrange	14	Des Moines	13	Bourbon	10	Russell	11	Jackson	10
Lake	13	Dickinson	15	Brown	11	Saline	11	Jefferson	10
Lawrence	11	Dubuque	14	Butler	.10	Scott	12	Jessamine	10
Madison	13	Emmet	15	Chase	10	Sedgwick	10	Johnson	11
Marion	12	Fayette	15	Chautauqua	9	Seward	10	Kenton	11
Marshall	13	Floyd	15	Cherokee	9	Shawnee	11	Knott	10
Martin	11	Franklin	15	Cheyenne	13	Sheridan	12	Knox	10
Miami	14	Fremont	13	Clark	10	Sherman	13	Larue	9
Monroe	11	Greene	14	Clay	11	Smith	12	Laurel	10
Montgomery	12	Grundy	15	Cloud	12	Stafford	11	Lawrence	11
Morgan	12	Guthrie	14	Coffey	10	Stanton	11	Lee	10
Newton	13	Hamilton	15	Comanche	9	Stevens	10	Leslie	10
Noble	14	Hancock	15	Cowley	9	Sumner	9	Letcher	10
Ohio	11	Hardin	15	Crawford	9	Thomas	13	Lewis	11
Orange	11	Harrison	14	Decatur	13	Trego	12	Lincoln	10
Owen	12	Henry	13	Dickinson	11	Wabaunsee	11	Livingston	9
Parke	12	Howard	15	Doniphan	11	Wallace	12	Logan	9
Perry	10	Humboldt	15	Douglas	10	Washington	12	Lyon	9
Pike	11	Ida	15	Edwards	11	Wichita	12	Madison	10
Porter	13	Iowa	14	Elk	9	Wilson	9	Magoffin	10
Posey	10	Jackson	14	Ellis	12	Woodson	10	Marion	10
Pulaski	13	Jasper	14	Ellsworth	11	Wyandotte	11	Marshall	9
Putnam	12	Jefferson	13	Finney	11			Martin	11
Randolph	13	Johnson	13	Ford	11	**KENTUCKY**		Mason	11

C-3

County	Zone	County	Zone	County	Zone	County	Zone	County	Zone
Mccracken	9	Ouachita	6	Hampden	14	Presque Isle	15	Rice	15
Mccreary	10	Plaquemines	3	Hampshire	14	Roscommon	15	Rock	15
Mclean	9	Pointe Coupee	4	Middlesex	13	Saginaw	14	Roseau	17
Meade	9	Rapides	5	Nantucket	12	Sanilac	14	Scott	15
Menifee	10	Red River	5	Norfolk	13	Schoolcraft	16	Sherburne	16
Mercer	10	Richland	6	Plymouth	12	Shiawassee	14	Sibley	15
Metcalfe	9	Sabine	5	Suffolk	13	St Clair	14	St Louis	17
Monroe	9	St Bernard	3	Worcester	14	St Joseph	14	Stearns	16
Montgomery	10	St Charles	3			Tuscola	14	Steele	15
Morgan	10	St Helena	4	**MICHIGAN**		Van Buren	14	Stevens	16
Muhlenberg	9	St James	3	Alcona	15	Washtenaw	13	Swift	16
Nelson	10	St John The Baptist	3	Alger	16	Wayne	13	Todd	16
Nicholas	11	St Landry	4	Allegan	14	Wexford	15	Traverse	16
Ohio	9	St Martin	4	Alpena	15			Wabasha	15
Oldham	10	St Mary	3	Antrim	15	**MINNESOTA**		Wadena	17
Owen	10	St Tammany	4	Arenac	15	Aitkin	17	Waseca	15
Owsley	10	Tangipahoa	4	Baraga	17	Anoka	16	Washington	15
Pendleton	11	Tensas	5	Barry	14	Becker	17	Watonwan	15
Perry	10	Terrebonne	3	Bay	15	Beltrami	17	Wilkin	17
Pike	10	Union	6	Benzie	15	Benton	16	Winona	15
Powell	10	Vermilion	4	Berrien	14	Big Stone	16	Wright	16
Pulaski	10	Vernon	5	Branch	14	Blue Earth	15	Yellow Medicine	15
Robertson	11	Washington	4	Calhoun	14	Brown	15		
Rockcastle	10	Webster	6	Cass	14	Carlton	17	**MISSISSIPPI**	
Rowan	11	West Baton Rouge	4	Charlevoix	15	Carver	15	Adams	5
Russell	10	West Carroll	6	Cheboygan	15	Cass	17	Alcorn	7
Scott	11	West Feliciana	4	Chippewa	16	Chippewa	16	Amite	4
Shelby	10	Winn	5	Clare	15	Chisago	16	Attala	6
Simpson	9			Clinton	14	Clay	17	Benton	7
Spencer	10	**MAINE**		Crawford	15	Clearwater	17	Bolivar	6
Taylor	9	Androscoggin	15	Delta	16	Cook	17	Calhoun	6
Todd	9	Aroostook	17	Dickinson	16	Cottonwood	15	Carroll	6
Trigg	9	Cumberland	15	Eaton	14	Crow Wing	17	Chickasaw	6
Trimble	10	Franklin	16	Emmet	15	Dakota	15	Choctaw	6
Union	9	Hancock	15	Genesee	14	Dodge	15	Claiborne	5
Warren	9	Kennebec	15	Gladwin	15	Douglas	16	Clarke	5
Washington	10	Knox	15	Gogebic	17	Faribault	15	Clay	6
Wayne	10	Lincoln	15	Grand Traverse	15	Fillmore	15	Coahoma	7
Webster	9	Oxford	16	Gratiot	14	Freeborn	15	Copiah	5
Whitley	10	Penobscot	15	Hillsdale	14	Goodhue	15	Covington	5
Wolfe	10	Piscataquis	17	Houghton	17	Grant	16	De Soto	7
Woodford	10	Sagadahoc	15	Huron	14	Hennepin	15	Forrest	5
		Somerset	17	Ingham	14	Houston	15	Franklin	5
LOUISIANA		Waldo	15	Ionia	14	Hubbard	17	George	4
Acadia	4	Washington	15	Iosco	15	Isanti	16	Greene	5
Allen	4	York	15	Iron	17	Itasca	17	Grenada	6
Ascension	4			Isabella	15	Jackson	15	Hancock	4
Assumption	3	**MARYLAND**		Jackson	14	Kanabec	16	Harrison	4
Avoyelles	5	Allegany	12	Kalamazoo	14	Kandiyohi	16	Hinds	6
Beauregard	4	Anne Arundel	9	Kalkaska	15	Kittson	17	Holmes	6
Bienville	6	Baltimore	10	Kent	14	Koochiching	17	Humphreys	6
Bossier	6	Baltimore City	9	Keweenaw	17	Lac Qui Parle	15	Issaquena	6
Caddo	6	Calvert	9	Lake	15	Lake	17	Itawamba	7
Calcasieu	4	Caroline	9	Lapeer	14	Lake Of The Woods	7	Jackson	4
Caldwell	6	Carroll	11	Leelanau	15	Le Sueur	15	Jasper	5
Cameron	4	Cecil	10	Lenawee	14	Lincoln	15	Jefferson	5
Catahoula	5	Charles	9	Livingston	14	Lyon	15	Jefferson Davis	5
Claiborne	6	Dorchester	9	Luce	16	Mahnomen	17	Jones	5
Concordia	5	Frederick	11	Mackinac	16	Marshall	17	Kemper	6
De Soto	5	Garrett	13	Macomb	14	Martin	15	Lafayette	7
East Baton Rouge	6	Harford	10	Manistee	15	Mcleod	15	Lamar	4
East Carroll	6	Howard	10	Marquette	16	Meeker	16	Lauderdale	6
East Feliciana	4	Kent	10	Mason	15	Mille Lacs	16	Lawrence	5
Evangeline	4	Montgomery	10	Mecosta	15	Morrison	16	Leake	6
Franklin	6	Prince Georges	10	Menominee	16	Mower	15	Lee	7
Grant	5	Queen Annes	9	Midland	15	Murray	15	Leflore	6
Iberia	4	Somerset	9	Missaukee	15	Nicollet	15	Lincoln	5
Iberville	4	St Marys	9	Monroe	13	Nobles	15	Lowndes	6
Jackson	6	Talbot	9	Montcalm	14	Norman	17	Madison	6
Jefferson	3	Washington	11	Montmorency	15	Olmsted	15	Marion	4
Jefferson Davis	4	Wicomico	9	Muskegon	14	Otter Tail	17	Marshall	7
La Salle	5	Worcester	9	Newaygo	15	Pennington	17	Monroe	6
Lafayette	4			Oakland	14	Pine	16	Montgomery	6
Lafourche	3	**MASSACHUSETTS**		Oceana	15	Pipestone	15	Neshoba	6
Lincoln	6	Barnstable	12	Ogemaw	15	Polk	17	Newton	6
Livingston	4	Berkshire	14	Ontonagon	17	Pope	16	Noxubee	6
Madison	6	Bristol	12	Osceola	15	Ramsey	15	Oktibbeha	6
Morehouse	6	Dukes	12	Oscoda	15	Red Lake	17	Panola	7
Natchitoches	5	Essex	13	Otsego	15	Redwood	15	Pearl River	4
Orleans	3	Franklin	14	Ottawa	14	Renville	15	Perry	5

County	Zone	County	Zone	County	Zone	County	Zone	County	Zone
Pike	4	Johnson	11	Deer Lodge	16	Fillmore	13	Nye	12
Pontotoc	7	Knox	12	Fallon	15	Franklin	13	Pershing	12
Prentiss	7	Laclede	10	Fergus	15	Frontier	13	Storey	12
Quitman	7	Lafayette	11	Flathead	16	Furnas	13	Washoe	12
Rankin	6	Lawrence	10	Gallatin	15	Gage	13	White Pine	15
Scott	6	Lewis	12	Garfield	15	Garden	14		
Sharkey	6	Lincoln	11	Glacier	16	Garfield	14	**NEW HAMPSHIRE**	
Simpson	5	Linn	12	Golden Valley	15	Gosper	13	Belknap	15
Smith	5	Livingston	12	Granite	16	Grant	14	Carroll	15
Stone	4	Macon	12	Hill	16	Greeley	14	Cheshire	15
Sunflower	6	Madison	10	Jefferson	15	Hall	13	Coos	16
Tallahatchie	7	Maries	11	Judith Basin	15	Hamilton	13	Grafton	15
Tate	7	Marion	12	Lake	15	Harlan	13	Hillsborough	15
Tippah	7	Mcdonald	9	Lewis And Clark	15	Hayes	13	Merrimack	15
Tishomingo	7	Mercer	13	Liberty	16	Hitchcock	13	Rockingham	15
Tunica	7	Miller	11	Lincoln	15	Holt	14	Strafford	15
Union	7	Mississippi	9	Madison	15	Hooker	14	Sullivan	15
Walthall	4	Moniteau	11	Mccone	15	Howard	14		
Warren	6	Monroe	12	Meagher	15	Jefferson	13	**NEW JERSEY**	
Washington	6	Montgomery	11	Mineral	15	Johnson	13	Atlantic	10
Wayne	5	Morgan	11	Missoula	15	Kearney	13	Bergen	12
Webster	6	New Madrid	9	Musselshell	15	Keith	14	Burlington	11
Wilkinson	4	Newton	9	Park	15	Keya Paha	14	Camden	10
Winston	6	Nodaway	13	Petroleum	15	Kimball	14	Cape May	10
Yalobusha	7	Oregon	9	Phillips	16	Knox	14	Cumberland	10
Yazoo	6	Osage	11	Pondera	15	Lancaster	13	Essex	11
		Ozark	9	Powder River	15	Lincoln	14	Gloucester	10
MISSOURI		Pemiscot	9	Powell	16	Logan	14	Hudson	11
Adair	12	Perry	10	Prairie	15	Loup	14	Hunterdon	12
Andrew	12	Pettis	11	Ravalli	15	Madison	14	Mercer	11
Atchison	13	Phelps	10	Richland	15	Mcpherson	14	Middlesex	11
Audrain	12	Pike	12	Roosevelt	16	Merrick	13	Monmouth	11
Barry	9	Platte	11	Rosebud	15	Morrill	14	Morris	12
Barton	10	Polk	10	Sanders	15	Nance	13	Ocean	11
Bates	11	Pulaski	10	Sheridan	16	Nemaha	13	Passaic	12
Benton	11	Putnam	13	Silver Bow	16	Nuckolls	13	Salem	10
Bollinger	10	Ralls	12	Stillwater	15	Otoe	13	Somerset	12
Boone	11	Randolph	12	Sweet Grass	15	Pawnee	13	Sussex	13
Buchanan	12	Ray	11	Teton	15	Perkins	13	Union	11
Butler	9	Reynolds	10	Toole	16	Phelps	13	Warren	12
Caldwell	12	Ripley	9	Treasure	15	Pierce	14		
Callaway	11	Saline	11	Valley	16	Platte	13	**NEW MEXICO**	
Camden	11	Schuyler	13	Wheatland	15	Polk	13	Bernalillo	9
Cape Girardeau	9	Scotland	13	Wibaux	15	Red Willow	13	Catron	11
Carroll	12	Scott	9	Yellowstone	15	Richardson	13	Chaves	7
Carter	10	Shannon	10	Yellowstone National		Rock	14	Cibola	12
Cass	11	Shelby	12	Park	15	Saline	13	Colfax	13
Cedar	11	St Charles	10			Sarpy	13	Curry	9
Chariton	12	St Clair	11	**NEBRASKA**		Saunders	13	De Baca	9
Christian	10	St Francois	10	Adams	13	Scotts Bluff	14	Dona Ana	7
Clark	13	St Louis	10	Antelope	14	Seward	13	Eddy	7
Clay	11	St Louis City	10	Arthur	14	Sheridan	15	Grant	9
Clinton	12	Ste Genevieve	10	Banner	14	Sherman	14	Guadalupe	9
Cole	11	Stoddard	9	Blaine	14	Sioux	15	Harding	11
Cooper	11	Stone	9	Boone	14	Stanton	14	Hidalgo	7
Crawford	10	Sullivan	12	Box Butte	15	Thayer	13	Lea	7
Dade	10	Taney	9	Boyd	14	Thomas	14	Lincoln	9
Dallas	10	Texas	10	Brown	14	Thurston	14	Los Alamos	13
Daviess	12	Vernon	11	Buffalo	13	Valley	14	Luna	7
De Kalb	12	Warren	11	Burt	14	Washington	13	Mckinley	13
Dent	10	Washington	10	Butler	13	Wayne	14	Mora	15
Douglas	10	Wayne	10	Cass	13	Webster	13	Otero	7
Dunklin	9	Webster	10	Cedar	14	Wheeler	14	Quay	8
Franklin	10	Worth	13	Chase	13	York	13	Rio Arriba	12
Gasconade	11	Wright	10	Cherry	14			Roosevelt	8
Gentry	13			Cheyenne	14	**NEVADA**		San Juan	12
Greene	10	**MONTANA**		Clay	13	Carson City	12	San Miguel	12
Grundy	12	Beaverhead	15	Colfax	13	Churchill	12	Sandoval	13
Harrison	13	Big Horn	15	Cuming	14	Clark	5	Santa Fe	13
Henry	11	Blaine	16	Custer	14	Douglas	13	Sierra	8
Hickory	11	Broadwater	15	Dakota	14	Elko	15	Socorro	9
Holt	12	Carbon	15	Dawes	15	Esmeralda	12	Taos	15
Howard	11	Carter	15	Dawson	13	Eureka	15	Torrance	11
Howell	9	Cascade	15	Deuel	14	Humboldt	13	Union	11
Iron	10	Chouteau	15	Dixon	14	Lander	13	Valencia	10
Jackson	11	Custer	15	Dodge	13	Lincoln	12		
Jasper	9	Daniels	16	Douglas	13	Lyon	13	**NEW YORK**	
Jefferson	10	Dawson	15	Dundy	13	Mineral	12	Albany	14

County	Zone	County	Zone	County	Zone	County	Zone	County	Zone
Allegany	15	Catawba	8	Wilson	7	Delaware	13	Cherokee	8
Bronx	11	Chatham	8	Yadkin	8	Erie	13	Choctaw	6
Broome	15	Cherokee	9	Yancey	11	Fairfield	12	Cimarron	10
Cattaraugus	15	Chowan	7			Fayette	12	Cleveland	7
Cayuga	14	Clay	9	**NORTH DAKOTA**		Franklin	12	Coal	7
Chautauqua	13	Cleveland	7	Adams	16	Fulton	14	Comanche	7
Chemung	15	Columbus	6	Barnes	17	Gallia	11	Cotton	7
Chenango	15	Craven	6	Benson	17	Geauga	13	Craig	9
Clinton	15	Cumberland	7	Billings	16	Greene	12	Creek	8
Columbia	13	Currituck	7	Bottineau	17	Guernsey	12	Custer	8
Cortland	15	Dare	6	Bowman	16	Hamilton	11	Delaware	8
Delaware	15	Davidson	8	Burke	17	Hancock	13	Dewey	9
Dutchess	13	Davie	8	Burleigh	16	Hardin	13	Ellis	9
Erie	14	Duplin	6	Cass	17	Harrison	13	Garfield	8
Essex	16	Durham	8	Cavalier	17	Henry	14	Garvin	7
Franklin	16	Edgecombe	7	Dickey	16	Highland	11	Grady	7
Fulton	15	Forsyth	8	Divide	17	Hocking	12	Grant	9
Genesee	14	Franklin	8	Dunn	16	Holmes	13	Greer	7
Greene	14	Gaston	7	Eddy	17	Huron	13	Harmon	7
Hamilton	16	Gates	7	Emmons	16	Jackson	11	Harper	9
Herkimer	15	Graham	9	Foster	17	Jefferson	13	Haskell	7
Jefferson	15	Granville	8	Golden Valley	16	Knox	13	Hughes	7
Kings	10	Greene	7	Grand Forks	17	Lake	13	Jackson	7
Lewis	15	Guilford	8	Grant	16	Lawrence	11	Jefferson	6
Livingston	14	Halifax	7	Griggs	17	Licking	12	Johnston	6
Madison	14	Harnett	7	Hettinger	16	Logan	13	Kay	9
Monroe	14	Haywood	9	Kidder	17	Lorain	13	Kingfisher	8
Montgomery	14	Henderson	9	La Moure	16	Lucas	14	Kiowa	7
Nassau	11	Hertford	7	Logan	16	Madison	12	Latimer	7
New York	10	Hoke	7	Mchenry	17	Mahoning	13	Le Flore	7
Niagara	14	Hyde	6	Mcintosh	16	Marion	13	Lincoln	7
Oneida	15	Iredell	8	Mckenzie	16	Medina	13	Logan	8
Onondaga	14	Jackson	9	Mclean	17	Meigs	11	Love	6
Ontario	14	Johnston	7	Mercer	16	Mercer	13	Major	9
Orange	12	Jones	6	Morton	16	Miami	13	Marshall	6
Orleans	14	Lee	7	Mountrail	17	Monroe	12	Mayes	8
Oswego	14	Lenoir	7	Nelson	17	Montgomery	12	Mcclain	7
Otsego	15	Lincoln	7	Oliver	16	Morgan	12	Mccurtain	7
Putnam	12	Macon	9	Pembina	17	Morrow	13	Mcintosh	7
Queens	10	Madison	9	Pierce	17	Muskingum	12	Murray	7
Rensselaer	14	Martin	7	Ramsey	17	Noble	12	Muskogee	7
Richmond	11	Mcdowell	8	Ransom	16	Ottawa	13	Noble	8
Rockland	12	Mecklenburg	7	Renville	17	Paulding	14	Nowata	9
Saratoga	14	Mitchell	11	Richland	16	Perry	12	Okfuskee	7
Schenectady	14	Montgomery	7	Rolette	17	Pickaway	12	Oklahoma	8
Schoharie	15	Moore	7	Sargent	16	Pike	11	Okmulgee	8
Schuyler	15	Nash	7	Sheridan	17	Portage	13	Osage	8
Seneca	14	New Hanover	6	Sioux	16	Preble	12	Ottawa	9
St Lawrence	15	Northampton	7	Slope	16	Putnam	13	Pawnee	8
Steuben	15	Onslow	6	Stark	16	Richland	13	Payne	8
Suffolk	11	Orange	8	Steele	17	Ross	12	Pittsburg	7
Sullivan	15	Pamlico	6	Stutsman	17	Sandusky	13	Pontotoc	7
Tioga	15	Pasquotank	7	Towner	17	Scioto	11	Pottawatomie	7
Tompkins	15	Pender	6	Traill	17	Seneca	13	Pushmataha	6
Ulster	15	Perquimans	7	Walsh	17	Shelby	13	Roger Mills	9
Warren	15	Person	8	Ward	17	Stark	13	Rogers	9
Washington	15	Pitt	7	Wells	17	Summit	13	Seminole	7
Wayne	14	Polk	7	Williams	17	Trumbull	13	Sequoyah	7
Westchester	12	Randolph	8			Tuscarawas	13	Stephens	7
Wyoming	14	Richmond	7	**OHIO**		Union	13	Texas	10
Yates	14	Robeson	7	Adams	11	Van Wert	13	Tillman	7
		Rockingham	8	Allen	13	Vinton	11	Tulsa	8
NORTH CAROLINA		Rowan	7	Ashland	13	Warren	12	Wagoner	8
Alamance	8	Rutherford	7	Ashtabula	13	Washington	11	Washington	9
Alexander	8	Sampson	6	Athens	11	Wayne	13	Washita	8
Alleghany	11	Scotland	7	Auglaize	13	Williams	14	Woods	9
Anson	7	Stanly	7	Belmont	12	Wood	14	Woodward	9
Ashe	11	Stokes	9	Brown	11	Wyandot	13		
Avery	11	Surry	9	Butler	12			**OREGON**	
Beaufort	6	Swain	9	Carroll	13	**OKLAHOMA**		Baker	15
Bertie	7	Transylvania	9	Champaign	13	Adair	8	Benton	10
Bladen	6	Tyrrell	6	Clark	13	Alfalfa	9	Clackamas	10
Brunswick	6	Union	7	Clermont	11	Atoka	7	Clatsop	11
Buncombe	9	Vance	8	Clinton	12	Beaver	10	Columbia	11
Burke	8	Wake	7	Columbiana	13	Beckham	8	Coos	9
Cabarrus	7	Warren	8	Coshocton	12	Blaine	8	Crook	14
Caldwell	8	Washington	7	Crawford	13	Bryan	7	Curry	9
Camden	7	Watauga	11	Cuyahoga	13	Caddo	8	Deschutes	14
Carteret	6	Wayne	7	Darke	13	Canadian	8	Douglas	9
Caswell	8	Wilkes	9	Defiance	14	Carter	6	Gilliam	12

County	Zone	County	Zone	County	Zone	County	Zone	County	Zone
Grant	15	Pike	13	Brown	16	Coffee	8	Williamson	8
Harney	15	Potter	15	Brule	15	Crockett	8	Wilson	9
Hood River	12	Schuylkill	13	Buffalo	15	Cumberland	9		
Jackson	11	Snyder	13	Butte	15	Davidson	8	**TEXAS**	
Jefferson	13	Somerset	13	Campbell	15	De Kalb	9	Anderson	5
Josephine	9	Sullivan	14	Charles Mix	14	Decatur	8	Andrews	6
Klamath	14	Susquehanna	15	Clark	16	Dickson	9	Angelina	4
Lake	15	Tioga	15	Clay	14	Dyer	8	Aransas	3
Lane	10	Union	13	Codington	16	Fayette	7	Archer	6
Lincoln	11	Venango	14	Corson	15	Fentress	10	Armstrong	9
Linn	10	Warren	14	Custer	15	Franklin	8	Atascosa	3
Malheur	12	Washington	12	Davison	15	Gibson	9	Austin	4
Marion	10	Wayne	15	Day	16	Giles	8	Bailey	9
Morrow	12	Westmoreland	13	Deuel	16	Grainger	9	Bandera	4
Multnomah	10	Wyoming	14	Dewey	15	Greene	9	Bastrop	4
Polk	10	York	11	Douglas	14	Grundy	9	Baylor	7
Sherman	13			Edmunds	15	Hamblen	9	Bee	3
Tillamook	11	**RHODE ISLAND**		Fall River	15	Hamilton	8	Bell	5
Umatilla	12	Bristol	12	Faulk	15	Hancock	10	Bexar	4
Union	13	Kent	12	Grant	16	Hardeman	8	Blanco	5
Wallowa	15	Newport	12	Gregory	14	Hardin	8	Borden	7
Wasco	13	Providence	14	Haakon	15	Hawkins	9	Bosque	5
Washington	10	Washington	12	Hamlin	16	Haywood	8	Bowie	6
Wheeler	13			Hand	15	Henderson	8	Brazoria	3
Yamhill	10	**SOUTH CAROLINA**		Hanson	15	Henry	9	Brazos	4
		Abbeville	7	Harding	15	Hickman	9	Brewster	5
PENNSYLVANIA		Aiken	6	Hughes	15	Houston	9	Briscoe	8
Adams	11	Allendale	5	Hutchinson	14	Humphreys	9	Brooks	3
Allegheny	12	Anderson	7	Hyde	15	Jackson	9	Brown	5
Armstrong	13	Bamberg	5	Jackson	14	Jefferson	9	Burleson	4
Beaver	12	Barnwell	5	Jerauld	15	Johnson	10	Burnet	5
Bedford	13	Beaufort	5	Jones	15	Knox	8	Caldwell	4
Berks	12	Berkeley	5	Kingsbury	15	Lake	9	Calhoun	3
Blair	13	Calhoun	6	Lake	15	Lauderdale	8	Callahan	6
Bradford	15	Charleston	5	Lawrence	15	Lawrence	8	Cameron	2
Bucks	11	Cherokee	7	Lincoln	15	Lewis	8	Camp	6
Butler	14	Chester	7	Lyman	15	Lincoln	8	Carson	9
Cambria	13	Chesterfield	7	Marshall	16	Loudon	8	Cass	6
Cameron	15	Clarendon	6	Mccook	15	Macon	9	Castro	9
Carbon	13	Colleton	5	Mcpherson	16	Madison	8	Chambers	4
Centre	13	Darlington	6	Meade	15	Marion	8	Cherokee	5
Chester	11	Dillon	6	Mellette	14	Marshall	8	Childress	7
Clarion	14	Dorchester	5	Miner	15	Maury	9	Clay	6
Clearfield	15	Edgefield	6	Minnehaha	15	Mcminn	8	Cochran	8
Clinton	13	Fairfield	7	Moody	15	Mcnairy	8	Coke	6
Columbia	13	Florence	6	Pennington	15	Meigs	8	Coleman	5
Crawford	14	Georgetown	5	Perkins	15	Monroe	8	Collin	5
Cumberland	12	Greenville	7	Potter	15	Montgomery	9	Collingsworth	7
Dauphin	12	Greenwood	7	Roberts	16	Moore	8	Colorado	4
Delaware	10	Hampton	5	Sanborn	15	Morgan	10	Comal	4
Elk	15	Horry	5	Shannon	15	Obion	9	Comanche	5
Erie	14	Jasper	5	Spink	15	Overton	9	Concho	5
Fayette	12	Kershaw	7	Stanley	15	Perry	8	Cooke	6
Forest	15	Lancaster	7	Sully	15	Pickett	10	Coryell	5
Franklin	11	Laurens	7	Todd	14	Polk	8	Cottle	7
Fulton	12	Lee	6	Tripp	14	Putnam	9	Crane	5
Greene	12	Lexington	6	Turner	15	Rhea	8	Crockett	5
Huntingdon	12	Marion	6	Union	14	Roane	9	Crosby	7
Indiana	13	Marlboro	6	Walworth	15	Robertson	9	Culberson	6
Jefferson	15	Mccormick	6	Yankton	14	Rutherford	8	Dallam	9
Juniata	12	Newberry	6	Ziebach	15	Scott	10	Dallas	5
Lackawanna	14	Oconee	7			Sequatchie	8	Dawson	7
Lancaster	11	Orangeburg	6	**TENNESSEE**		Sevier	9	De Witt	3
Lawrence	14	Pickens	7	Anderson	9	Shelby	7	Deaf Smith	9
Lebanon	12	Richland	6	Bedford	8	Smith	9	Delta	6
Lehigh	12	Saluda	6	Benton	9	Stewart	9	Denton	5
Luzerne	13	Spartanburg	7	Bledsoe	8	Sullivan	9	Dickens	7
Lycoming	13	Sumter	6	Blount	8	Sumner	9	Dimmit	3
Mckean	15	Union	7	Bradley	8	Tipton	8	Donley	8
Mercer	14	Williamsburg	6	Campbell	10	Trousdale	9	Duval	3
Mifflin	12	York	7	Cannon	9	Unicoi	10	Eastland	6
Monroe	13			Carroll	9	Union	9	Ector	6
Montgomery	11	**SOUTH DAKOTA**		Carter	10	Van Buren	9	Edwards	4
Montour	13	Aurora	15	Cheatham	9	Warren	9	El Paso	6
Northampton	12	Beadle	15	Chester	8	Washington	9	Ellis	5
Northumberland	13	Bennett	14	Claiborne	10	Wayne	8	Erath	5
Perry	12	Bon Homme	14	Clay	9	Weakley	9	Falls	5
Philadelphia	10	Brookings	16	Cocke	9	White	9	Fannin	6

County	Zone	County	Zone	County	Zone	County	Zone	County	Zone
Fayette	4	Marion	6	Victoria	3	Botetourt	9	Sussex	8
Fisher	6	Martin	6	Walker	4	Brunswick	8	Tazewell	11
Floyd	8	Mason	5	Waller	4	Buchanan	10	Virginia Beach	8
Foard	7	Matagorda	3	Ward	6	Buckingham	9	Warren	11
Fort Bend	4	Maverick	3	Washington	4	Campbell	9	Washington	11
Franklin	6	Mcculloch	5	Webb	3	Caroline	9	Westmoreland	8
Freestone	5	Mclennan	5	Wharton	3	Carroll	11	Wise	10
Frio	3	Mcmullen	3	Wheeler	8	Charles City	8	Wythe	11
Gaines	7	Medina	4	Wichita	7	Charlotte	9	York	8
Galveston	3	Menard	5	Wilbarger	7	Chesterfield	9		
Garza	7	Midland	6	Willacy	2	Clarke	11	**VIRGINIA INDEPENDENT CITIES**	
Gillespie	5	Milam	4	Williamson	4	Craig	10	Alexandria	10
Glasscock	6	Mills	5	Wilson	4	Culpeper	10	Bedford	9
Goliad	3	Mitchell	6	Winkler	6	Cumberland	9	Bristol	11
Gonzales	4	Montague	6	Wise	5	Dickenson	10	Buena Vista	9
Gray	9	Montgomery	4	Wood	6	Dinwiddie	8	Charlottesville	9
Grayson	6	Moore	9	Yoakum	8	Essex	9	Chesapeake	8
Gregg	6	Morris	6	Young	6	Fairfax	10	Clifton Forge	10
Grimes	4	Motley	7	Zapata	2	Fauquier	10	Colonial Hts	9
Guadalupe	4	Nacogdoches	5	Zavala	3	Floyd	11	Covington	10
Hale	8	Navarro	5			Fluvanna	9	Danville	9
Hall	8	Newton	4	**UTAH**		Franklin	10	Emporia	8
Hamilton	5	Nolan	6	Beaver	14	Frederick	11	Fairfax	10
Hansford	9	Nueces	3	Box Elder	12	Fredericksburg Giles	10	Falls Church	10
Hardeman	7	Ochiltree	9	Cache	15	Gloucester	8	Franklin	8
Hardin	4	Oldham	9	Carbon	.14	Goochland	9	Fredericksburg	10
Harris	4	Orange	4	Daggett	15	Grayson	11	Galax	11
Harrison	6	Palo Pinto	6	Davis	12	Greene	10	Hampton	8
Hartley	9	Panola	5	Duchesne	15	Greensville	8	Harrisonburg	11
Haskell	6	Parker	5	Emery	14	Halifax	9	Hopewell	8
Hays	4	Parmer	9	Garfield	14	Hampton	8	Lexington	9
Hemphill	8	Pecos	5	Grand	10	Hanover	9	Lynchburg	9
Henderson	5	Polk	4	Iron	12	Henrico	8	Manassas	10
Hidalgo	2	Potter	9	Juab	12	Henry	10	Manassas Park	10
Hill	5	Presidio	5	Kane	10	Highland	11	Martinsville	10
Hockley	8	Rains	6	Millard	13	Isle Of Wight	8	Newport News	8
Hood	5	Randall	9	Morgan	15	James City	8	Norfolk	8
Hopkins	6	Reagan	5	Piute	13	King And Queen	9	Norton	10
Houston	5	Real	4	Rich	15	King George	9	Petersburg	8
Howard	6	Red River	6	Salt Lake	12	King William	9	Poquoson	8
Hudspeth	6	Reeves	6	San Juan	13	Lancaster	8	Portsmouth	8
Hunt	6	Refugio	3	Sanpete	14	Lee	10	Radford	11
Hutchinson	9	Roberts	9	Sevier	13	Loudoun	10	Richmond	8
Irion	5	Robertson	4	Summit	15	Louisa	9	Roanoke	9
Jack	6	Rockwall	5	Tooele	12	Lunenburg	9	Salem	9
Jackson	3	Runnels	5	Uintah	15	Madison	11	South Boston	9
Jasper	4	Rusk	5	Utah	12	Mathews	8	Staunton	11
Jeff Davis	6	Sabine	5	Wasatch	15	Mecklenburg	9	Suffolk	8
Jefferson	4	San Augustine	5	Washington	10	Middlesex	8	Virginia Beach	8
Jim Hogg	3	San Jacinto	4	Wayne	14	Montgomery	11	Waynesboro	11
Jim Wells	3	San Patricio	3	Weber	12	Nansemond	8	Williamsburg	8
Johnson	5	San Saba	5			Nelson	9	Winchester	11
Jones	6	Schleicher	5	**VERMONT**		New Kent	8		
Karnes	3	Scurry	7	Addison	15	Newport News	8	**WASHINGTON**	
Kaufman	5	Shackelford	6	Bennington	15	Norfolk	8	Adams	12
Kendall	5	Shelby	5	Caledonia	16	Northampton	8	Asotin	12
Kenedy	2	Sherman	9	Chittenden	15	Northumberland	8	Benton	11
Kent	7	Smith	5	Essex	16	Nottoway	9	Chelan	12
Kerr	5	Somervell	5	Franklin	15	Orange	10	Clallam	12
Kimble	5	Starr	2	Grand Isle	15	Page	11	Clark	11
King	7	Stephens	6	Lamoille	16	Patrick	10	Columbia	12
Kinney	4	Sterling	6	Orange	16	Pittsylvania	9	Cowlitz	11
Kleberg	2	Stonewall	7	Orleans	16	Powhatan	9	Douglas	14
Knox	7	Sutton	5	Rutland	15	Prince Edward	9	Ferry	15
La Salle	3	Swisher	9	Washington	16	Prince George	8	Franklin	11
Lamar	6	Tarrant	5	Windham	15	Prince William	10	Garfield	12
Lamb	8	Taylor	6	Windsor	15	Pulaski	11	Grant	12
Lampasas	5	Terrell	5			Rappahannock	11	Grays Harbor	11
Lavaca	4	Terry	7	**VIRGINIA**		Richmond	8	Island	12
Lee	4	Throckmorton	6	Accomack	8	Roanoke	9	Jefferson	11
Leon	5	Titus	6	Albemarle	9	Rockbridge	9	King	10
Liberty	4	Tom Green	5	Alleghany	10	Rockingham	11	Kitsap	11
Limestone	5	Travis	4	Amelia	9	Russell	10	Kittitas	14
Lipscomb	9	Trinity	4	Amherst	9	Scott	10	Klickitat	12
Live Oak	3	Tyler	4	Appomattox	9	Shenandoah	11	Lewis	11
Llano	5	Upshur	6	Arlington	10	Smyth	11	Lincoln	15
Loving	6	Upton	5	Augusta	11	Southampton	8	Mason	11
Lubbock	7	Uvalde	4	Bath	11	Spotsylvania	10	Okanogan	15
Lynn	7	Val Verde	4	Bedford	9	Stafford	10	Pacific	11
Madison	4	Van Zandt	5	Bland	11	Surry	8		

County	Zone	County	Zone	County	Zone
Pend Oreille	15	Buffalo	15	Hot Springs	15
Pierce	11	Burnett	17	Johnson	15
San Juan	12	Calumet	15	Laramie	15
Skagit	11	Chippewa	15	Lincoln	17
Skamania	11	Clark	15	Natrona	15
Snohomish	11	Columbia	15	Niobrara	15
Spokane	14	Crawford	15	Park	15
Stevens	15	Dane	15	Platte	14
Thurston	11	Dodge	15	Sheridan	15
Wahkiakum	11	Door	15	Sublette	17
Walla Walla	11	Douglas	17	Sweetwater	16
Whatcom	12	Dunn	15	Teton	17
Whitman	14	Eau Claire	15	Uinta	16
Yakima	12	Florence	17	Washakie	15
		Fond Du Lac	15	Weston	15
WEST VIRGINIA		Forest	17		
Barbour	13	Grant	15		
Berkeley	11	Green	15		
Boone	10	Green Lake	15		
Braxton	11	Iowa	15		
Brooke	12	Iron	17		
Cabell	10	Jackson	15		
Calhoun	11	Jefferson	15		
Clay	11	Juneau	15		
Doddridge	12	Kenosha	15		
Fayette	12	Kewaunee	15		
Gilmer	11	La Crosse	15		
Grant	13	Lafayette	15		
Greenbrier	12	Langlade	17		
Hampshire	11	Lincoln	17		
Hancock	12	Manitowoc	15		
Hardy	12	Marathon	15		
Harrison	12	Marinette	15		
Jackson	11	Marquette	15		
Jefferson	11	Menominee	15		
Kanawha	10	Milwaukee	15		
Lewis	12	Monroe	15		
Lincoln	10	Oconto	15		
Logan	10	Oneida	17		
Marion	12	Outagamie	15		
Marshall	12	Ozaukee	15		
Mason	11	Pepin	15		
Mcdowell	11	Pierce	15		
Mercer	11	Polk	16		
Mineral	12	Portage	15		
Mingo	10	Price	17		
Monongalia	12	Racine	15		
Monroe	11	Richland	15		
Morgan	11	Rock	15		
Nicholas	12	Rusk	16		
Ohio	12	Sauk	15		
Pendleton	13	Sawyer	17		
Pleasants	11	Shawano	15		
Pocahontas	13	Sheboygan	15		
Preston	13	St Croix	15		
Putnam	10	Taylor	17		
Raleigh	12	Trempealeau	15		
Randolph	13	Vernon	15		
Ritchie	11	Vilas	17		
Roane	11	Walworth	15		
Summers	12	Washburn	17		
Taylor	12	Washington	15		
Tucker	13	Waukesha	15		
Tyler	11	Waupaca	15		
Upshur	12	Waushara	15		
Wayne	10	Winnebago	15		
Webster	12	Wood	15		
Wetzel	12				
Wirt	11	**WYOMING**			
Wood	11	Albany	16		
Wyoming	11	Big Horn	15		
		Campbell	15		
WISCONSIN					
Adams	15	Carbon	16		
Ashland	17	Converse	15		
Barron	16	Crook	15		
Bayfield	17	Fremont	15		
Brown	15	Goshen	14		

1995 Model Energy Code
Summary of Basic Requirements

MECcheck

Air Leakage	Joints, penetrations, and all other such openings in the building envelope that are sources of air leakage must be caulked, gasketed, weatherstripped, or otherwise sealed. The maximum leakage rates for manufactured windows and doors are shown on the reverse side. Recessed lights must be type IC rated and installed with no penetrations or installed inside an appropriate air-tight assembly with a 0.5-in. clearance from combustible materials and 3-in. clearance from insulation.
Vapor Retarder	Vapor retarders must be installed on the warm-in-winter side of all non-vented framed ceilings, walls, and floors. This requirement does not apply to the following locations nor where moisture or its freezing will not damage the materials. • Texas — Zones 2-5 • Alabama, Georgia, N. Carolina, Oklahoma, S. Carolina — Zones 4-6 • Arkansas, Tennessee — Zones 6-7 • Florida, Hawaii, Louisiana, Mississippi — All Zones
Materials and Insulation Information	Materials and equipment must be identified so that compliance can be determined. Manufacturer manuals for all installed heating and cooling equipment and service water heating equipment must be provided. Insulation R-values, glazing and door U-values, and heating and cooling equipment efficiency (if high-efficiency credit is taken) must be clearly marked on the building plans or specifications.
Duct Insulation	Supply and return ducts for heating and cooling systems located in unconditioned spaces must be insulated to the levels shown on the reverse side of this sheet. Exceptions: Insulation is not required for exhaust air ducts, ducts within HVAC equipment, and when the design temperature difference between the air in the duct and the surrounding air is 15°F or less.
Duct Construction	Ducts must be sealed using mastic with fibrous backing tape. For fibrous ducts, pressure-sensitive tape may be used. Other sealants may be approved by the building official. Duct tape is not permitted. The HVAC system must provide a means for balancing air and water systems.
Temperature Controls	Thermostats are required for each separate HVAC system in single-family buildings and each dwelling unit in multifamily buildings (non-dwelling portions of multifamily buildings must have one thermostat for each system or zone). Thermostats must have the following ranges: Heating Only — 55°F - 75°F Cooling Only — 70°F - 85°F Heating and Cooling — 55°F - 85°F A manual or automatic means to partially restrict or shut off the heating and/or cooling input to each zone or floor shall be provided for single-family homes and to each room for multifamily buildings.
HVAC Piping Insulation	HVAC piping in unconditioned spaces conveying fluids at temperatures above 120°F or chilled fluids at less than 55°F must be insulated to the levels shown on the reverse side of this sheet.
Swimming Pools	All heated swimming pools must have an on/off pool heater switch. Heated pools require a pool cover unless over 20% of the heating energy is from non-depletable sources. All swimming pool pumps must be equipped with a time clock.
Circulating Hot Water	Circulating hot water systems must have automatic or manual controls and pipes must be insulated to the levels shown on the reverse side of this sheet.
Electric Systems	Each multifamily dwelling unit must be equipped with separate electric meters.

Version 2.0/Dec. 1995/U.S. Dept. of Housing and Urban Development/Rural Economic and Community Development/U.S. Dept. of Energy/Pacific Northwest National Laboratory

Duct Insulation R-Value Requirements

Zone Number	Ducts in Unconditioned Spaces (i.e. Attics, Crawl Spaces, Unheated Basements and Garages, and Exterior Cavities)	Ducts Outside the Building
Zones 1-4	R-5	R-8
Zones 5-14	R-5	R-6.5
Zone 15-19	R-5	R-8

Maximum Leakage Rates for Manufactured Windows and Doors

Frame Type	Windows (cfm/ft of Operable Sash Crack)	Doors (cfm per ft^2 of Door Area)	
		Sliders	Swinging
Wood	0.34	0.35	0.5
Aluminum	0.37	0.37	0.5
PVC	0.37	0.37	0.5

Minimum Insulation Thickness for HVAC Pipes[a]

Piping System Types	Fluid Temp Range (°F)	Insulation Thickness in Inches by Pipe Sizes[b]			
		Runouts 2 in.[c]	1 in. and Less	1.25 in. to 2 in.	2.5 in. to 4 in.
Heating Systems					
Low Pressure/Temperature	201-250	1.0	1.5	1.5	2.0
Low Temperature	120-200	0.5	1.0	1.0	1.5
Steam Condensate (for feed water)	Any	1.0	1.0	1.5	2.0
Cooling Systems					
Chilled Water	40-55	0.5	0.5	0.75	1.0

(a) The pipe insulation thicknesses specified in this table are based on insulation R-values ranging from R-4 to R-4.6 per inch of thickness. For materials with an R-value greater than R-4.6, the insulation thickness specified in this table may be reduced as follows:

$$\text{New Minimum Thickness} = \frac{4.6 \times \text{Table Thickness}}{\text{Actual R-Value}}$$

For materials with an R-value less than R-4, the minimum insulation thickness must be increased as follows:

$$\text{New Minimum Thickness} = \frac{4.0 \times \text{Table Thickness}}{\text{Actual R-Value}}$$

(b) For piping exposed to outdoor air, increase thickness by 0.5 in.
(c) Applies to runouts not exceeding 12 ft in length to individual terminal units.

Minimum Insulation Thickness for Circulating Hot Water Pipes

Heated Water Temperature (°F)	Insulation Thickness in Inches by Pipe Sizes[a]			
	Non-Circulating Runouts	Circulating Mains and Runouts		
	Up to 1 in.	Up to 1.25 in.	1.5 - 2.0 in.	Over 2 in.
170-180	0.5	1.0	1.5	2.0
140-160	0.5	0.5	1.0	1.5
100-130	0.5	0.5	0.5	1.0

(a) Nominal pipe size and insulation thickness.

Trade-Off Worksheet

Builder Name _____   Date _____

Builder Address _____

Building Address _____ Zone # _____

Submitted By _____ Phone Number _____

PROPOSED

U-values and F-values can be found in Tables 4-1 through 4-10.

Ceilings, Skylights, and Floors Over Outside Air

Description	Insulation R-Value	U-Value	x Area	= UA
Ceiling			ft²	
Floor Over Outside Air			ft²	
Skylight	—		ft²	
			ft²	
			ft²	
Ceilings: Total Area			ft²	

Walls, Windows, and Doors

Description	Insulation R-Value	U-Value	x Area	= UA
Wall			ft²	
Window	—		ft²	
Door	—		ft²	
Sliding Glass Door	—		ft²	
			ft²	
			ft²	
			ft²	
Walls: Total Area			ft²	

Floors and Foundations

Description	Insulation Depth	Insulation R-Value	U-Value or F-Value	x Area or Perimeter	= UA
Floor Over Unconditioned				ft²	
Basement Wall				ft²	
Unheated Slab	in.			ft	
Heated Slab	in.			ft	
Crawl Wall	in.			ft²	

Total Proposed UA [ ]

REQUIRED

Required U-values can be found in Table 4-11.

Required U-Value	x Area	= UA
	ft²	

Required U-Value	x Area	= UA
	ft²	

Required U-Value or F-Value	x Area or Perimeter	= UA
	ft²	
	ft²	
	ft	
	ft	
	ft²	

Total Required UA [ ]

Total Proposed UA must be less than or equal to the *Total Required UA*.

Statement of Compliance: The proposed building design represented in these documents is consistent with the building plans, specifications, and other calculations submitted with the permit application. The proposed building has been designed to meet the requirements of the 1995 CABO Model Energy Code.

Builder/Designer _____   Company Name _____   Date _____

Version 2.0 / December 1995 / U.S. Dept. of Housing and Urban Development / Rural and Economic Community Development / U.S. Dept. of Energy / Pacific Northwest National Laboratory

Energy Label

MEC*check*

Street Address _____

This home includes the following energy features:

Insulation R-Values   ## Glazing/Door U-Values

Insulating
Sheathing R-Value   U-Value

_____ _____ Ceiling   _____ Windows

_____ _____ Ceiling   _____ Windows

_____ _____ Wall   _____ Sliding Glass Doors

_____ _____ Wall   _____ Doors

_____ Floor   _____ Doors

_____ Basement Wall

_____ Crawl Space Wall

_____ Slab

_____ Duct

Mechanical System   Type and Fuel   Efficiency

Heating System   _____   _____

Cooling System   _____   _____ SEER

Water Heater   _____   _____ EF

Other Energy Features

_____

_____

_____

_____

Builder _____ Date _____

Version 2.0; December 1995
U.S. Dept. of Housing and Urban Development / Rural Economic and Community Development
U.S. Dept. of Energy / Pacific Northwest National Laboratory

Field Inspection Checklist

	Requirement	Installed (Y/N)	Comments

Pre-Inspection

* Approved Building Plans on Site (104.1)

Foundation Inspection   Inspection Date _____ Approved: Yes ___ No ___ Init._____

* Slab-Edge Insulation (502.2.1.4)   Depth: _____
* Basement Wall Exterior Insulation (502.2.1.6)   Depth: _____
* Crawl Space Wall Insulation (502.2.1.5)   Depth: _____

Framing Inspection   Inspection Date _____ Approved: Yes ___ No ___ Init._____

* Floor Insulation (502.2.1.3)
* Glazing and Door Area (502.2.1.1)
* Mass Walls (502.1.2)
* Caulking/Sealing Penetrations (502.4.3)
* Duct Insulation (503.9.1)
* Duct Construction (503.10.2)
* HVAC Piping Insulation (503.11)
* Circulating Hot-Water Piping Insulation (504.7)

Insulation Inspection   Inspection Date _____ Approved: Yes ___ No ___ Init._____

* Wall Insulation (502.2.1.1)
* Basement Wall Interior Insulation (502.2.1.6)   Depth: _____
* Ceiling Insulation (502.2.1.2)
* Glazing and Door U-Values (502.2.1.1)
* Vapor Retarder (502.1.4)

Final Inspection   Inspection Date _____ Approved: Yes ___ No ___ Init._____

* Heating Equipment (102.1)
    Make and Model Number
    Efficiency (AFUE or HSPF)
* Cooling Equipment (102.1)
    Make and Model Number
    Efficiency (SEER)
* Multifamily Units Separately Metered (505.2)
* Thermostats for Each System (503.8.3)
* Heat Pump Thermostat (503.4.2.3)
* Window and Door Air Leakage (502.4.2)
* Weatherstripping at Doors/Windows (502.4.3)
* Equipment Maintenance Information (102.2)

Version 2.0 / December 1995 / U.S. Dept. of Housing and Urban Development / Rural and Economic Development / U.S. Dept. of Energy / Pacific Northwest National Laboratory

MEC_check_

Take-Off Worksheet

Builder Name _____ Date _____

Builder Address _____

Building Address _____

Submitted By _____ Phone Number _____

Ceilings, Skylights, and Floors Over Outside Air

Description	Area	Insulation R-Value	Skylight U-Value
Ceiling	ft²		—
Floor Over Outside Air	ft²		—
Skyulight	ft²	—	

Walls, Windows, and Doors

Description	Area	Insulation R-Value	Glazing/Door U-Value
Wall	ft²		—
Window	ft²	—	
Door	ft²	—	
Sliding Glass Door	ft²	—	

Floors and Foundations

Description	Area or Perimeter	Insulation R-Value	Insulation Depth
Floor Over Unconditioned Space	ft²		—
Basement Wall	ft²		
Unheated Slab	ft		
Heated Slab	ft		
Crawl Space Wall	ft²		

Equipment Efficiency (This section may be left blank if no credit will be taken for high-efficiency equipment.)

Heating _____   AFUE/HSPF _____

Cooling _____   SEER _____
               Efficiency   Make & Model Number

Version 2.0 / December 1995 / U.S. Dept. of Housing and Urban Development / Rural and Economic Community Development / U.S. Dept. of Energy / Pacific Northwest National Laboratory

APPENDIX

B

Prescriptive Method for Residential Cold-Formed Steel Framing

PRESCRIPTIVE METHOD FOR RESIDENTIAL COLD-FORMED STEEL FRAMING

First Edition

for
The U.S. Department of Housing and Urban Development
Office of Policy, Development and Research
Washington, D.C.

*Co-Sponsored by the American Iron and Steel Institute
and the National Association of Home Builders*

by

*NAHB Research Center
400 Prince George's Boulevard
Upper Marlboro, MD 20774-8731*

May 1996

I

Notice

The U.S. Government does not endorse products or manufacturers. Trade or manufacturer's names appear herein solely because they are considered essential to the object of this report.

Disclaimer

While the material presented in this document has been prepared in accordance with recognized engineering principles and sound judgement, this document should not be used without first securing competent advice to its suitability with respect for any given application. **The use of this document is subject to approval by the local code authority**. Although the information in this document is believed to be accurate, neither the authors, nor reviewers, nor the U.S. Department of Housing and Urban Development of the U.S. Government, nor the NAHB Research Center, nor any of their employees or representatives makes any warranty, guarantee, or representation, expressed or implied, with respect to the accuracy, effectiveness, or usefulness of any information, method, or material in this document, nor assumes any liability for the use of any information, methods, or materials disclosed herein, or for damages arising from such use.

IV

Foreword

For centuries home builders in the United States have made wood their material of choice because of its satisfactory performance, abundant supply, and relatively low cost. However, increases and unpredictable fluctuations in the price of framing lumber, as well as problems with its quality, are causing builders and other providers of affordable housing to seek alternative building products.

Use of cold-formed steel framing in the residential market has increased over the past years. Its price stability, consistent quality, similarity to conventional framing, successes in the commercial market, and resistance to fire, rot, and termites have attracted the attention of many builders. However, lack of prescriptive construction requirements has prevented this alternative material from gaining wider acceptance among home builders and code officials.

Prescriptive Method for Residential Cold-Formed Steel Framing is the result of a 3-year research and development program sponsored by the U.S. Department of Housing and Urban Development (HUD), with co-funding from the National Association of Home Builders (NAHB) and the American Iron and Steel Institute (AISI). The program was conducted by the NAHB Research Center with assistance and input from steering, advisory, and engineering committees. These committees represented steel manufacturers, steel producers, code officials, academics, researchers, professional engineers, and builders experienced in residential and cold-formed steel framing.

Prescriptive Method for Residential Cold-Formed Steel Framing facilitates the construction of steel-framed housing and thereby expands housing affordability through competition from new methods and materials. It also provides cold-form steel suppliers and consumers with standardized requirements for steel framing materials that will enhance market acceptance and consistent user application. Finally, this report provides code officials and inspectors with the guidance necessary to perform their duties in the home construction process when cold-formed steel is one of the materials.

Michael A. Stegman
Assistant Secretary for Policy
Development and Research

VI

Acknowledgments

This report was prepared by the NAHB Research Center under sponsorship of the U.S. Department of Housing and Urban Development (HUD). We wish to recognize the National Association of Home Builders (NAHB) and the American Iron and Steel Institute (AISI) whose co-funding made this project possible. Special appreciation is extended to William Freeborne of HUD for guidance throughout the project

The preparation of this report required the talents of many dedicated professionals. The principal authors were Nader R. Elhajj, P.E. and Kevin Bielat with review by Jay Crandell, P.E. and Mark Nowak.

Appreciation is extended to members of the steering committee (listed below) who met several times to review and provide guidance on this document and who, with their input, made this work much more complete.

Joe Bertoni
Kevin Bielat; NAHB Research Center
Delbert F. Boring, P.E.; AISI
Roger Brockenbrough; RL Brockenbrough & Assoc. Inc
Jay Crandell, P.E.; NAHB Research Center
Ed DiGirolamo; The Steel Network
John D. Ewing; USS - POSCO
Nader Elhajj, P.E.; NAHB Research Center
William Freeborne; HUD
Richard Haws; AISI
Bill Knorr; Knorr Steel Framing Systems
Dr. Roger LaBoube; University of Missouri-Rolla

Mike Meyers; U.S. Steel
Mark Nowak; NAHB Research Center
Dr. Teoman Pekoz; Cornell University
Neal L. Peterson, P.E.; Devco Engineering
Greg Ralph; Dietrich, Inc.
Dr. Paul Seaburg; Pennsylvania State University
David Smith, P.E.
Kurt Stochlia; ICBO Evaluation Service, Inc.
Mark Tipton; Met Homes
Ken Vought; USS - POSCO
Tim Waite, P.E.; NAHB Research Center
Steve Walker, P.E.
Julian White, Jr.; Dale Incor, Inc.
Riley Williams; Mitchell and Best
Roger Wildt; Bethlehem Steel

Appreciation is also extended to the many other individuals who provided input throughout the development of this document. Special thanks to Robert L. Madsen, P.E. of Devco Engineering for his technical support throughout this project.

The NAHB Research Center appreciates and recognizes the following individuals and companies who donated steel, tools, computer programs, and other materials to support the development and testing of the provisions in this document.

Julian White, Jr. of Dale Incor for donating steel.
Ed Daingeau of Clark-Cincinnati Steel Framing Systems for donating steel.
Neal Peterson, P.E. of Devco Engineering for donating the AISIWIN computer program.
Bob Glauz of RSG Software for donating the CFS computer program.
Martin Williams & Brian Mucha, P.E. of Atlantic Framing Design, Inc. for donating the C-Stud Analyzer II computer program.
Black and Decker/Dewalt for donating screw guns and other tools.

VIII

PRESCRIPTIVE METHOD FOR RESIDENTIAL COLD-FORMED STEEL FRAMING

CONTENTS

PRESCRIPTIVE METHOD FOR RESIDENTIAL COLD-FORMED STEEL FRAMING

X

PRESCRIPTIVE METHOD FOR RESIDENTIAL COLD-FORMED STEEL FRAMING

LIST OF TABLES

PRESCRIPTIVE METHOD FOR RESIDENTIAL COLD-FORMED STEEL FRAMING

LIST OF FIGURES

PRESCRIPTIVE METHOD FOR RESIDENTIAL COLD-FORMED STEEL FRAMING

Executive Summary

The *Prescriptive Method* was developed as an interim guideline for the construction of one- and two-family residential dwellings using cold-formed steel framing. It provides a strictly prescriptive method to stick-frame typical homes with cold-formed steel framing. This document standardizes the basic cold-formed steel members, provides an identification system for labelling, and gives minimum corrosion protection requirements. It also includes floor joist span tables, ceiling joist span tables, rafters span tables, wall stud tables, wall bracing requirements, and connection requirements. The requirements are supplemented with appropriate construction details in an easy-to-read format.

The *Prescriptive Method* is consistent with current U.S. building code provisions, engineering standards, and industry specifications, but it is not written as a regulatory document. The *Prescriptive Method* is written in a building code compatible style to facilitate future code adoption.

PRESCRIPTIVE METHOD FOR RESIDENTIAL COLD-FORMED STEEL FRAMING

PRESCRIPTIVE METHOD FOR RESIDENTIAL COLD-FORMED STEEL FRAMING

INTRODUCTION

The *Prescriptive Method* was developed as an interim guideline to facilitate the use of cold-formed steel framing in the construction of one- and two-family residential dwellings. It provides a strictly prescriptive method to stick-frame typical homes with cold-formed steel framing. The provisions in this document were developed by applying accepted engineering practices. However, builders using this document should verify its compliance with local code requirements.

This document is designed to be as compatible as possible with building code provisions, but it is not written as a regulatory document. The user shall refer to the applicable building code requirements where the provisions of this document are not applicable, in conflict with the building code, or where engineered design is called out.

1.0 GENERAL

1.1 Purpose

The purpose of this document is to provide prescriptive requirements for the construction of residential buildings framed with cold-formed steel. These provisions include definitions, span tables, fastener schedules, and other related information appropriate for use by home builders and building code officials.

1.2 Approach

These requirements are based primarily on the AISI *Specification for the Design of Cold-Formed Steel Structural Members* (AISI-86 with 1989 addendum) and ASCE *Minimum Design Loads for Buildings and Other Structures* (ASCE 7-93).

These provisions are intended to represent sound engineering and construction practice. A commentary is available that documents the assumptions and the derivation of the requirements contained in this document.

1.3 Scope

These provisions apply to the construction of detached one- or two-family dwellings, townhouses, and other attached single-family dwellings not more than two stories in height using in-line framing practices. Steel-framed construction in accordance with this prescriptive method shall be limited by the applicability limits set forth in Table 1. Intermixing of these provisions with other construction materials, such as wood, in a single structure shall be in accordance with the applicable building code requirements for that material and the applicability limits set forth in Table 1.

PRESCRIPTIVE METHOD FOR RESIDENTIAL COLD-FORMED STEEL FRAMING

Table 1
Applicability Limits[1]

House Geometry, Site, and Loading Conditions	LIMITATION
House Width	36 feet (9.14 m) maximum
Number of Stories	2 story
Design Wind Speed[2]	110 mph maximum (49 m/sec) fastest-mile wind speed (Exposure C) [except as noted][2]
Ground Snow Load	70 psf (3.353 kN/m^2) max. ground snow load
Seismic Zone[2]	0, 1, 2, 3, and 4 [except as noted][2]
Floors	
Floor dead load	10 psf (0.479 kN/m^2) maximum
Floor live load 　1st floor 　2nd floor (sleeping rooms)	40 psf (1.92 kN/m^2) maximum 30 psf (1.437 kN/m^2) maximum
Cantilever (supporting one story wall + roof)	24" (610 mm) maximum
Walls	
Wall dead load	10 psf (0.479 kN/m^2) maximum
Interior partitions lateral load	5 psf (0.2395 kN/m^2) maximum
Bearing Wall height	10 feet (3.048 m) maximum
Roofs	
Roof dead load	12 psf (0.479 kN/m^2) maximum total load. 7 psf (0.335 kN/m^2) maximum roof covering
Roof live load	Up to 70 psf (3.353 kN/m^2) ground snow load 16 psf (0.7664 kN/m^2) minimum.
Ceiling dead load	5 psf (0.2395 kN/m^2) maximum
Roof slope	3:12 to 12:12
Rake overhang	12" (305 mm) maximum
Soffit overhang	24" (609.6 mm) maximum
Attic live load with storage	20 psf (0.958 kN/m^2) maximum
Attic live load without storage	10 psf (0.479 kN/m^2) maximum

For SI:　1" = 25.4 mm, 1 psf = 0.0479 kN/m^2, 1 mph = 1.609 km/hr = .447 m/sec, 1 foot = 304.8 mm.
[1]　The limitations cover the majority of one-and two-family dwelling construction.
[2]　For design wind speeds > 90 mph (40.23 m/sec) Exposure C and Seismic Zones 3 & 4 engineered design is required for wall bracing, hold-downs, and/or uplift straps.

PRESCRIPTIVE METHOD FOR RESIDENTIAL COLD-FORMED STEEL FRAMING

1.4 Definitions

Accepted Engineering Practice: That which conforms to accepted principles, tests, or standards of nationally recognized technical or scientific authorities.

Approved: Refers to approval by the code official or other authority having jurisdiction as the result of investigation and tests conducted by him, or by reason of accepted principles or tests by nationally recognized organizations.

Axial Load: The longitudinal force acting on a member. Examples are the gravity loads carried by columns or studs.

Buckling: The bending, warping or crumpling of a member (such as a wall stud) subjected to axial, bending, bearing, or shear loads.

Ceiling Joist: A horizontal structural framing member which supports a ceiling and/or attic loads. Refer to Fig. 2.

C-section: Used for structural framing members (such as studs, joists, headers, beams, girders, and rafters). The name comes from the member's "C" shaped cross-sectional configuration consisting of a web, flange and lip. Figure 1 shows this cross-section and defines the different parts of the C-section. C-section web depth measurements are taken to the outside of the flanges. C-section flange width measurements also use outside dimensions.

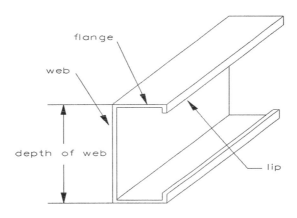

Figure 1 C-Section Configuration

Clip Angle: An L shaped short piece of metal (normally with a 90 degree bend). It is typically used for connections. Refer to Fig. 6.

Cold-forming: A process where light-gauge steel members are manufactured by (1) press-braking blanks sheared from sheets or cut length of coils or plates, or by (2) roll forming cold- or hot-rolled coils of sheet steel; both forming operations are performed at ambient room temperature, that is, without manifest addition of heat such as would be required for hot forming.

PRESCRIPTIVE METHOD FOR RESIDENTIAL COLD-FORMED STEEL FRAMING

Flange: The part of a C-section or track that is perpendicular to the web. Refer to Figure 1.

Flat Strap: Sheet steel cut to a specified width without any bends (Typically used for bracing and other flat applications). Refer to Fig. 2.

Floor Joist: A horizontal structural framing member that supports floor loads. Refer to Fig. 2.

Galvanized Steel: Steel that has a zinc protective coating for resistance against corrosion. The level of protection provided is measured by the weight of the galvanized coating applied to the surface area of the steel (eg. G-40 or G-60).

Header: A horizontal built-up structural framing member used over wall or roof openings to carry loads across the opening. Refer to Fig. 2.

In-Line Framing: Framing method where all vertical and horizontal load carrying members are aligned. Refer to Fig. 2.

Jack Stud: A vertical structural member that does not span the full height of the wall and supports vertical loads and/or transfers lateral loads. Jack studs are used to support headers. Refer to Fig. 2.

King Stud: A vertical structural member that spans the full height of the wall and supports vertical loads and lateral loads. Usually located at both ends of a header adjacent to the jack studs to resist lateral loads. Refer to Fig. 2.

Lip: The part of a C-section which extends from the flange at the open end. The lip increases the strength characteristics of the member and acts as a stiffener to the flange. Refer to Figure 1.

Loads, Live and Dead: Dead loads are the weight of the walls, partitions, framing, floors, ceilings, roofs, and all other permanent stationary construction entering into and becoming a part of a building. Live loads are all loads except dead and lateral loads.

Material Properties: The chemical, mechanical, and physical properties of steel before or after the cold-forming process.

Material Thickness: The base metal thickness excluding any protective coatings. Thickness is expressed in mils (1/1000 of an inch).
Mil: A unit of measurement typically used in measuring the thickness of thin elements. One mil equals 1/1000 of an inch.

Multiple Span: The span made by a continuous member having intermediate supports.

Non-Load Bearing Walls *(Non-structural walls)*: Refer to Walls.

Punchout: A hole in the web of a steel framing member allowing for the installation of plumbing, electrical, and other trade installation. Refer to Fig. 3.

Rafter: A structural framing member (sloped) which supports roof loads. Refer to Fig. 2
Seismic Zone: Seismic Zones designate areas with varying degrees of seismic risk and associated seismic design parameters (i.e. effective peak ground acceleration). Seismic Zones 1, 2, 3, and 4 correspond to effective peak ground acceleration of 0.1g, 0.2g, 0.3g, and 0.4g, respectively (1g is the acceleration of the earth's gravity at sea level). Refer to ASCE 7-93 for more details.

PRESCRIPTIVE METHOD FOR RESIDENTIAL COLD-FORMED STEEL FRAMING

Seismic Zone: Seismic Zones designate areas with varying degrees of seismic risk and associated seismic design parameters (i.e. effective peak ground acceleration).   Seismic Zones 1, 2, 3, and 4 correspond to effective peak ground acceleration of 0.1g, 0.2g, 0.3g, and 0.4g, respectively (1g is the acceleration of the earth's gravity at sea level).   Refer to ASCE 7-93 for more details.

Shearwall: A wall assembly capable of resisting lateral forces to prevent racking from wind or seismic loads acting parallel to the plane of the wall.

Single Span: The span made by one continuous structural member without any intermediate supports.

Span: The clear horizontal distance between bearing supports.

Structural Sheathing: The covering (eg. plywood) used directly over structural members (eg. studs or joists) to distribute loads, brace walls, and strengthen the assembly.   Refer to Fig. 2.

Stud: Vertical structural element of a wall assembly which supports vertical loads and/or transfers lateral loads.   Refer to Fig. 2.

Track: Used for applications such as top and bottom plate for walls and band joists for flooring systems. The track has a web and flanges, but no lips.   Track web depth measurements are taken to the inside of the flanges. Refer to Fig. 2.

Walls:

> *Structural or load bearing*: (i.e. transverse and/or axial load bearing). Steel framing systems that exceed the limits for a non-structural system (eg. wall studs).

> *Non-Structural*: *(i.e. non-load bearing)*. Steel framing systems that are limited to 10 psf (0.479 kN/m^2) maximum lateral (transverse) load and/or limited, exclusive of sheathing materials, to 100 pounds (450 N) per lineal foot (305 mm) or 200 pounds (900 N) maximum superimposed vertical load per member (eg. partitions).

Web: The part of a C-section or track that connects the two flanges. Refer to Fig. 1.

Web Crippling: The bending, warping or crumpling of the web of a member subjected to concentrated load.

Web Stiffener: Additional material that is attached to the web to strengthen the member against web crippling. Also called bearing stiffener.

Wind Exposure: Wind exposure is determined by site conditions that affect the ground level wind speeds experienced at a given site. For the purpose of this document, Exposure B represents a suburban or wooded terrain and Exposure C represents open terrain with scattered obstructions. Refer to ASCE 7-93 for more details.

Wind Speed: Wind speed is the design wind speed related to winds that could be expected to occur once every 50 years at a given site (i.e. 50 year return period). Wind speeds in this document are given in units of miles per hour by "fastest-mile" measurements. Refer to ASCE 7-93 for more details.

Yield Strength: A characteristic of the basic strength of the steel material.   It is the highest unit stress that the material can endure before permanent deformation occurs.

PRESCRIPTIVE METHOD FOR RESIDENTIAL COLD-FORMED STEEL FRAMING

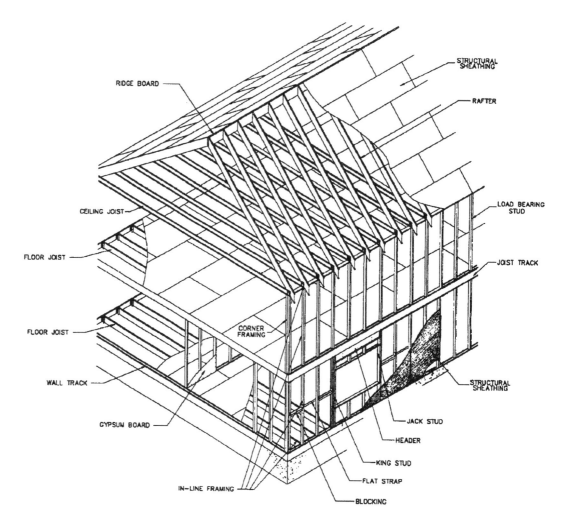

FIGURE 2
STEEL FRAMED BUILDING

PRESCRIPTIVE METHOD FOR RESIDENTIAL COLD-FORMED STEEL FRAMING

2.0 MATERIALS, SHAPES, AND STANDARD SIZES

2.1 Types of Cold-Formed Steel

2.1.1 Structural Members

Load bearing steel framing members shall be cold-formed to shape from structural quality sheet steel complying with the requirements of one of the following:

1   ASTM A 653: Grades 33, 37, 40, & 50 (Class 1 and 3); or

2.   ASTM A 792: Grades 33, 37, 40, & 50A; or

3.   ASTM A 875: Grades 33, 37, 40, & 50 (Class 1 and 3); or

4.   Steels that comply with ASTM A 653, except for tensile and elongation, shall be permitted provided the ratio of tensile strength to yield point is at least 1.08 and the total elongation is at least is at least 10 percent for a two-inch gage length or 7 percent for an eight-inch gauge length.

2.1.2 Non-Structural Members

Non-structural members shall comply with ASTM C-645.

2.2 Physical Dimensions

Cold-formed structural steel members shall comply with the dimensional requirements specified in Table 2. In addition, tracks shall have a minimum of 1-1/4 inch (32 mm) flanges. Members with different geometrical shapes shall not be used with these provisions without the approval of a design professional.

Table 2
C-Section Dimensions

Nominal Member Size	Industry Designator[2]	Web Depth inches (mm)	Minimum Flange Width[1] inches (mm)	Minimum Lip Size inches (mm)
2 x 4	350S162-t	3.5 (88.9)	1.625 (41.3)	0.5 (12.7)
2 x 6	550S162-t	5.5 (139.7)	1.625 (41.3)	0.5 (12.7)
2 x 8	800S162-t	8 (203.2)	1.625 (41.3)	0.5 (12.7)
2 x 10	1000S162-t	10 (254)	1.625 (41.3)	0.5 (12.7)
2 x 12	1200S162-t	12 (304.8)	1.625 (41.3)	0.5 (12.7)

For SI: 1" = 25.4 mm.

[1]   Maximum flange size permitted is 2" (50.8 mm)

[2]   "t" is the uncoated material thickness in mils. "S" indicates studs and joist sections with lips. Track sections shall use the designator "T" instead of "S".

PRESCRIPTIVE METHOD FOR RESIDENTIAL COLD-FORMED STEEL FRAMING

2.3   Uncoated Material Thickness

The steel thicknesses of steel framing members in their end-use shall meet or exceed the minimum uncoated thickness values given in Table 3.

Table 3
Minimum Uncoated Material Thickness

Designation (mils)	Minimum Uncoated Thickness inches (mm)	Reference Gauge Number
18	0.018 (0.455)	25
27	0.027 (0.683)	22
33	0.033 (0.836)	20
43	0.043 (1.087)	18
54	0.054 (1.367)	16
68	0.068 (1.720)	14
97	0.097 (2.454)	12

For SI: 1" = 25.4 mm,

2.4   Bend Radius

The maximum inside bend radius shall be the greater of 3/32" (2.38 mm) or two times the material thickness (2t).

2.5   Yield Strength

The minimum yield strength (or yield point) of steel for C-sections, tracks, flat straps, and other members shall be 33 ksi (227,700 kPa) as determined by the 0.2 percent offset method, or by other methods in accordance with ASTM A370.

2.6   Corrosion Protection

Cold-formed steel framing components identified in accordance with this document shall have a metallic coating complying with Table 4. Coatings shall not flake when the steel is cold-formed to produce a framing component.

PRESCRIPTIVE METHOD FOR RESIDENTIAL COLD-FORMED STEEL FRAMING

Table 4
Minimum Coating Requirements

Steel Component	Reference ASTM Standard		
	A 653 / A 653M	A 792 / A 792M	A 875 / A 875M
Structural	G60	AZ50	GF60
Non-structural	G40	AZ50	GF45

Other approved metallic coatings shall be permitted provided the alternate coatings can be demonstrated to have a corrosion resistance that is equal to or greater than the corresponding hot-dipped galvanized coatings (i.e. G40 and G60) and provide protection at cut edges, scratches, etc., by cathodic or sacrificial protection.

The minimum coating designations shown in Table 4 assume normal exposure conditions and construction practices. Cold formed steel members located in harsh environments (such as coastal areas) shall have a minimum coating of G-90.

Steel framing members shall be located within the building envelope and adequately shielded from direct contact with moisture from the ground or the outdoor climate.

2.6.1   Compatibility With Other Metals

- Copper materials shall not be used in direct contact with galvanized steel members or components.

- Iodized aluminum members are compatible with cold-formed steel members.

- Zinc coated or zinc-aluminum coated steel shall not be embedded in concrete, unless approved   for that purp

2.7   Web Punchouts

Unreinforced punchouts in webs of structural members shall not exceed 1-1/2 inches wide x 4 inches long (38 mm wide x 102 mm long), located along the centerline of the web at a minimum center to center spacing of 24 inches (600 mm). Punchout locations and sizes shall comply with Figure 3.

PRESCRIPTIVE METHOD FOR RESIDENTIAL COLD-FORMED STEEL FRAMING

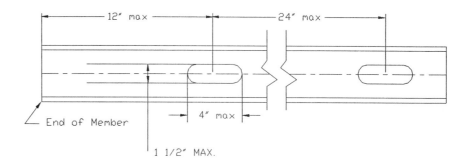

Fig. 3   Web Punchout

2.8   Cutting, Notching, and Hole Stiffening

Flanges of joists, studs, headers, and other structural members shall not be cut or notched without an approved design. Web holes closer than 12 inches (305 mm) from center of hole to joist bearings, or that exceed the size limits specified in Section 2.7, shall be reinforced with a solid steel plate, stud, joist, or track section of an equivalent thickness to the member it reinforces, provided that the following limitations are met:

  a.   The depth of the hole (across the web) does not exceed half the depth of the web, and

  b.   The length of the hole (along the web) does not exceed 4" (102 mm) or the depth of
          the web, whichever is greater.

Reinforcement shall extend at least 1 inch (25.4 mm) beyond all edges of the hole and shall be attached to the web with #8 screws (minimum) spaced no greater than one inch (25.4 mm) on center along the edges of the reinforcement, as shown in Figure 4.

2.9   Bearing Stiffeners

Bearing stiffeners (also called web stiffeners) shall be at least as thick as the members to which they are attached. A stiffener shall be fabricated from a track or a C-section. Stiffeners shall be screwed to the web of the member it is stiffening by a minimum of two #8 screws on each end of the stiffener (four screws total). Stiffeners shall extend across the full depth of the web. Stiffeners shall be seated between the flanges of the C-sections or installed across the web opposite to the flanges. Where required by these provisions, stiffeners shall be installed in accordance with Figure 5.

PRESCRIPTIVE METHOD FOR RESIDENTIAL COLD-FORMED STEEL FRAMING

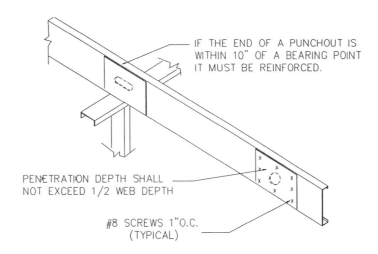

IF THE END OF A PUNCHOUT IS
WITHIN 10" OF A BEARING POINT
IT MUST BE REINFORCED.

PENETRATION DEPTH SHALL
NOT EXCEED 1/2 WEB DEPTH

#8 SCREWS 1"O.C.
(TYPICAL)

Fig. 4   Hole Stiffener

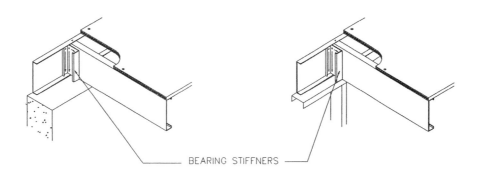

BEARING STIFFNERS

FLOOR JOISTS BEARING ON FOUNDATION   FLOOR FRAMING AT EXTERIOR WALL

Fig. 5   Bearing Stiffener

PRESCRIPTIVE METHOD FOR RESIDENTIAL COLD-FORMED STEEL FRAMING

2.10   Fasteners

Fasteners shall comply with Sections 2.10.1 and 2.10.2. Other fastening techniques, such as the use of pneumatically driven fasteners, powder actuated fasteners, crimping, welding, or other approved fasteners may be used provided that they are designed in accordance with accepted engineering practice.

2.10.1   Screws

All screws shall be a minimum size of #8 unless otherwise specified. All screws shall be corrosion resistant and shall be spaced such that the minimum center-to-center or edge distance of three screw diameters is maintained. Self-drilling tapping screws shall conform to SAE J78. A minimum of three exposed threads shall extend through the steel, for steel to steel connections, and when attaching other materials to steel framing. Screws shall penetrate individual components of connections without causing permanent separation between the components. Screws shall be installed in a manner such that the threads or holes are not stripped. Head styles, threads, and point types for screws shall be selected based on application conditions and manufacturer recommendations. Refer to Section 10.0 for additional guidelines on screws.

2.10.2   Bolts

Bolts shall meet or exceed the requirements of ASTM A307. Bolts shall be installed with nuts and washers. Center to center spacing of bolt holes, connecting sheet metal material to concrete, shall be a minimum of three bolt diameters. Distance from the center of the bolt hole to the edge of the connecting member shall not be less than 1-1/2 bolt diameters. Bolt hole diameter shall not exceed the bolt diameter by more than 1/16" (1.6 mm).

3.0   FIELD IDENTIFICATION SYSTEM

Load bearing steel framing members shall have a legible label, stamp, stencil, or embossment, at a minimum of 48 inches on center, with the following information as a minimum:

a.   The manufacturer's identification.
b.   The minimum uncoated steel thickness in mils (such as 43 mils).
c.   "ST" for structural members.
d.   Metallic coating weight (mass).
e.   Minimum yield strength in kips per square inch.

An example of an acceptabel label : ABC 43 mils ST G60 33 ksi

4.0   FOUNDATION

The foundation shall comply with applicable building codes. Steel framing shall be attached to the foundation structure according to the requirements of Sections 5 and 6 of this document.

Foundation Anchor bolts shall be located no more than 12 inches (305 mm) from corners and the termination of bottom tracks (i.e. at door openings). Anchor bolts shall have a minimum embedment of 6" (152.4 mm) in concrete or 15" (381 mm) in masonry.

PRESCRIPTIVE METHOD FOR RESIDENTIAL COLD-FORMED STEEL FRAMING

5.0 FLOORS

5.1 Floor Construction

Cold-formed steel floors shall be constructed in accordance with this section and Figure 6. Cold-formed steel framing members shall comply with the provisions of Section 2.0. Steel floors constructed per this section are limited to design wind speeds of 110 mph (177 km/hr) fastest-mile or design seismic conditions of Seismic Zones 0, 1, 2, 3, and 4. All floor joists shall be aligned with other horizontal or vertical load carrying members above or below the joists.

5.2 Floor to Foundation or Bearing Wall Connection

Cold-formed steel floors shall be anchored to the foundation or bearing walls in accordance with Table 5 and Figure 7. Cold-formed steel floors supported by interior load-bearing walls shall be constructed per Figures 8a or 8b.

5.3 Allowable Joist Spans

The clear span of cold-formed steel floor joists shall not exceed the limits set forth in Tables 6 and 7. Bearing stiffeners, in accordance with Section 2.9, shall be installed at all bearing locations for floor joists. Floor joists shall have a bearing length of not less than 1.5" (38.1 mm) and 3.5" (88.9 mm) for end and interior supports respectively.

5.4 Joist Bracing

The top flanges of floor joists shall be laterally braced by the application of floor sheathing (plywood or oriented strand board) in accordance with the applicable building code and fastened to the joists according to the fastening requirements in Table 8.

Floor joists spans that exceed 12 feet (3658 mm) shall have the bottom flanges laterally braced in accordance with one of the following:

1. Gypsum board installed in accordance with the applicable building code (such as ceilings).

2. Steel strapping installed as shown in Figure 6. Steel straps shall be at least 1-1/2" in width and 33 mils in thickness (38.1 mm x 0.836 mm) and shall be installed without slack. Straps shall be attached to the bottom flange at each joist with at least one #8 screw. In line blocking or bridging (X-bracing) shall be installed between joists at the termination of all straps and at a maximum spacing of 12 feet (3657 mm), measured perpendicular to the joists. The ends of straps shall be fastened to blocking or to the exterior walls with at least two #8 screws.

5.5 Floor Cantilevers

Floor overhangs shall not exceed 24" (609.6 mm). Bearing stiffeners shall be installed at bearing points and at the end of cantilevered section. All punchouts in the cantilevered section of the joist shall be reinforced as described in Section 2.8. Floor cantilevers shall support loads from one story only.

Approved design is required for floor overhangs for balconies, decks, and other extensions that have higher live loads than 40 psf (1.92 kN/m^2).

PRESCRIPTIVE METHOD FOR RESIDENTIAL COLD-FORMED STEEL FRAMING

5.6   Splicing

Floor Joists shall not be spliced without an approved design.

Splicing of joist track members only shall conform with Figure 9. Joist tracks shall not be spliced within 3 inches (76.2 mm) of joist connections.

5.7   Framing of Openings

Openings in floor framing shall be framed with built-up header and trimmer joists. Header joist spans shall not exceed 8 feet (2438 mm). Header and trimmer joists shall be fabricated from joist and track sections, which shall be of equivalent size and thickness as the floor joists, in accordance with Figure 6. Header joists shall be connected to trimmer joists with a minimum of two 2 inch x 2 inch (51 x 51 mm) clip angles installed on opposite sides of each end of the header. Four #10 screws shall be installed through each leg of the clip angles. The clip angles shall be at least as thick as the floor joists or 54 mils (0.054 mm), whichever is greater.

5.8   Floor Trusses

Cold-formed steel floor trusses shall be designed, braced, and installed in accordance with an approved design. Truss members shall not be notched, cut, or altered in any manner unless so designed.

PRESCRIPTIVE METHOD FOR RESIDENTIAL COLD-FORMED STEEL FRAMING

Table 5
Floor to Foundation or Bearing Wall Connection Requirements

Framing Condition	Wind Speed (mph), Exposure, and Seismic Zones [1]			
	Up to 70 B or Zones 0, 1, 2	Up to 90 B or 70 C or Zone 3	Up to 100 B or 90 C or Zone 4	Up to 110 C
One-Story or Second Floor of Two-Story Building				
Floor Joist to Bearing Wall Top Track	2 - #10 screws	2- #10 screws	2- #10 screws	Approved Design Required
Joist Track or End Joist to Bearing Wall Top Track	1-#10 screw at 24" oc	1-#10 screw at 24" oc	1-#10 screw at 24" oc	
Joist Track to Foundation Wall (concrete or masonry)	1/2" anchor bolt, clip angle, and 8 - #10 screws at 6' oc	1/2" anchor bolt, clip angle, and 8 - #10 screws at 6' oc	1/2" anchor bolt, clip angle, and 8 - #10 screws at 4' oc	1/2" anchor bolt, clip angle, and 8 - #10 screws at 2' oc
Floor Joist to Wood Sill on Foundation Wall[2]	2- #10 screws through joist flange[3]	2- #10 screws through joist flange[3]	2- #10 screws through joist flange[3]	Approved Design Required
Joist Track to Wood Sill on Foundation Wall[2]	1-#10 screw at 24" oc	1-#10 screw at 24" oc	1-#10 screw at 24" oc	
First Floor of Two-Story Building				
Joist Track to Foundation Wall (concrete or masonry)	1/2" anchor bolt, clip angle, and 8 - #10 screws at 6' oc	1/2" anchor bolt, clip angle, and 8 - #10 screws at 6' oc	5/8" anchor bolt, clip angle, and 8 - #10 screws at 4' oc	5/8" anchor bolt, clip angle, and 8 - #10 screws at 2' oc
Floor Joist to Wood Sill on Foundation Wall[2]	2- #10 screws through joist flange[3]	2- #10 screws through joist flange[3]	2- #10 screws through joist flange[3]	Approved Design Required
Joists Track to Wood Sill on Foundation Wall[2]	1-#10 screw at 24" oc	1-#10 screw at 24" oc	1-#10 screw at 12" oc	

For SI: 1" = 25.4 mm, 1 psf = 0.0479 kN/m^2, 1 mph = 1.609 km/hr, 1 foot = 304.8 mm.

[1] Use the highest of the wind speed and exposure or the seismic requirements.
[2] Screw threads shall penetrate a minimum depth of 1 inch (25.4 mm) into the wood sill.
[3] A suitable alternate connection shall be a 4"x 6" x 54 mil (102 x 152 x .054 mm) plate fastened to the joist track with two #8 screws and to the wood sill with two 10d nails. The plates shall be located on the outside of the joist track at the same spacing as the floor joists.

PRESCRIPTIVE METHOD FOR RESIDENTIAL COLD-FORMED STEEL FRAMING

Table 6
Allowable Spans For Cold-Formed Steel Floor Joists
Single Span

Nominal Joist Size[4]	FLOOR JOISTS ALLOWABLE SPANS [1,2,3]					
	10 psf D.L. + 30 psf L.L.			10 psf D.L. + 40 psf L.L.		
	Spacing (inches)			Spacing (inches)		
	12	16	24	12	16	24
2 x 6 x 33	11'-7"	10'-7"	9'-1"	10'-7"	9'-7"	8'-1"
2 x 6 x 43	12'-8"	11'-6"	10'-0"	11'-6"	10'-5"	9'-1"
2 x 6 x 54	13'-7"	12'-4"	10'-9"	12'-4"	11'-2"	9'-9"
2 x 6 x 68	14'-6"	13'-2"	11'-6"	13'-2"	12'-0"	10'-6"
2 x 6 x 97	16'-1"	14'-7"	12'-9"	14'-7"	13'-3"	11'-7"
2 x 8 x 33	15'-8"	13'-3"	8'-10"	14'-0"	10'-7"	7'-1"
2 x 8 x 43	17'-1"	15'-6"	13'-7"	15'-6"	14'-1"	12'-3"
2 x 8 x 54	18'-4"	16'-8"	14'-7"	16'-8"	15'-2"	13'-3"
2 x 8 x 68	19'-8"	17'-11"	15'-7"	17'-11"	16'-3"	14'-2"
2 x 8 x 97	21'-10"	19'-10"	17'-4"	19'-10"	18'-0"	15'-9"
2 x 10 x 43	20'-6"	18'-8"	15'-3"	18'-8"	16'-8"	13'-1"
2 x 10 x 54	22'-1"	20'-1"	17'-6"	20'-1"	18'-3"	15'-11"
2 x 10 x 68	23'-8"	21'-6"	18'-10"	21'-6"	19'-7"	17'-1"
2 x 10 x 97	26'-4"	23'-11"	20'-11"	23'-11"	21'-9"	19'-0"
2 x 12 x 43	23'-5"	20'-3"	14'-1"	20'-11"	16'-10"	11'-3"
2 x 12 x 54	25'-9"	23'-4"	19'-7"	23'-4"	21'-3"	17'-6"
2 x 12 x 68	27'-8"	25'-1"	21'-11"	25'-1"	22'-10"	19'-11"
2 x 12 x 97	30'-9"	27'-11"	24'-5"	27'-11"	25'-4"	22'-2"

For SI: 1" = 25.4 mm,   1 psf = 0.0479 kN/m², 1 foot = 304.8 mm.

[1]   Table provides the maximum clear span in feet and inches.
[2]   Bearing stiffeners shall be installed at all support points and concentrated loads.
[3]   Deflection criteria: L/480 for live loads; L/240 for total loads.
[4]   For actual size refer to Table 2.

PRESCRIPTIVE METHOD FOR RESIDENTIAL COLD-FORMED STEEL FRAMING

Table 7
Allowable Spans For Cold-Formed Steel Floor Joists
Multiple Spans

Nominal [5]	FLOOR JOISTS ALLOWABLE SPANS [1,2,3,4,5]					
	10 psf D.L. + 30 psf L.L.			10 psf D.L. + 40 psf L.L.		
	Spacing (inches)			Spacing (inches)		
	12	16	24	12	16	24
2 x 6 x 33	12'-10"	10'-6"	7'-10"	11'-0"	9'-0"	6'-7"
2 x 6 x 43	15'-8"	13'-6"	11'-0"	14'-0"	12'-1"	9'-10"
2 x 6 x 54	17'-7"	15'-3"	12'-5"	15'-9"	13'-8"	11'-2"
2 x 6 x 68	19'-6"	17'-2"	14'-0"	17'-8"	15'-4"	12'-6"
2 x 6 x 97	21'-7"	19'-7"	16'-8"	19'-7"	17'-10"	14'-11"
2 x 8 x 33	12'-9"	10'-2"	7'-1"	10'-9"	8'-6"	5'-8"
2 x 8 x 43	19'-5"	16'-8"	12'-6"	17'-5"	14'-3"	10'-8"
2 x 8 x 54	23'-0"	19'-11"	16'-3"	20'-6"	17'-9"	14'-6"
2 x 8 x 68	25'-10"	22'-5"	18'-3"	23'-2"	20'-0"	16'-4"
2 x 8 x 97	29'-4"	26'-7"	21'-11"	26'-7"	24'-0"	19'-7"
2 x 10 x 43	20'-3"	16'-5"	12'-1"	17'-3"	13'-11"	10'-2"
2 x 10 x 54	25'-6"	22'-1"	18'-0"	22'-10"	19'-9"	15'-6"
2 x 10 x 68	30'-6"	26'-5"	21'-7"	27'-4"	23'-8"	19'-3"
2 x 10 x 97	35'-4"	31'-9"	25'-11"	32'-1"	28'-5"	23'-2"
2 x 12 x 43	19'-8"	15'-9"	11'-3"	16'-7"	13'-3"	9'-0"
2 x 12 x 54	27'-8"	23'-9"	17'-10"	24'-9"	20'-4"	15'-2"
2 x 12 x 68	32'-7"	28'-3"	23'-0"	29'-2"	25'-3"	20'-7"
2 x 12 x 97	41'-3"	36'-7"	29'-10"	37'-5"	32'-9"	26'-9"

For SI: 1" = 25.4 mm,   1 psf = 0.0479 kN/m², 1 foot = 304.8 mm.

[1]   Table provides the maximum clear span in feet and inches, to either the right or left of the interior support.
[2]   Interior bearing supports for multiple span joists shall consist of structural (bearing) walls or beams.
[3]   Bearing stiffeners shall be installed at all support points and concentrated loads.
[4]   Deflection criteria: L/480 for live loads; L/240 for total loads.
[5]   The interior support shall be located within ∀ two feet (610 mm) of mid-span provided that each of the resulting spans do not exceed the maximum span shown in the table above.
[6]   For actual size refer to Table 2.

PRESCRIPTIVE METHOD FOR RESIDENTIAL COLD-FORMED STEEL FRAMING

Table 8
Floor Fastening Schedule

Building Element	Number & size of screws	Spacing of screws
Floor Joist to Joist Track (connect at joist flanges or web stiffener flange)	2 # 8 screws	N/A
Web Stiffener to Floor Joist	4 # 8 screws	N/A
Subfloor (oriented strand board or plywood) to floor joist and joist track	# 8 screws[1]	6" oc edges 12" oc in field of panel

For SI: 1" = 25.4 mm,   1 psf = 0.0479 kN/m^2, 1 foot = 304.8 mm.

[1]   Head styles shall be bugle-head, flat-head, or similar with a minimum head diameter of 0.315 inches (8 mm.)

PRESCRIPTIVE METHOD FOR RESIDENTIAL COLD-FORMED STEEL FRAMING

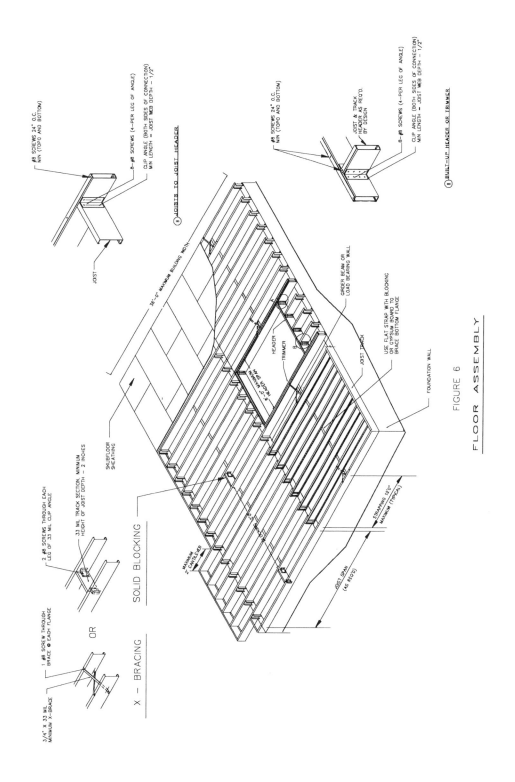

FIGURE 6

FLOOR ASSEMBLY

PRESCRIPTIVE METHOD FOR RESIDENTIAL COLD-FORMED STEEL FRAMING

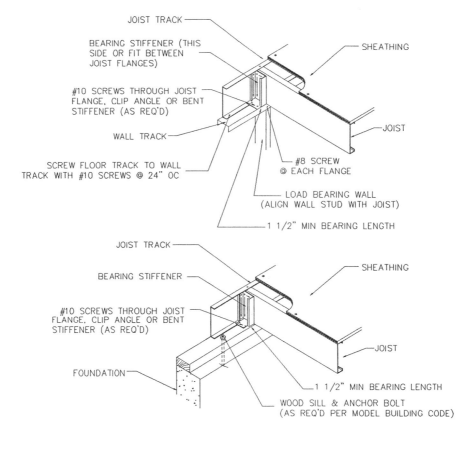

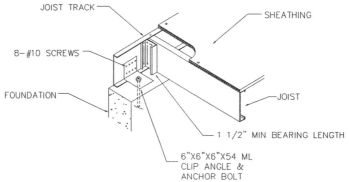

Fig. 7   Floor to Foundation or Bearing Wall Connection

PRESCRIPTIVE METHOD FOR RESIDENTIAL COLD-FORMED STEEL FRAMING

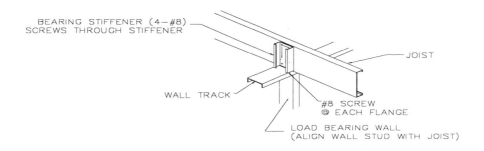

Fig. 8a Continuous Joist Span Supported on Stud

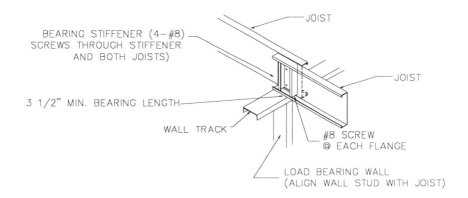

Fig. 8b Lapped Joists Supported on Studs

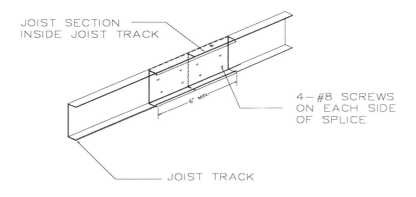

Fig. 9 Joist Track Splice

PRESCRIPTIVE METHOD FOR RESIDENTIAL COLD-FORMED STEEL FRAMING

6.0   STRUCTURAL WALLS
6.1   Wall Construction

Structural steel walls shall comply with this section and Figure 10. Cold-formed steel framing members shall comply with the provisions of Section 2.0. Structural steel walls in buildings constructed per this section shall be limited to design wind speeds up to 110 mph (177 km/hr) fastest-mile, Exposure C, design seismic conditions for zones 0, 1, 2, 3, and 4, or ground snow loads up to 70 psf (3.36 kN/m^2).

6.2   Wall to Foundation or Floor Connection

Cold-formed steel-framed walls shall be anchored to foundations or floors in accordance with Table 9 and Figure 11.

6.3   Load Bearing Walls

Framing of load bearing walls shall comply with Tables 10 through 16. Applied ground snow loads and wind speeds shall be established by local building code requirements. Wind exposure category, B or C, shall be established on an individual site basis according to the wind exposure definitions in Section 1.4. If wind exposure is unknown, exposure C shall be used to determine stud requirements.

All load bearing studs shall be aligned with trusses and other horizontal or vertical load carrying members above or below the wall stud. Top and bottom wall tracks shall be of equal or greater thickness than that of the studs.

6.4   Stud Bracing (Buckling Resistance)

The flanges of load bearing steel studs shall be laterally braced in accordance with one of the following:

1. Gypsum board or structural sheathing (i.e. APA rated OSB or plywood) fastened to wall studs in accordance with Table 16, or

2. Horizontal steel strapping installed at mid-height for 8-foot (2438 mm) walls, and 1/3 points for 9 and 10-foot (2743 and 3048 mm) walls (Refer to Fig. 10). Steel straps shall be at least 1-1/2 inches in width and 33 mils in thickness (38.1 mm x 0.836 mm) and shall be installed without slack. Straps shall be attached to the flanges of studs with at least one #8 screw. In line blocking shall be installed between studs at the termination of all straps and at 12-foot (3658 mm) intervals in between. Straps shall be fastened to the blocking with at least two #8 screws (refer to Fig. 10).

Adequate temporary or permanent stud bracing shall be provided to resist construction loads.

6.5   Splicing

Load bearing steel studs shall not be spliced without an approved design. Splicing of wall tracks only shall conform with Figure 12. Wall tracks shall not be spliced within 3 inches (76.2 mm) of studs or anchor bolt locations.

6.6   Corner Framing

Corner studs and the top track shall be installed in accordance with Figure 13. Other approved corner framing details may be used.

PRESCRIPTIVE METHOD FOR RESIDENTIAL COLD-FORMED STEEL FRAMING

6.7   Headers

Headers shall be installed above wall openings in accordance with Figures 10 and 14 and Tables 17 through 22. Header spans for house widths between those tabulated may be determined by interpolation. Headers shall be formed from two, equal sized C-shapes in a back-to-back or box type configuration. Cripple studs , installed below the headers, shall be the same size and thickness (or thicker) than the adjacent full height wall studs. The number of jack and king studs, on each side of the header, shall comply with Table 17. King and jack studs shall be of the same size and thickness as wall studs. Headers shall be connected to king studs in accordance with Table 22. Jack and king studs shall be interconnected with structural sheathing in accordance with Figure 14. Headers are not required for openings in interior non-load bearing walls.

6.8   Wall Bracing

Exterior steel-framed walls shall be braced (shearwall bracing) with diagonal steel straps or structural sheathing in accordance with Sections 6.8.1 or 6.8.2.

6.8.1   Strap Bracing (X-brace)

Diagonal steel straps or 'X-braces' and their connections shall be designed and installed in accordance with an approved design.

6.8.2   Structural Sheathing

Structural sheathing shall be installed on all exterior wall surfaces in accordance with Table 23 and Figure 10. Structural sheathing shall be a minimum of 7/16" (11.1 mm) APA rated oriented strand board or 15/32" (12 mm) APA rated plywood. Structural sheathing shall cover the full-height of the wall from the bottom track to the top track. All edges and interior areas of the structural sheathing panels shall be fastened to a framing member and tracks in accordance with Table 16. The percentage of wall length with full-height sheathing (i.e. no openings) shall comply with the minimum percentages of Table 23 for all exterior walls.

In conditions where design wind speeds are in excess of 90 mph (Exposure C) or 100 mph (Exposure B) and in Seismic Zone 3 or greater, the amount and type of structural sheathing shall be determined by accepted engineering practices.

6.9   Hold-down Requirements in High Wind or Seismic Conditions

In conditions where wind speeds are in excess of 90 mph (144.81 km/hr) Exposure C or 100 mph (160.9 km/hr) Exposure B and in Seismic Zone 3 or greater, hold-down brackets shall be provided in accordance with accepted engineering practices. At a minimum, hold-down brackets of adequate size shall be provided at the ends of each wall line.

6.10   Uplift Straps

At design wind speeds greater than 90 mph (144.81 km/hr) Exposure C or 100 mph (160.9 km/hr) Exposure B, uplift straps shall be provided at king and wall studs in accordance with accepted engineering practices.

PRESCRIPTIVE METHOD FOR RESIDENTIAL COLD-FORMED STEEL FRAMING

Table 9
Wall to Foundation or Floor Connection Requirements

Framing Condition	Wind Speed (mph), Exposure, & Seismic Zones[1]			
	Up to 70 B or Zones 0, 1, 2	Up to 90 B 70 C or Zone 3	Up to 100 B or 90 C or Zone 4	Up to 110 C
One-Story or Second Floor of Two-Story Building				
Wall Bottom Track to Floor Framing	2 - #10 screws at 24" oc	2 - #10 screws at 24"oc	2 - #10 screws at 24"oc	Approved Design Required
Wall Bottom Track to Foundation (Slab-on-grade)	1/2" anchor bolt at 6' oc	1/2" anchor bolt at 6' oc	1/2" anchor bolt at 4' oc	Approved Design Required
First Floor of Two-Story Building				
Wall Bottom Track to Floor Framing	2 - #10 screws at 24" oc	2 - #10 screws at 24"oc	2 - #10 screws at 12"oc	Approved Design Required
Wall Bottom Track to Foundation (Slab-on-grade)	1/2" anchor bolt at 6' oc	1/2" anchor bolt at 6' oc	1/2" anchor bolt at 4' oc	Approved Design Required

For SI: 1" = 25.4 mm,   1 psf = 0.0479 kN/m^2,   1 mph = 1.609 km/hr, 1 foot = 304.8 mm.

[1]   Use the greater of the wind speed and exposure or the seismic requirements.

PRESCRIPTIVE METHOD FOR RESIDENTIAL COLD-FORMED STEEL FRAMING

Table 10
Steel Stud Thickness for 8' Walls Supporting Roof and Ceiling Only
(One Story or Second Floor of a Two Story Building)

Wind Exposure		Member Size[2]	Member Spacing (inches)	Required Stud Thickness (mils)[1]															
				Building Width (feet)[3]															
				24				28				32				36			
Exp. B	Exp. C			Ground Snow Load (psf)				Ground Snow Load (psf)				Ground Snow Load (psf)				Ground Snow Load (psf)			
				20	30	50	70	20	30	50	70	20	30	50	70	20	30	50	70
70 mph		2x4	16	33	33	33	33	33	33	33	33	33	33	33	33	33	33	33	33
			24	33	33	33	33	33	33	33	33	33	33	33	33	33	33	33	43
		2x6	16	33	33	33	33	33	33	33	33	33	33	33	33	33	33	33	33
			24	33	33	33	33	33	33	33	33	33	33	33	33	33	33	33	33
80 mph	70 mph	2x4	16	33	33	33	33	33	33	33	33	33	33	33	33	33	33	33	33
			24	33	33	33	33	33	33	33	33	33	33	33	43	33	33	43	43
		2x6	16	33	33	33	33	33	33	33	33	33	33	33	33	33	33	33	33
			24	33	33	33	33	33	33	33	33	33	33	33	33	33	33	33	33
90 mph	80 mph	2x4	16	33	33	33	33	33	33	33	33	33	33	33	33	33	33	33	33
			24	33	33	43	43	33	33	43	43	33	43	43	43	43	43	43	43
		2x6	16	33	33	33	33	33	33	33	33	33	33	33	33	33	33	33	33
			24	33	33	33	33	33	33	33	33	33	33	33	33	33	33	33	33
100 mph	90 mph	2x4	16	33	33	33	33	33	33	33	33	33	33	33	33	33	33	33	33
			24	43	43	43	43	43	43	43	54	43	43	54	54	43	43	54	54
		2x6	16	33	33	33	33	33	33	33	33	33	33	33	33	33	33	33	33
			24	33	33	33	33	33	33	33	33	33	33	33	33	33	33	33	33
110 mph	100 mph	2x4	16	33	33	43	43	33	43	43	43	43	43	43	43	43	43	43	43
			24	54	54	54	54	54	54	54	68	54	54	68	68	54	54	68	68
		2x6	16	33	33	33	33	33	33	33	33	33	33	33	33	33	33	33	33
			24	33	33	33	33	33	33	33	43	33	33	33	43	33	33	43	43
	110 mph	2x4	16	43	43	43	43	43	43	43	43	43	43	43	54	43	43	43	54
			24	68	68	68	68	68	68	68	68	68	68	68	68	68	68	68	97
		2x6	16	33	33	33	33	33	33	33	33	33	33	33	33	33	33	33	33
			24	33	43	43	43	43	43	43	43	43	43	43	43	43	43	43	43

For SI: 1" = 25.4 mm, 1 psf = 0.0479 kN/m², 1 mph = 1.609 km/hr, 1 foot = 304.8 mm.

[1] Exterior load bearing walls with a minimum of 1/2" (12.7 mm) gypsum board on the inside and 7/16" (11.1 mm) oriented strand board or plywood (APA rated 24/16) on the outside, and interior load bearing walls with a minimum of 1/2" (12.7 mm) gypsum board on both sides may use the next thinner stud but not less than 33 mils (0.836 mm). [2] For actual sizes of members, refer to Table 2. [3] Building width is in the direction of horizontal framing members supported by the wall studs.

PRESCRIPTIVE METHOD FOR RESIDENTIAL COLD-FORMED STEEL FRAMING

Table 11
Steel Stud Thickness for 8' Walls Supporting One Floor, Roof and Ceiling
(First Story of a Two Story Building)

Wind Exposure		Member Size[2]	Member Spacing (inches)	Required Stud Thickness (mils)[1] — Building Width (feet)[3]															
				24 Ground Snow Load (psf)				28 Ground Snow Load (psf)				32 Ground Snow Load (psf)				36 Ground Snow Load (psf)			
Exp. B	Exp. C			20	30	50	70	20	30	50	70	20	30	50	70	20	30	50	70
70 mph		2x4	16	33	33	33	33	33	33	33	33	33	33	33	43	33	33	33	43
			24	43	43	43	43	43	43	43	43	43	43	43	54	43	43	54	54
		2x6	16	33	33	33	33	33	33	33	33	33	33	33	33	33	33	33	33
			24	33	33	33	33	33	33	33	43	33	33	43	43	33	43	43	54
80 mph	70 mph	2x4	16	33	33	33	33	33	33	33	33	33	33	33	43	33	33	43	43
			24	43	43	43	54	43	43	54	54	54	54	54	54	54	54	54	54
		2x6	16	33	33	33	33	33	33	33	33	33	33	33	33	33	33	33	43
			24	33	33	33	33	33	33	33	43	33	33	43	43	43	43	43	54
90 mph	80 mph	2x4	16	33	33	33	33	33	33	43	43	43	43	43	43	43	43	43	43
			24	54	54	54	54	54	54	54	54	54	54	54	68	54	54	68	68
		2x6	16	33	33	33	33	33	33	33	33	33	33	33	33	33	33	33	33
			24	33	33	33	33	33	33	43	43	43	43	43	43	43	43	43	54
100 mph	90 mph	2x4	16	43	43	43	43	43	43	43	43	43	43	43	43	43	43	43	54
			24	54	54	54	68	54	68	68	68	68	68	68	68	68	68	68	68
		2x6	16	33	33	33	33	33	33	33	33	33	33	33	33	33	33	33	33
			24	33	43	43	43	43	43	43	43	43	43	43	43	43	43	43	54
110 mph	100 mph	2x4	16	43	43	43	54	43	54	54	54	54	54	54	54	54	54	54	54
			24	68	68	68	68	68	68	97	97	97	97	97	97	97	97	97	97
		2x6	16	33	33	33	33	33	33	33	33	33	33	33	33	33	33	33	43
			24	43	43	43	43	43	43	43	54	43	43	54	54	54	54	54	54
	110 mph	2x4	16	54	54	54	54	54	54	54	54	54	54	54	68	54	54	68	68
			24	97	97	97	97	97	97	97	97	97	97	97	97	97	97	97	97
		2x6	16	33	33	33	33	33	33	33	43	33	33	43	43	43	43	43	43
			24	43	43	54	54	54	54	54	54	54	54	54	54	54	54	54	54

For SI: 1" = 25.4 mm,   1 psf = 0.0479 kN/m², 1 mph = 1.609 km/hr, 1 foot = 304.8 mm.

[1]   Exterior load bearing walls with a minimum of 1/2" (12.7 mm) gypsum board on the inside and 7/16" (11.1 mm) oriented strand board or plywood (APA rated 24/16) on the outside, and interior load bearing walls with a minimum of 1/2" (12.7 mm) gypsum board on both sides may use the next thinner stud but not less than 33 mils (0.836 mm).[2]   For actual sizes of members, refer to Table 2[3] Building width is in the direction of horizontal framing members supported by the wall studs.

PRESCRIPTIVE METHOD FOR RESIDENTIAL COLD-FORMED STEEL FRAMING

Table 12
Steel Stud Thickness for 9' Walls Supporting Roof and Ceiling Only
(One Story or Second Floor of a Two Story Building)

Wind Exposure		Member Size[2]	Member Spacing (inches)	Required Stud Thickness (mils)[1] Building Width (feet)[3]															
				24				28				32				36			
Exp. B	Exp. C			Ground Snow Load (psf)				Ground Snow Load (psf)				Ground Snow Load (psf)				Ground Snow Load (psf)			
				20	30	50	70	20	30	50	70	20	30	50	70	20	30	50	70
70 mph		2x4	16	33	33	33	33	33	33	33	33	33	33	33	33	33	33	33	33
			24	33	33	33	33	33	33	33	33	33	33	33	33	33	33	33	43
		2x6	16	33	33	33	33	33	33	33	33	33	33	33	33	33	33	33	33
			24	33	33	33	33	33	33	33	33	33	33	33	33	33	33	33	33
80 mph	70 mph	2x4	16	33	33	33	33	33	33	33	33	33	33	33	33	33	33	33	33
			24	33	33	43	43	33	33	43	43	33	43	43	43	43	43	43	43
		2x6	16	33	33	33	33	33	33	33	33	33	33	33	33	33	33	33	33
			24	33	33	33	33	33	33	33	33	33	33	33	33	33	33	33	33
90 mph	80 mph	2x4	16	33	33	33	33	33	33	33	33	33	33	33	33	33	33	33	33
			24	43	43	43	43	43	43	43	43	43	43	43	54	43	43	43	54
		2x6	16	33	33	33	33	33	33	33	33	33	33	33	33	33	33	33	33
			24	33	33	33	33	33	33	33	33	33	33	33	33	33	33	33	33
100 mph	90 mph	2x4	16	33	33	33	33	33	33	33	43	33	33	43	43	33	33	43	43
			24	54	54	54	54	54	54	54	54	54	54	54	54	54	54	54	68
		2x6	16	33	33	33	33	33	33	33	33	33	33	33	33	33	33	33	33
			24	33	33	33	33	33	33	33	33	33	33	33	33	33	33	33	43
110 mph	100 mph	2x4	16	43	43	43	43	43	43	43	43	43	43	43	54	43	43	43	54
			24	68	68	68	68	68	68	68	68	68	68	68	68	68	68	68	97
		2x6	16	33	33	33	33	33	33	33	33	33	33	33	33	33	33	33	33
			24	33	43	43	43	43	43	43	43	43	43	43	43	43	43	43	43
	110 mph	2x4	16	54	54	54	54	54	54	54	54	54	54	54	54	54	54	54	54
			24	97	97	97	97	97	97	97	97	97	97	97	97	97	97	97	97
		2x6	16	33	33	33	33	33	33	33	33	33	33	33	33	33	33	33	33
			24	43	43	43	43	43	43	43	43	43	43	43	54	43	43	43	54

For SI: 1" = 25.4 mm, 1 psf = 0.0479 kN/m², 1 mph = 1.609 km/hr, 1 foot = 304.8 mm.

[1] Exterior load bearing walls with a minimum of 1/2" (12.7 mm) gypsum board on the inside and 7/16" (11.1 mm) oriented strand board or plywood (APA rated 24/16) on the outside, and interior load bearing walls with a minimum of 1/2" (12.7 mm) gypsum board on both sides may use the next thinner stud but not less than 33 mils (0.836 mm).

[2] For actual sizes of members, refer to Table 2.

[3] Building width is in the direction of horizontal framing members supported by the wall studs.

PRESCRIPTIVE METHOD FOR RESIDENTIAL COLD-FORMED STEEL FRAMING

Table 13
Steel Stud Thickness for 9' Walls Supporting One Floor, Roof and Ceiling
(First Story of a Two Story Building)

Wind Exposure		Member Size[2]	Member Spacing (inches)	Required Stud Thickness (mils)[1]															
				Building Width (feet)[3]															
				24				28				32				36			
Exp. B	Exp. C			Ground Snow Load (psf)				Ground Snow Load (psf)				Ground Snow Load (psf)				Ground Snow Load (psf)			
				20	30	50	70	20	30	50	70	20	30	50	70	20	30	50	70
70 mph		2x4	16	33	33	33	33	33	33	33	33	33	33	33	33	33	33	33	43
			24	43	43	43	43	43	43	43	54	43	43	54	54	54	54	54	54
		2x6	16	33	33	33	33	33	33	33	33	33	33	33	33	33	33	33	33
			24	33	33	33	33	33	33	33	43	33	33	43	43	33	33	43	43
80 mph	70 mph	2x4	16	33	33	33	33	33	33	33	43	33	43	43	43	43	43	43	43
			24	43	54	54	54	54	54	54	54	54	54	54	54	54	54	68	68
		2x6	16	33	33	33	33	33	33	33	33	33	33	33	33	33	33	33	33
			24	33	33	33	33	33	33	33	43	33	33	43	43	43	43	43	43
90 mph	80 mph	2x4	16	43	43	43	43	43	43	43	43	43	43	43	43	43	43	43	43
			24	54	54	54	54	54	54	54	54	54	54	54	54	54	54	68	68
		2x6	16	33	33	33	33	33	33	33	33	33	33	33	33	33	33	33	33
			24	33	33	43	43	43	43	43	43	43	43	43	43	43	43	43	54
100 mph	90 mph	2x4	16	43	43	43	43	43	43	43	54	43	43	54	54	54	54	54	54
			24	68	68	68	68	68	68	68	68	68	68	68	97	68	68	97	97
		2x6	16	33	33	33	33	33	33	33	33	33	33	33	33	33	33	33	33
			24	43	43	43	43	43	43	43	43	43	43	43	54	43	43	54	54
110 mph	100 mph	2x4	16	54	54	54	54	54	54	54	54	54	54	54	68	54	54	68	68
			24	97	97	97	97	97	97	97	97	97	97	97	97	97	97	97	97
		2x6	16	33	33	33	33	33	33	33	43	33	33	43	43	33	43	43	43
			24	43	43	54	54	54	54	54	54	54	54	54	54	54	54	54	54
	110 mph	2x4	16	68	68	68	68	68	68	68	68	68	68	68	68	68	68	68	68
			24	97	97	97	97	97	97	97	--	97	97	--	--	--	--	--	--
		2x6	16	33	33	43	43	43	43	43	43	43	43	43	43	43	43	43	43
			24	54	54	54	54	54	54	54	54	54	54	54	68	54	54	68	68

For SI: 1" = 25.4 mm, 1 psf = 0.0479 kN/m², 1 mph = 1.609 km/hr, 1 foot = 304.8 mm.

[1] Exterior load bearing walls with a minimum of 1/2" (12.7 mm) gypsum board on the inside and 7/16" (11.1 mm) oriented strand board or plywood (APA rated 24/16) on the outside, and interior load bearing walls with a minimum of 1/2" (12.7 mm) gypsum board on both sides may use the next thinner stud but not less than 33 mils (0.836 mm). [2] For actual sizes of members, refer to Table 2. [3] Building width is in the direction of horizontal framing members supported by the wall studs.

PRESCRIPTIVE METHOD FOR RESIDENTIAL COLD-FORMED STEEL FRAMING

Table 14
Steel Stud Thickness for 10' Walls Supporting Roof and Ceiling Only
(One Story or Second Floor of a Two Story Building)

Wind Exposure		Member Size[2]	Member Spacing (inches)	Required Stud Thickness (mils)[1]																
				Building Width (feet)[3]																
				24				28				32				36				
Exp. B	Exp. C			Ground Snow Load (psf)				Ground Snow Load (psf)				Ground Snow Load (psf)				Ground Snow Load (psf)				
				20	30	50	70	20	30	50	70	20	30	50	70	20	30	50	70
70 mph		2x4	16	33	33	33	33	33	33	33	33	33	33	33	33	33	33	33	33
			24	33	33	33	43	33	33	33	43	33	33	43	43	33	43	43	43
		2x6	16	33	33	33	33	33	33	33	33	33	33	33	33	33	33	33	33
			24	33	33	33	33	33	33	33	33	33	33	33	33	33	33	33	33
80 mph	70 mph	2x4	16	33	33	33	33	33	33	33	33	33	33	33	33	33	33	33	33
			24	43	43	43	43	43	43	43	54	43	43	43	54	43	43	54	54
		2x6	16	33	33	33	33	33	33	33	33	33	33	33	33	33	33	33	33
			24	33	33	33	33	33	33	33	33	33	33	33	33	33	33	33	33
90 mph	80 mph	2x4	16	33	33	33	43	33	33	43	43	33	33	43	43	33	33	43	43
			24	54	54	54	54	54	54	54	54	54	54	54	54	54	54	54	68
		2x6	16	33	33	33	33	33	33	33	33	33	33	33	33	33	33	33	33
			24	33	33	33	33	33	33	33	33	33	33	33	33	33	33	33	43
100 mph	90 mph	2x4	16	43	43	43	43	43	43	43	43	43	43	43	43	43	43	43	54
			24	68	68	68	68	68	68	68	68	68	68	68	68	68	68	68	68
		2x6	16	33	33	33	33	33	33	33	33	33	33	33	33	33	33	33	33
			24	33	33	43	43	33	33	43	43	33	43	43	43	43	43	43	43
110 mph	100 mph	2x4	16	54	54	54	54	54	54	54	54	54	54	54	54	54	54	54	68
			24	97	97	97	97	97	97	97	97	97	97	97	97	97	97	97	97
		2x6	16	33	33	33	33	33	33	33	33	33	33	33	33	33	33	33	33
			24	43	43	43	43	43	43	43	54	43	43	54	54	43	43	54	54
	110 mph	2x4	16	68	68	68	68	68	68	68	68	68	68	68	68	68	68	68	68
			24	97	97	97	--	97	97	--	--	97	97	--	--	97	--	--	--
		2x6	16	33	33	33	43	33	33	43	43	33	33	43	43	33	43	43	43
			24	54	54	54	54	54	54	54	54	54	54	54	54	54	54	54	68

For SI: 1" = 25.4 mm,   1 psf = 0.0479 kN/m²,   1 mph = 1.609 km/hr,   1 foot = 304.8 mm.

[1]   Exterior load bearing walls with a minimum of 1/2" (12.7 mm) gypsum board on the inside and 7/16" (11.1 mm) oriented strand board or plywood (APA rated 24/16) on the outside, and interior load bearing walls with a minimum of 1/2" (12.7 mm) gypsum board on both sides may use the next thinner stud but not less than 33 mils (0.836 mm).

[2]   For actual sizes of members, refer to Table 2.

[3]   Building width is in the direction of horizontal framing members supported by the wall studs.

PRESCRIPTIVE METHOD FOR RESIDENTIAL COLD-FORMED STEEL FRAMING

Table 15
Steel Stud Thickness for 10' Walls Supporting One Floor, Roof and Ceiling
(First Story of a Two Story Building)

Wind Exposure		Member Size[2]	Member Spacing (inches)	Required Stud Thickness (mils)[1]															
				Building Width (feet)[3]															
				24				28				32				36			
Exp. B	Exp. C			Ground Snow Load (psf)				Ground Snow Load (psf)				Ground Snow Load (psf)				Ground Snow Load (psf)			
				20	30	50	70	20	30	50	70	20	30	50	70	20	30	50	70
70 mph		2x4	16	33	33	33	43	33	33	43	43	43	43	43	43	43	43	43	43
			24	54	54	54	54	54	54	54	54	54	54	54	68	54	54	68	68
		2x6	16	33	33	33	33	33	33	33	33	33	33	33	33	33	33	33	33
			24	33	33	33	33	33	33	33	43	33	33	43	43	43	43	43	54
80 mph	70 mph	2x4	16	43	43	43	43	43	43	43	43	43	43	43	43	43	43	43	54
			24	54	54	68	68	68	68	68	68	68	68	68	68	68	68	68	97
		2x6	16	33	33	33	33	33	33	33	33	33	33	33	33	33	33	33	33
			24	33	33	43	43	43	43	43	43	43	43	43	43	43	43	43	54
90 mph	80 mph	2x4	16	43	43	43	43	43	43	54	54	43	54	54	54	54	54	54	54
			24	68	68	68	68	68	68	68	97	68	68	97	97	97	97	97	97
		2x6	16	33	33	33	33	33	33	33	33	33	33	33	33	33	33	33	33
			24	43	43	43	43	43	43	43	43	43	43	43	54	43	43	54	54
100 mph	90 mph	2x4	16	54	54	54	54	54	54	54	54	54	54	54	54	54	54	54	54
			24	97	97	97	97	97	97	97	97	97	97	97	97	97	97	97	97
		2x6	16	33	33	33	33	33	33	33	33	33	33	33	43	33	33	43	43
			24	43	43	43	54	43	43	54	54	54	54	54	54	54	54	54	54
110 mph	100 mph	2x4	16	68	68	68	68	68	68	68	68	68	68	68	68	68	68	68	97
			24	97	97	--	--	--	--	--	--	--	--	--	--	--	--	--	--
		2x6	16	43	43	43	43	43	43	43	43	43	43	43	43	43	43	43	43
			24	54	54	54	54	54	54	54	68	54	68	68	68	68	68	68	68
	110 mph	2x4	16	68	97	97	97	97	97	97	97	97	97	97	97	97	97	97	97
			24	--	--	--	--	--	--	--	--	--	--	--	--	--	--	--	--
		2x6	16	43	43	43	43	43	43	43	43	43	43	43	54	43	43	54	54
			24	68	68	68	68	68	68	68	68	68	68	68	68	68	68	68	97

For SI: 1" = 25.4 mm,   1 psf = 0.0479 kN/m², 1 mph = 1.609 km/hr, 1 foot = 304.8 mm.
[1]   Exterior load bearing walls with a minimum of 1/2" (12.7 mm) gypsum board on the inside and 7/16" (11.1 mm) oriented strand board or plywood (APA rated 24/16) on the outside, and interior load bearing walls with a minimum of 1/2" (12.7 mm) gypsum board on both sides may use the next thinner stud but not less than 33 mils (0.836 mm).[2]   For actual sizes of members, refer to Table 2.[3]   Building width is in the direction of horizontal framing members supported by the wall studs.

PRESCRIPTIVE METHOD FOR RESIDENTIAL COLD-FORMED STEEL FRAMING

Table 16
Wall Fastening Schedule

Building Element	Number & Type of Fasteners	Spacing of Fasteners
Stud to Top or Bottom Track	2 # 8 screws	Each end of stud, one per flange
Structural Sheathing	# 8 screws[1]	6" on edges 12" on intermediate supports
1/2" Gypsum Board	# 6 screws	12" oc.

For SI: 1" = 25.4 mm

[1]   Head styles shall be bugle-head, flat-head, or similar with a minimum head diameter of 0.315 inches (8 mm.)

Table 17
Jack and King Stud Table

Size of Opening	24" o.c. stud spacing		16" o.c. stud spacing	
	No. of Jack studs[1]	No. of King studs[2]	No. of Jack studs[1]	No. of King studs[2]
Up to 3'-6"	1	1	1	1
> 3'-6" to 5'-0"	1	2	1	2
> 5'-0" to 5'-6"	1	2	2	2
> 5'-6" to 8'-0"	1	2	2	2
> 8'-0" to 10'-6"	2	2	2	3
> 10'-6" to 12'-0"	2	2	3	3
> 12'-0" to 13'-0"	2	3	3	3
> 13'-0" to 14'-0"	2	3	3	4
> 14'-0" to 16'-0"	2	3	3	4
> 16'-0" to 17'-0"	3	3	4	4
> 17'-0" to 18'-0"	3	3	4	4

For SI: 1" = 25.4 mm,   1 foot = 304.8 mm.

[1]   Total number of jack studs required at each end of the header.
[2]   Total number of king studs required at each end of the header.

PRESCRIPTIVE METHOD FOR RESIDENTIAL COLD-FORMED STEEL FRAMING

Table 18
Allowable Header Spans for
Headers Supporting Roof and Ceiling Only[1,2]

Nominal Member Size[3]	70 psf Ground Snow Load				50 psf Ground Snow Load			
	Building Width				Building Width			
	24'	28'	32'	36'	24'	28'	32'	36'
2-2 x 4 x 33	2'-4"	2'-1"	--	--	3'-0"	2'-7"	2'-4"	2'-1"
2-2 x 4 x 43	3'-5"	3'-2"	3'-0"	2'-9"	3'-10"	3'-7"	3'-4"	3'-2"
2-2 x 4 x 54	3'-10"	3'-7"	3'-4"	3'-2"	4'-3"	4'-0"	3'-9"	3'-7"
2-2 x 4 x 68	4'-3"	4'-0"	3'-9"	3'-7"	4'-10"	4'-6"	4'-3"	4'-0"
2-2x 4 x 97	5'-1"	4'-9"	4'-5"	4'-3"	5'-8"	5'-4"	5'-0"	4'-9"
2-2x 6 x 33	--	--			2'-6"	2'-2"	--	--
2-2 x 6 x 43	4'-5"	3'-10"	3'-5"	3'-1"	5'-2"	4'-10"	4'-4"	3'-11"
2-2x 6 x 54	5'-2"	4'-10"	4'-7"	4'-4"	5'-10"	5'-5"	5'-1"	4'-10"
2-2 x 6 x 68	5'-10"	5'-5"	5'-1"	4'-10"	6'-6"	6'-1"	5'-9"	5'-6"
2-2 x 6 x 97	6'-11"	6'-6"	6'-1"	5'-9"	7'-10"	7'-3"	6'-10"	6'-5"
2-2 x 8 x 33	--	--	--	--	--	--	--	--
2-2 x 8 x 43	3'-4"	2'-11"	2'-7"	2'-4"	4'-3"	3'-9"	3'-4"	3'-0"
2-2x 8 x 54	6'-9"	5'-10"	5'-3"	4'-8"	7'-7"	7'-1"	6'-7"	5'-11"
2-2 x 8 x 68	7'-7"	7'-1"	6'-8"	6'-4"	8'-6"	8'-0"	7'-6"	7'-2"
2-2 x 8 x 97	9'-1"	8'-6"	8'-0"	7'-7"	10'-3"	9'-7"	9'-0"	8'-7"
2-2 x 10 x 43	2'-10"	2'-6"	2'-2"	--	3'-7"	3'-1"	2'-9"	2'-6"
2-2 x 10 x 54	5'-7"	4'-11"	4'-4"	3'-11"	7'-1"	6'-2"	5'-6"	4'-11"
2-2 x 10 x 68	8'-11"	8'-4"	7'-11"	7'-6"	10'-1"	9'-5"	8'-10"	8'-5"
2-2 x 10 x 97	10'-9"	10'-1"	9'-6"	9'-0"	12'-1"	11'-4"	10'-8"	10'-1"
2-2 x 12 x 43	2'-5"	2'-1"	--	--	3'-1"	2'-8"	2'-4"	2'-2"
2-2 x 12 x 54	4'-10"	4'-2"	3'-9"	3'-4"	6'-1"	5'-4"	4'-9"	4'-3"
2-2 x 12 x 68	9'-7"	8'-5"	7'-6"	6'-9"	10'-9"	10'-1"	9'-6"	8'-6"
2-2 x 12 x 97	2'-5"	11'-7"	10'-11"	10'-4"	13'-11"	13'-0"	12'-3"	11'-8"

For SI: 1" = 25.4 mm, 1 psf = 0.0479 kN/m^2, 1 foot = 304.8 mm.

[1] Table provides the maximum header clear span in feet and inches.
[2] Deflection criteria: L/360 for live loads; L/240 for total loads.

PRESCRIPTIVE METHOD FOR RESIDENTIAL COLD-FORMED STEEL FRAMING

Table 19
Allowable Header Spans for
Headers Supporting Roof and Ceiling Only[1,2]

| Nominal member Size[3] | 30 psf Ground Snow Load | | | | 20 psf Ground Snow Load | | | |
| | Building Width | | | | Building Width | | | |
	24'	28'	32'	36'	24'	28'	32'	36'
2-2 x 4 x 33	3'-8"	3'-5"	3'-2"	2'-10"	3'-11"	3'-8"	3'-5"	3'-3"
2-2 x 4 x 43	4'-5"	4'-2"	3'-11"	3'-9"	4'-9"	4'-5"	4'-2"	4'-0"
2-2 x 4 x 54	5'-0"	4'-8"	4'-5"	4'-2"	5'-4"	5'-0"	4'-9"	4'-6"
2-2 x 4 x 68	5'-7"	5'-3	4'-11"	4'-8"	6'-0"	5'-7"	5'-3"	5'-0"
2-2x 4 x 97	6'-8"	6'-2"	5'-10"	5'-7"	7'-1"	6'-8"	6'-3"	5'-11"
2-2x 6 x 33	3'-5"	3'-0"	2'-8"	2'-5"	3'-11"	3'-5"	3'-0"	2'-9"
2-2 x 6 x 43	6'-0"	5'-8"	5'-4"	5'-0"	6'-5"	6'-0"	5'-8"	5'-5"
2-2x 6 x 54	6'-9"	6'-4"	6'-0"	5'-8"	7'-3"	6'-10"	6'-5"	6'-1"
2-2 x 6 x 68	7'-7"	7'-2"	6'-9"	6'-4"	8'-2"	7'-8"	7'-2"	6'-10"
2-2 x 6 x 97	9'-1"	8'-6"	8'-0"	7'-7"	9'-9"	9'-1"	8'-7"	8'-2"
2-2 x 8 x 33	2'-7"	2'-3"	--	--	3'-0"	2'-8"	2'-4"	2'-1"
2-2 x 8 x 43	5'-10"	5'-1"	4'-6"	4'-1"	6'-8"	5'-10"	5'-2"	4'-8"
2-2x 8 x 54	8'-10"	8'-3"	7'-9"	7'-5"	9'-6"	8'-10"	8'-4"	7'-11"
2-2 x 8 x 68	9'-11"	9'-4"	8'-9"	8'-4"	10'-8"	10'-0"	9'-5"	8'-11"
2-2 x 8 x 97	11'-11"	11'-2"	10'-6"	10'-0"	12'-10"	11'-11"	11'-3"	10'-8"
2-2 x 10 x 43	4'-10"	4'-3"	3'-9"	3'-5"	5'-7"	4'-10"	4'-4"	3'-11"
2-2 x 10 x 54	9'-8"	8'-5"	7'-6"	6'-9"	10'-6"	9'-8"	8'-7"	7'-9"
2-2 x 10 x 68	11'-9"	10'-12"	10'-4"	9'-10"	12'-7"	11'-9"	11'-1"	10'-6"
2-2 x 10 x 97	14'-1"	13'-2"	12'-5"	11'-10"	15'-2"	14'-2"	13'-4"	12'-8"
2-2 x 12 x 43	4'-2"	3'-8"	3'-3"	2'-11"	4'-9"	4'-2"	3'-8"	3'-4"
2-2 x 12 x 54	8'-3"	7'-3"	6'-5"	5'-9"	9'-6"	8'-3"	7'-4"	6'-7"
2-2 x 12 x 68	12'-6"	11'-9"	11'-1"	10'-6"	13'-5"	12'-7"	11'-10"	11'-3"
2-2 x 12 x 97	16'-3"	15'-2"	14'-4"	13'-7"	17'-5"	16'-3"	15'-4"	14'-7"

For SI: 1" = 25.4 mm, 1 psf = 0.0479 kN/m², 1 foot = 304.8 mm.

[1] Table provides the maximum header clear span in feet and inches.
[2] Deflection criteria: L/360 for live loads; L/240 for total loads.
[3] Refer to Table 2 for actual size.

PRESCRIPTIVE METHOD FOR RESIDENTIAL COLD-FORMED STEEL FRAMING

Table 20
Allowable Header Spans for
Headers Supporting One Floor, Roof and Ceiling[1,2]

Nominal Member Size[3]	70 psf Ground Snow Load				50 psf Ground Snow Load			
	Building Width				Building Width			
	24'	28'	32'	36'	24'	28'	32'	36'
2-2 x 4 x 33	-	-	-	-	-	-	-	-
2-2 x 4 x 43	2'-10"	2'-6"	2'-3"	2'-1"	3'-1"	2'-11"	2'-8"	2'-5"
2-2 x 4 x 54	3'-3"	3'-0"	2'-10"	2'-9"	3'-6"	3'-3"	3'-1"	3'-0"
2-2 x 4 x 68	3'-7"	3'-5"	3'-2"	3'-1"	3'-11"	3'-8"	3'-6"	3'-4"
2-2x 4 x 97	4'-3"	4'-0"	3'-10"	3'-7"	4'-8"	4'-4"	4'-2"	3'-11"
2-2x 6 x 33	-	-	-	-	-	-	-	-
2-2 x 6 x 43	3'-2"	2'-9"	2'-6"	2'-3"	3'-8"	3'-3"	2'-11"	2'-8"
2-2x 6 x 54	4'-4"	4'-1"	3'-11"	3'-8"	4'-9"	4'-6"	4'-3"	4'-0"
2-2 x 6 x 68	4'-11"	4'-7"	4'-4"	4'-2"	5'-4"	5'-0"	4'-9"	4'-6"
2-2 x 6 x 97	5'-10"	5'-6"	5'-3"	4'-11"	6'-4"	6'-0"	5'-8"	5'-5"
2-2 x 8 x 33	--	--	--	--	--	--	--	--
2-2 x 8 x 43	2'-5"	2'-2"	--	--	2'-10"	2'-6"	2'-3"	--
2-2x 8 x 54	4'-10"	4'-3"	3'-10"	3'-5"	5'-8"	5'-0"	4'-6"	4'-1"
2-2 x 8 x 68	6'-5"	6'-0"	5'-8"	5'-5"	7'-0"	6'-7"	6'-2"	5'-11"
2-2 x 8 x 97	7'-8"	7'-3"	6'-10"	6'-6"	8'-4"	7'-10"	7'-5"	7'-1"
2-2 x 10 x 43	--	--	--	--	2'-4"	2'-1"	--	--
2-2 x 10 x 54	4'-0"	3'-6"	3'-2"	2'-10"	4'-9"	4'-2"	3'-9"	3'-5"
2-2 x 10 x 68	7'-7"	7'-1"	6'-5"	5'-9"	8'-3"	7'-9"	7'-4"	6'-10"
2-2 x 10 x 97	9'-1"	8'-7"	8'-1"	7'-8"	9'-10"	9'-3"	8'-9"	8'-4"
2-2 x 12 x 43	--	--	--	--	--	--	--	--
2-2 x 12 x 54	3'-5"	3'-0"	2'-9"	2'-5"	4'-0"	3'-7"	3'-2"	2'-11"
2-2 x 12 x 68	6'-11"	6'-1"	5'-6"	4'-11"	8'-2"	7'-2"	6'-5"	5'-10"
2-2 x 12 x 97	10'-6"	9'-0"	9'-4"	8'-10"	11'-4"	10'-8"	10'-1"	9'-8"

For SI: 1" = 25.4 mm,   1 psf = 0.0479 kN/m², 1 foot = 304.8 mm.

[1] Table provides the maximum header clear span in feet and inches.
[2] Deflection criteria: L/360 for live loads; L/240 for total loads.
[3] Refer to Table 2 for actual size.

Table 21
Allowable Header Spans for
Headers Supporting One Floor, Roof and Ceiling [1,2]

Nominal Member Size[3]	30 psf Ground Snow Load				20 psf Ground Snow Load			
	Building Width				Building Width			
	24'	28'	32'	36'	24'	28'	32'	36'
2-2 x 4 x 33	2'-2"	--	--	--	2'-3"	--	--	--
2-2 x 4 x 43	3'-3"	3'-1"	2'-11"	2'-8"	3'-4"	3'-1"	2'-11"	2'-9"
2-2 x 4 x 54	3'-8"	3'-6"	3'-3"	3'-1"	3'-9"	3'-6"	3'-4"	3'-2"
2-2 x 4 x 68	4'-1"	4'-0"	3'-8"	3'-6"	4'-2"	3'-11"	3'-9"	3'-6"
2-2x 4 x 97	4'-11"	4'-7"	4'-4"	4'-2"	4'-11"	4'-8"	4'-5"	4'-2"
2-2x 6 x 33	--	--	--	--	--	--	--	--
2-2 x 6 x 43	4'-1"	3'-8"	3'-3"	3'-0"	4'-2"	3'-9"	3'-4"	3'-0"
2-2x 6 x 54	5'-0"	4'-8"	4'-5"	4'-3"	5'-1"	4'-9"	4'-6"	4'-3"
2-2 x 6 x 68	5'-7"	5'-3"	5'-0"	4'-9"	5'-8"	5'-4"	5'-1"	4'-10"
2-2 x 6 x 97	6'-8"	6'-4"	6'-0"	5'-8"	6'-9"	6'-4"	6'-0"	5'-9"
2-2 x 8 x 33	--	--	--	--	--	--	--	--
2-2 x 8 x 43	3'-2"	2'-9"	2'-6"	2'-3"	3'-3"	2'-10"	2'-7"	2'-4"
2-2x 8 x 54	6'-3"	5'-7"	5'-0"	4'-6"	6'-5"	5'-8"	5'-1"	4'-7"
2-2 x 8 x 68	7'-4"	6'-11"	6'-6"	6'-3"	7'-5"	7'-0"	6'-7"	6'-4"
2-2 x 8 x 97	8'-9"	8'-3"	7'-10"	7'-5"	8'-11"	8'-4"	7'-11"	7'-7"
2-2 x 10 x 43	2'-8"	2'-4"	2'-1"	2'-0"	2'-8"	2'-5"	2'-2"	2'-0"
2-2 x 10 x 54	5'-3"	4'-8"	4'-2"	3'-9"	5'-4"	4'-9"	4'-3"	3'-10"
2-2 x 10 x 68	8'-8"	8'-2"	7'-8"	7'-4"	8'-9"	8'-3"	7'-10"	7'-5"
2-2 x 10 x 97	10'-5"	9'-9"	9'-3"	8'-10"	10'-6"	9'-11"	9'-4"	8'-11"
2-2 x 12 x 43	2'-3"	--	--	--	2'-4"	2'-1"	--	--
2-2 x 12 x 54	4'-6"	4'-0"	3'-7"	3'-3"	4'-7"	4'-1"	3'-8"	3'-4"
2-2 x 12 x 68	9'-0"	8'-0"	7'-2"	6'-6"	9'-3"	8'-2"	7'-4"	6'-8"
2-2 x 12 x 97	12'-0"	11'-3"	10'-8"	10'-2"	12'-1"	11'-5"	10'-9"	10'-3"

For SI: 1" = 25.4 mm,   1 psf = 0.0479 kN/m^2, 1 foot = 304.8 mm.

[1] Table provides the maximum header clear span in feet and inches.
[2] Deflection criteria: L/360 for live loads; L/240 for total loads.
[3] Refer to Table 2 for actual size.

PRESCRIPTIVE METHOD FOR RESIDENTIAL COLD-FORMED STEEL FRAMING

Table 22
Header to King Stud Connection Requirements

Header Span	Wind Speed (mph), Exposure & Seismic Zones [1,2,3]			
	Up to 70 B or Zones 0, 1, 2	Up to 90 B or 70 C or Zone 3	Up to 100 B or 90 C or Zone 4	Up to 110 C
# 4'	4 - #10 screws	4 - #10 screws	6 - #10 screws	8 - #10 screws
> 4' to 8'	4 - #10 screws	4 - #10 screws	8 - #10 screws	12 - #10 screws
> 8' to 12'	4 - #10 screws	6 - #10 screws	10 - #10 screws	16 - #10 screws
> 12' to 16'	4 - #10 screws	8 - #10 screws	12 - #10 screws	20 - #10 screws

For SI: 1" = 25.4 mm, 1 foot = 304.8 mm, 1 mph = 1.609 km/hr.

[1]   One-half of the total number of screws shall be applied to the header and one half to the king stud by use of an angle or suitable connector of thickness equivalent to the header member or king stud, whichever is higher.

[2]   Total number of screws shall be reduced by 2 screws when headers are located on the 1st floor of a two-story building, but the total number of screws shall be no less than 4.

[3]   For roof slopes of 6:12 or greater, the required number of screws shall be reduced by 1/2, but the total number of screws shall be no less than 4.

PRESCRIPTIVE METHOD FOR RESIDENTIAL COLD-FORMED STEEL FRAMING

Table 23
Minimum Percent Length of Full-Height Sheathed Wall
Required for Wall Bracing[1,2,3,4,5]

Wall Condition	Roof Slope[7]	Wind Speed (mph) and Exposure			
		less than or equal to 90 B (or 70 C)		less than or equal to 100 B (or 90 C)	
		endwall[6]	sidewall[6]	endwall[6]	sidewall[6]
One-Story Construction or Second Floor of Two-Story	3:12	30%	30%	30%	30%
	6:12	30%	30%	40%	30%
	9:12	45%	30%	75%	50%
	12:12	60%	40%	100%	70%
1st Floor of Two-Story Construction	3:12	50%	35%	80%	55%
	6:12	55%	40%	90%	60%
	9:12	75%	50%	Approved Design Required	Approved Design Required
	12:12	95%	65%	Approved Design Required	Approved Design Required

For **SI**: 1 mph = 49 m/sec, 1" = 25.4 mm, 1 foot = 304.8 mm.

[1]   Percentages are given as a percentage of the total wall length. For example, a 48-foot (14630.4 mm) long wall that requires 25% of full-height sheathing would result in 0.25x48 ft. = 12 ft (3657.6 mm) of wall with full-height sheathing. Areas above or below openings would also be sheathed.

[2]   The total length of full-height sheathing provided for the sidewalls shall not be less than that required for the endwalls, even if the tabulated percentages would require less for a given sidewall length.

[3]   A 48-inch-wide (1219.2 mm) panel of structural sheathing shall be located at each end of a wall or as near thereto as possible. Individual segments of wall (i.e. between openings) with full-height sheathing shall be at least 48" (1219.2 mm) in length to count toward the length of full-height sheathing required by the tabulated percentages.

[4]   These requirements are applicable to Seismic Zones 0,1, and 2.

[5]   Required amounts of full-height sheathing shall be increased for 9' (2743.2 mm) and 10' (3048 mm) walls by multiplying the percentages by 1.10 and 1.20 respectively.

[6]   Sidewalls are parallel to the ridge and endwalls are perpendicular to the ridge.

[7]   Interpolation may be used for intermediate roof slopes or wind conditions.

PRESCRIPTIVE METHOD FOR RESIDENTIAL COLD-FORMED STEEL FRAMING

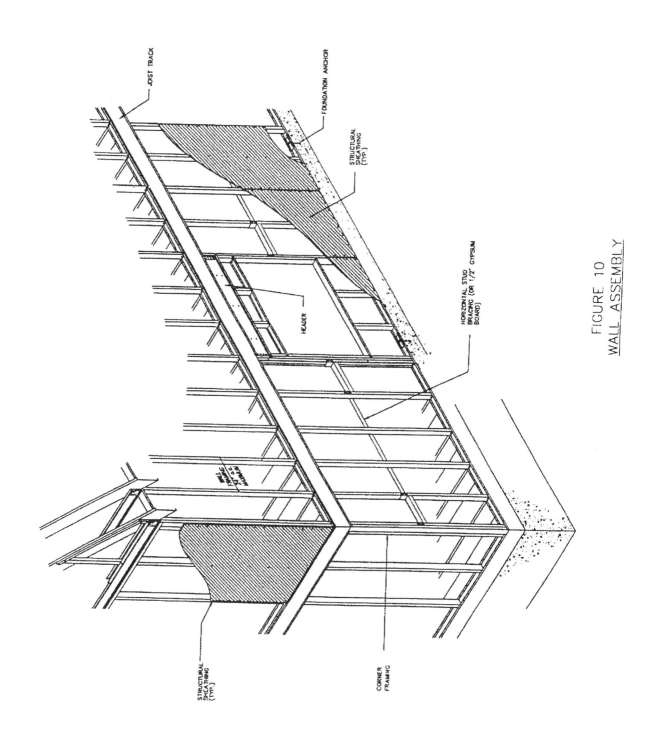

FIGURE 10
WALL ASSEMBLY

PRESCRIPTIVE METHOD FOR RESIDENTIAL COLD-FORMED STEEL FRAMING

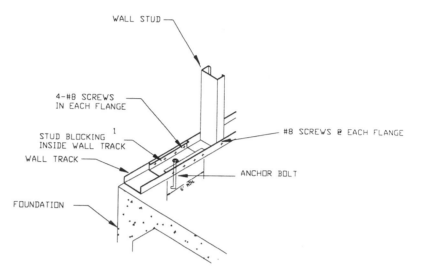

1 STUD BLOCKING THICKNESS EQUIVALENT TO WALL STUDS

Fig. 11 Wall to Foundation Connection

PRESCRIPTIVE METHOD FOR RESIDENTIAL COLD-FORMED STEEL FRAMING

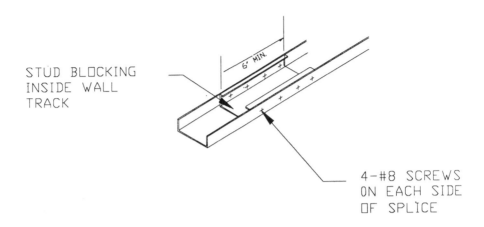

Fig. 12 Wall Track Splice

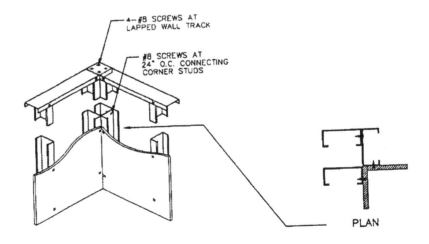

Fig. 13 Corner Framing

PRESCRIPTIVE METHOD FOR RESIDENTIAL COLD-FORMED STEEL FRAMING

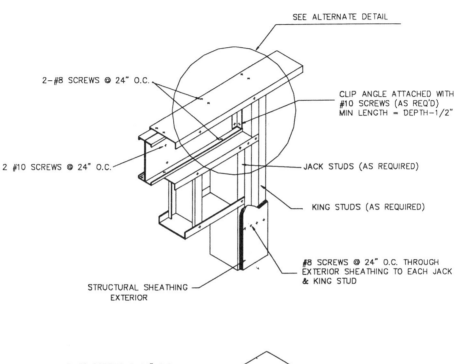

SEE ALTERNATE DETAIL

2-#8 SCREWS @ 24" O.C.

CLIP ANGLE ATTACHED WITH
#10 SCREWS (AS REQ'D)
MIN LENGTH = DEPTH-1/2"

2 #10 SCREWS @ 24" O.C.

JACK STUDS (AS REQUIRED)

KING STUDS (AS REQUIRED)

#8 SCREWS @ 24" O.C. THROUGH
EXTERIOR SHEATHING TO EACH JACK
& KING STUD

STRUCTURAL SHEATHING
EXTERIOR

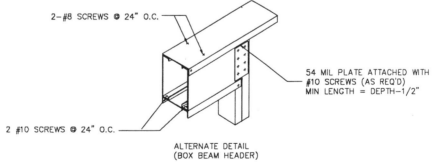

2-#8 SCREWS @ 24" O.C.

54 MIL PLATE ATTACHED WITH
#10 SCREWS (AS REQ'D)
MIN LENGTH = DEPTH-1/2"

2 #10 SCREWS @ 24" O.C.

ALTERNATE DETAIL
(BOX BEAM HEADER)

Fig. 14 Header Detail

PRESCRIPTIVE METHOD FOR RESIDENTIAL COLD-FORMED STEEL FRAMING

7.0   NON-STRUCTURAL WALLS

7.1   Non-Load Bearing Studs

Non-load bearing studs shall comply with ASTM C 645 and shall have a minimum base metal thickness of 18 mils (0.455 mm). Table 24 provides limiting heights of selected non-load bearing studs. The values in Table 24 are provided for information only. Manufacturer's shapes, sizes, tables, and recommendations shall be followed where applicable.

Table 24
Limiting Heights for Selected
Non-Load Bearing Steel Studs

Actual Stud Size (Flange x Web) (inches)	Stud Spacing (inches)	Maximum Allowable Clear Height	
		Stud Thickness	
		18 mil	27 mil
1.25 x 3.5	16" o.c.	11'-11"	16'-3"
	24" o.c.	9'-10"	13'-3"
1.25 x 3.625	16" o.c.	12'-3"	14'-3"
	24" o.c.	10'-0"	12'-5"
1.25 x 4.0	16" o.c.	12'-11"	15'-5"
	24" o.c.	10'-7"	13'-5"

For SI: 1" = 25.4 mm,   1 psf = 0.0479 kN/m^2, 1 foot = 304.8 mm, 1 mil = 1/1000 inch..

Notes:
1.   Stud height based on stud capacity alone.
2.   Maximum lateral load is 5 psf (0.2395 kN/m^2).
3.   Yield strength is 33 ksi (227.7 kPa)
4.   All sections have a minimum lip size of 1/8" (3.2 mm)

PRESCRIPTIVE METHOD FOR RESIDENTIAL COLD-FORMED STEEL FRAMING

8.0   ROOF SYSTEMS

8.1   Roof Construction

Cold-formed steel roof systems shall be constructed in accordance with this section, Figures 15 through 17, and Tables 25 through 32. Cold-formed steel framing members shall comply with the provisions of Section 2.0. Steel roofs constructed per this section are limited to design wind speeds of 110 mph (177 km/hr) fastest-mile or design seismic conditions of Seismic Zones 0, 1, 2, 3, and 4. Roof systems constructed in accordance with this section shall consist of both ceiling joist (rafter ties) and rafters. The provisions of this section do not apply to cathedral ceilings

All roof framing members (i.e. ceiling joists and rafters) shall be aligned with load carrying members below.

8.2   Allowable Ceiling Joist Span

The clear span of cold-formed steel ceiling joists shall not exceed the limits set forth in Tables 25 through 28. Ceiling joists shall have a bearing length of not less than 1.5" (38.1 mm).

8.3   Ceiling Joist Bracing

The bottom flanges of ceiling joists shall be laterally braced by sheathing materials such as gypsum boards installed in accordance with the applicable building code requirements.

Bracing of the top flanges shall be provided in accordance with Tables 25 through 28. When required, bracing shall be accomplished by the installation of 1-1/2" x 33 mil (38.1 mm x 0.836 mm) continuous strap at mid span or 1/3 span as indicated in the tables. The strap shall be installed without slack and shall be attached to the top flanges of the ceiling joists with a minimum of one #8 screw at each intersection. Solid blocking (or X-bracing) shall be provided between the ceiling joists at intervals not exceeding 12 feet (3.66 m) in-line with the strapping. The ends of the straps shall be fastened to blocking or a stable component of the building (i.e. the end-wall or side walls) with a minimum of two #8 screws.

The third point bracing span values shall apply for fully sheathed ceiling joists (eg. an attic floor)

8.4   Rafters

Steel rafters spans shall be installed in accordance with Figures 15 through 17 and Tables 29 through 31. Wood rafters may be substituted provided that they are designed and installed in accordance with accepted engineering practices or applicable building code requirements. Wind speeds (load effects) shall be converted to equivalent snow loads in accordance with Table 30 for use in selecting rafters from Table 29.

Rafters shall be connected to a ridge board. The ridge board connection shall be constructed with a 2 inch x 2 inch (50.8 mm x 50.8 mm) clip angle of equivalent size and thickness as the rafter member extending the depth of the rafter. The clip angle shall connect the rafter to the ridge board with #10 screws applied to each leg of the clip angle in accordance with Table 31. The heel joint connection, which connects the ceiling joist to the rafter, shall comply with Table 31. Lapped ceiling joists (i.e. lapped at bearing supports) shall be connected with the same number and size of screws as the heel joint connection. Ceiling joists shall run parallel to the rafters to form a continuous tie between the rafter heel joints.

PRESCRIPTIVE METHOD FOR RESIDENTIAL COLD-FORMED STEEL FRAMING

8.5 Rafter Bottom Bracing

Steel rafters shall be braced in accordance with Figure 15 with steel straps or a 2 x 4 x 33 mil C-section at a spacing of 8 feet (2438 mm) as measured parallel to the rafters. Steel straps shall be at least 1-1/2" inches (38 mm) wide and 33 mils (0.836 mm) thick and shall be installed without slack. Straps shall be attached to the bottom flange at each rafter with at least one #8 screw. In line blocking or bridging (X-bracing) shall be installed between rafters at the termination of all straps and at a maximum spacing of 12 feet (3657 mm), measured perpendicular to the rafters. The ends of straps shall be fastened to blocking or anchored to a stable building component with at least two #8 screws.

8.6 Splicing

Ceiling joists and rafters shall not be spliced without an approved design.

8.7 Roof Trusses

Steel or wood trusses shall be engineered and installed in accordance with the applicable building code requirements. Roof bracing shall be installed in accordance with the engineered truss system design. All trusses shall be aligned with load carrying members below.

8.8 Uplift Resistance and Gable Endwall Bracing

At design wind speeds greater than 90 mph Exposure C or 100 mph (1609 km/hr) Exposure B, uplift straps and gable endwall bracing shall be provided in accordance with an approved design.

PRESCRIPTIVE METHOD FOR RESIDENTIAL COLD-FORMED STEEL FRAMING

Table 25
Ceiling Joists Allowable Spans
Single Span, Without Attic Storage [1,2]

Nominal Joist [3]	Lateral Support of Top (Compression) Flange					
	Unbraced		Mid-Span Bracing		Third-Point Bracing	
	Spacing (inches)		Spacing (inches)		Spacing (inches)	
	16	24	16	24	16	24
2 x 4 x 33	9'-2"	8'-3"	11'-9"	10'-1"	11'-9"	10'-4"
2 x 4 x 43	9'-11"	8'-10"	12'-10"	11'-2"	12'-10"	11'-2"
2 x 4 x 54	10'-8"	9'-6"	13'-9"	12'-0"	13'-9"	12'-0"
2 x 4 x 68	11'-7"	10'-4"	14'-8"	12'-10"	14'-8"	12'-10"
2 x 4 x 97	13'-7"	12'-0"	16'-2"	14'-1"	16'-2"	14'-1"
2 x 6 x 33	10'-5"	9'-5"	14'-5"	12'-8"	16'-4"	13'-10"
2 x 6 x 43	11'-2"	10'-1"	15'-7"	13'-10"	18'-0"	15'-5"
2 x 6 x 54	12'-0"	10'-9"	16'-7"	14'-9"	19'-5"	16'-8"
2 x 6 x 68	12'-11"	11'-7"	17'-8"	15'-10"	20'-11"	18'-1"
2 x 6 x 97	14'-11"	13'-2"	19'-10"	17'-8"	23'-2"	20'-3"
2 x 8 x 33	11'-8"	10'-6"	16'-5"	14'-9"	19'-5"	16'-7"
2 x 8 x 43	12'-6"	11'-3"	17'-6"	15'-10"	21'-2"	18'-7"
2 x 8 x 54	13'-4"	11'-11"	18'-7"	16'-9"	22'-7"	20'-0"
2 x 8 x 68	14'-3"	12'-9"	19'-8"	17'-8"	23'-11"	21'-4"
2 x 8 x 97	16'-2"	14'-5"	21'-10"	19'-6"	26'-3"	23'-6"
2 x 10 x 43	13'-4"	12'-1"	18'-9"	16'-11"	22'-11"	20'-6"
2 x 10 x 54	14'-2"	12'-9"	19'-10"	17'-10"	24'-2"	21'-9"
2 x 10 x 68	15'-2"	13'-7"	21'-0"	18'-11"	25'-6"	23'-0"
2 x 10 x 97	17'-1"	15'-2"	23'-2"	20'-9"	27'-11"	25'-1"
2 x 12 x 43	14'-1"	12'-8"	19'-10"	17'-11"	24'-3"	21'-6"
2 x 12 x 54	15'-0"	13'-5"	20'-11"	18'-11"	25'-7"	23'-1"
2 x 12 x 68	15'-11"	14'-4"	22'-2"	19'-11"	27'-0"	24'-4"
2 x 12 x 97	17'-10"	15'-11"	24'-4"	21'-10"	29'-4"	26'-5"

For SI: 1" = 25.4 mm,   1 foot = 304.8 mm,   1 psf = 0.0479 kN/m².

[1] Bearing stiffeners shall be installed at all bearing locations.
[2] Deflection criteria: L/240 for total loads.
[3] Refer to Table 2 for actual size.

PRESCRIPTIVE METHOD FOR RESIDENTIAL COLD-FORMED STEEL FRAMING

Table 26
Ceiling Joists Allowable Spans
Two Equal Spans, Without Attic Storage[1,2,3,4,5]

Nominal Joist[6]	Lateral Support of Top (Compression Flange)					
	Unbraced		Mid-Span Bracing		Third-Point Bracing	
	Spacing (inches)		Spacing (inches)		Spacing (inches)	
	16	24	16	24	16	24
2 x 4 x 33	12'-4"	10'-11"	13'-5"	10'-11"	13'-5"	10'-11"
2 x 4 x 43	13'-6"	12'-1"	16'-4"	13'-4"	16'-4"	13'-4"
2 x 4 x 54	14'-9"	13'-1"	18'-4"	15'-0"	18'-4"	15'-0"
2 x 4 x 68	16'-4"	14'-5"	19'-8"	16'-9"	19'-8"	16'-9"
2 x 4 x 97	19'-6"	17'-2"	21'-8"	18'-11"	21'-8"	18'-11"
2 x 6 x 33	14'-0"	12'-7"	18'-2"	14'-10"	18'-2"	14'-10"
2 x 6 x 43	15'-2"	13'-7"	20'-11"	18'-1"	22'-1"	18'-1"
2 x 6 x 54	16'-5"	14'-8"	22'-5"	19'-5"	24'-11"	20'-4"
2 x 6 x 68	17'-11"	15'-11"	24'-1"	21'-5"	28'-0"	22'-10"
2 x 6 x 97	21'-2"	18'-8"	27'-7"	24'-5"	31'-1"	27'-2"
2 x 8 x 33	15'-7"	14'-1"	21'-3"	15'-10"	21'-3"	15'-10"
2 x 8 x 43	16'-10"	15'-1"	23'-6"	21'-2"	27'-6"	22'-5"
2 x 8 x 54	18'-1"	16'-2"	24'-11"	22'-5"	30'-2"	26'-6"
2 x 8 x 68	19'-7"	17'-6"	26'-8"	23-'11"	32'-2"	28'-7"
2 x 8 x 97	22'-10"	20'-2"	30'-2"	26'-10"	35'-10"	31'-11"
2 x 10 x 43	17'-11"	16'-2"	25'-1"	22'-7"	30'-6"	24'-9"
2 x 10 x 54	19'-3"	17'-3"	26'-7"	23'-11"	32'-4"	29'-1"
2 x 10 x 68	20'-9"	18'-6"	28'-5"	25'-6"	34'-4"	30'-10"
2 x 10 x 97	23'-11"	21'-2"	31'-10"	28'-4"	38'-0"	34'-0"
2 x 12 x 43	18'-11"	17'-0"	26'-6"	23'-10"	32'-4"	24'-5"
2 x 12 x 54	20'-2"	18'-1"	28'-1"	25'-3"	34'-2"	30'-9"
2 x 12 x 68	21'-9"	19'-5"	29'-10"	26'-10"	36'-2"	32'-7"
2 x 12 x 97	24'-10"	22'-1"	33'-4"	29'-9"	39'-10"	35'-8"

For SI: 1" = 25.4 mm, 1 foot = 304.8 mm, 1 psf = 0.0479 kN/m².

[1] Table provides the maximum ceiling joist span in feet and inches, to either the right or left of   the interior s

[2] Bearing stiffeners shall be installed at all bearing locations.

[3] Deflection criteria: L/240 for total loads.

[4] Interior bearing supports for multiple span joists shall consist of structural walls or beams.

[5] The interior support may be located ∀ two feet from mid-span provided that each of the resulting spans do not exceed the span given in the table above.

[6] Refer to Table 2 for actual sizes.

PRESCRIPTIVE METHOD FOR RESIDENTIAL COLD-FORMED STEEL FRAMING

Table 27
Ceiling Joists Allowable Spans
Single Span, With Attic Storage [1,2]

Nominal Joist Size[3]	Lateral Support of Top (Compression Flange)					
	Unbraced		Mid-Span Bracing		Third-Point Bracing	
	Spacing (inches)		Spacing (inches)		Spacing (inches)	
	16	24	16	24	16	24
2 x 4 x 33	8'-0"	7'-0"	9'-8"	8'-1"	9'-11"	8'-3"
2 x 4 x 43	8'-8"	7'-8"	10'-9"	9'-1"	10'-10"	9'-5"
2 x 4 x 54	9'-3"	8'-3"	11'-7"	9'-11"	11'-7"	10'-1"
2 x 4 x 68	10'-0"	8'-11"	12'-5"	10'-10"	12'-5"	10'-10"
2 x 4 x 97	11'-7"	10'-3"	13'-7"	11'-11"	13'-7"	11'-11"
2 x 6 x 33	9'-2"	8'-3"	12'-2"	10'-5"	13'-3"	11'-0"
2 x 6 x 43	9'-10"	8'-10"	13'-4"	11'-6"	14'-9"	12'-5"
2 x 6 x 54	10'-5"	9'-5"	14'-4"	12'-6"	16'-1"	13'-7"
2 x 6 x 68	11'-3"	10'-0"	15'-4"	13'-5"	17'-5"	14'-10"
2 x 6 x 97	12'-9"	11'-4"	17'-1"	15'-1"	19'-7"	16'-9
2 x 8 x 33	10'-3"	9'-3"	14'-4"	12'-5"	15'-11"	13'-4"
2 x 8 x 43	10'-11"	9'-10"	15'-5"	13'-8"	17'-11"	15'-5"
2 x 8 x 54	11'-8"	10'-6"	16'-3"	14'-7"	19'-3"	16'-8"
2 x 8 x 68	12'-5"	11'-2"	17'-3"	15'-6"	20'-7"	18'-0"
2 x 8 x 97	13'-11"	12'-5"	18'-7"	17'-0"	22'-9"	20'-1"
2 x 10 x 43	11'-9"	10'-7"	16'-6"	14'-10"	19'-10"	17'-1"
2 x 10 x 54	12'-5"	11'-2"	17'-5"	15'-8"	21'-1"	18'-7"
2 x 10 x 68	13'-3"	11'-10"	18'-5"	16'-7"	22'-4"	19'-11"
2 x 10 x 97	14'-9"	13'-2"	20'-2"	18'-1"	24'-4"	21'-10"
2 x 12 x 43	12'-5"	11'-2"	17'-5"	15'-8"	20'-9"	18'-0"
2 x 12 x 54	13'-1"	11'-9"	18'-5"	16'-7"	22'-5"	20'-1"
2 x 12 x 68	13'-11"	12'-6"	19'-5"	17'-6"	23'-8"	21'-3"
2 x 12 x 97	15'-5"	13'-10"	21'-2"	19'-0"	25'-8"	23'-1"

For SI: 1" = 25.4 mm,   1 foot = 304.8 mm,   1 psf = 0.0479 kN/m^2.

[1]   Bearing stiffeners shall be installed at all bearing locations.
[2]   Deflection criteria: L/240 for total loads.
[3]   Refer to Table 2 for actual size.

PRESCRIPTIVE METHOD FOR RESIDENTIAL COLD-FORMED STEEL FRAMING

Table 28
Ceiling Joists Allowable Spans
Two Equal Spans, With Attic Storage [1,2,3,4,5]

Nominal Joist	Lateral Support of Top (Compression Flange)					
	Unbraced		Mid-Span Bracing		Third-Point Bracing	
	Spacing (inches)		Spacing (inches)		Spacing (inches)	
	16	24	16	24	16	24
2 x 4 x 33	10'-5"	8'-6"	10'-5"	8'-6"	10'-5"	8'-6"
2 x 4 x 43	11'-8"	10'-4"	12'-8"	10'-4"	12'-8"	10'-4"
2 x 4 x 54	12'-9"	11'-3"	14'-3"	11'-7"	14'-3"	11'-7"
2 x 4 x 68	14'-0"	12'-4"	15'-11"	13'-0"	15'-11"	13'-0"
2 x 4 x 97	16'-7"	14'-5"	18'-3"	15'-5"	18'-3"	15'-5"
2 x 6 x 33	12'-3"	11'-0"	14'-1"	11'-0"	14'-1"	11'-0"
2 x 6 x 43	13'-3"	11'-10"	17'-2"	14'-0"	17'-2"	14'-0"
2 x 6 x 54	14'-3"	12'-9"	19'-2"	15'-9"	19'-4"	15'-9"
2 x 6 x 68	15'-6"	13'-9"	20'-9"	17'-8"	21'-8"	17'-8"
2 x 6 x 97	18'-0"	15'-11"	23'-7"	20'-6"	25'-11"	21'-1"
2 x 8 x 33	13'-8"	10'-9"	14'-8"	10'-9"	14'-8"	10'-9"
2 x 8 x 43	14'-9"	13'-3"	20'-6"	17'-5"	21'-3"	17'-5"
2 x 8 x 54	15'-9"	14'-1"	21'-10"	19'-6"	25'-2"	20'-6"
2 x 8 x 68	17'-0"	15'-2"	23'-3"	20'-10"	27'-8"	23'-2"
2 x 8 x 97	19'-6"	17'-3"	26'-0"	23'-2"	30'-11"	27'-0"
2 x 10 x 43	15'-9"	14'-2"	22'-0"	17'-3"	23'-0"	17'-3"
2 x 10 x 54	16'-9"	15'-0"	23'-4"	21'-0"	27'-11"	22'-10"
2 x 10 x 68	18'-0"	16'-1"	24'-9"	22'-3"	30'-0"	26'-7"
2 x 10 x 97	20'-6"	18'-2"	27'-6"	24'-7"	33'-0"	29'-6"
2 x 12 x 43	16'-7"	14'-11"	22'-7"	16'-7"	22'-7"	16'-7"
2 x 12 x 54	17'-7"	15'-10"	24'-7"	22'-2"	30'-0"	24'-9"
2 x 12 x 68	18'-10"	16'-11"	26'-1"	23'-5"	31'-8"	28'-5"
2 x 12 x 97	21'-5"	19'-0"	28'-10"	25'-10"	34'-8"	34'-1"

For SI: 1" = 25.4 mm, 1 foot = 304.8 mm, 1 psf = 0.0479 kN/m².

[1] Table provides the maximum ceiling joist span in feet and inches, to either the right or left of the interior support.

[2] Bearing stiffeners shall be installed at all support locations.

[3] Deflection criteria: L/240 for total loads.

[4] Interior bearing supports for multiple span joists shall consist of structural walls or beams.

[5] The interior support may be located ∀ two feet from mid-span provided that each of the resulting spans do not exceed the span given in the table above.

[6] Refer to Table 2 for actual sizes.

PRESCRIPTIVE METHOD FOR RESIDENTIAL COLD-FORMED STEEL FRAMING

Table 29
Allowable Horizontal Rafter Spans[1,2]

Nominal	Ground Snow Loads[3]							
	20 psf Ground		30 psf Ground		50 psf Ground		70 psf Ground	
	Spacing (inches)		Spacing (inches)		Spacing (inches)		Spacing (inches)	
	16	24	16	24	16	24	16	24
2 x 6 x 33	12'-8"	10'-4"	11'-9"	9'-7"	9'-11"	8'-1"	8'-10"	7'-2"
2 x 6 x 43	15'-5"	12'-7"	14'-3"	11'-8"	12'-1"	9'-10"	10'-8"	8'-9"
2 x 6 x 54	13'-0"	14'-2"	16'-1"	13'-1"	13'-8"	11'-2"	12'-1"	9'-10"
2 x 6 x 68	18'-1"	15'-10"	17'-3"	14'-9"	15'-4"	12'-6"	13'-6"	11'-1"
2 x 6 x 97	20'-1"	17'-6"	19'-1"	16'-8"	17'-1"	14'	15'-7"	13'-2"
2 x 8 x 33	15'-5"	11'-5"	14'-4"	9'-10"	10'-7"	7'-1"	8'-3"	5'-6"
2 x 8 x 43	19'-1"	15'-7'	17'-9"	14'-6"	15'-1"	12'-3"	13'-3"	10'-9"
2 x 8 x 54	22'-7"	18'-5"	21'-0"	17'-1"	17'-9"	14'-6"	15'-9"	12-10"
2 x 8 x 68	24'-7"	20'-9"	23'-4"	19'-3"	20'-0"	16'-4"	17'-8"	14'-5"
2 x 8 x 97	27'-3"	23'-9"	26'-0"	22'-8"	23'-3"	19'-7"	21'-3"	17'-4"
2 x 10 x 43	21'-2"	17'-3"	19'-8"	16'-0"	16'-8"	13'-1"	14'-9"	10'-3"
2 x 10 x 54	25'-1"	20'-6"	23'-3"	19'-0"	19'-9"	16'-1"	17'-5"	14'-3"
2 x 10 x 68	29'-6"	24'-6"	27'-9"	22'-9"	23'-8"	19'-3"	21'-0"	17'-1"
2 x 10 x 97	32'-0"	28'-8"	31'-3"	27'-3"	28'-0"	23'-2"	25'-1"	20'-6"
2 x 12 x 43	23'-0"	18'-2"	21'-4"	15'-7"	16'-9"	11'-3"	13'-2"	8'-9"
2 x 12 x 54	27'-3"	22'-3"	25'-3"	20'-7"	21'-5"	17'-6"	18'-11	15'-5"
2 x 12 x 68	32'-1"	26'-2"	29'-9"	24'-3"	25'-3"	20'-7"	22'-4"	18'-2"
2 x 12 x 97	38'-4"	33'-6"	36'-6"	31'-6"	32'-8"	26'-9"	29'-0"	23'-7"

For SI: 1" = 25.4 mm,   1 foot = 304.8 mm,   1 psf = 0.0479 kN/m².

[1] Table provides the maximum horizontal rafter span in feet and inches for slopes from 3:12 to
     12:12.
[2] Deflection criteria: L/240 for live loads and L/180 for total loads.
[3] Use Table 30 to determine equivalent ground snow load when wind controls the design.
[4] Refer to Table 2 for actual member size.

PRESCRIPTIVE METHOD FOR RESIDENTIAL COLD-FORMED STEEL FRAMING

Table 30
Wind Speed to Equivalent Snow Load Conversion

Roof Slope	Equivalent Ground Snow Load (psf)				
	Wind Speed (mph)			Exposure C	
	70	80	90	100	110
3 / 12	20	20	20	30	50
4 / 12	20	20	30	50	50
5 / 12	20	20	30	50	50
6 / 12	20	20	30	50	70
7 / 12	30	30	50	70	70
8 / 12	30	30	50	70	
9 / 12	30	50	50	70	
10 / 12	30	50	50	Design Required	
11 / 12	30	50	70		
12 / 12	50	50	70		
	Wind Speed (mph)			Exposure C	
	70	80	90	100	110
3 / 12	20	20	20	30	50
4 / 12	20	20	20	30	50
5 / 12	20	20	20	30	50
6 / 12	20	20	20	50	50
7 / 12	20	30	30	5 0	70
8 / 12	20	30	50	50	70
9 / 12	30	30	50	70	70
10 / 12	30	30	50	70	Design Required
11 / 12	30	50	50	70	
12 / 12	30	50	50		

For SI: 1 mph = 1.609 km/hr, 1 psf = 0.0479 kN/m².

PRESCRIPTIVE METHOD FOR RESIDENTIAL COLD-FORMED STEEL FRAMING

Table 31
Ridge & Heel Joint Connections

Roof Slope	Number of Screws Required[1]															
	24' Building Width				28' Building Width				32' Building Width				36' Building Width			
	Ground Snow Load[2] (psf)				Ground Snow Load[2] (psf)				Ground Snow Load[2] (psf)				Ground Snow Load[2] (psf)			
	20	30	50	70	20	30	50	70	20	30	50	70	20	30	50	70
3/12	5	6	9	12	6	7	10	13	7	8	12	15	8	9	13	17
4/12	4	5	7	9	5	6	8	10	6	6	9	12	6	7	10	13
5/12	4	4	6	7	4	5	7	9	5	5	8	10	5	6	9	11
6/12	3	4	5	7	4	4	6	8	4	5	7	9	4	5	7	10
7/12	3	3	5	6	3	4	5	7	4	4	6	8	4	5	7	9
8/12	3	3	4	5	3	3	5	6	3	4	5	7	4	4	6	8
9/12	2	3	4	5	3	3	4	6	3	4	5	6	3	4	6	7
10/12	2	3	4	5	3	3	4	5	3	3	5	6	3	4	5	7
11/12	2	3	4	4	3	3	4	5	3	3	5	6	3	4	5	6
12/12	2	3	3	4	2	3	4	5	3	3	4	6	3	4	5	6
Ridge Board	2	2	3	3	2	2	3	4	2	2	3	4	2	3	4	5

For SI: 1" = 25.4 mm, 1 foot = 304.8 mm, 1 psf = 0.0479 kN/m².

[1] Screws shall be # 10 minimum.
[2] Use Table 30 to determine equivalent ground snow load for cases when wind controls the design.
[3] Ridge board connection applies to all roof slopes.

PRESCRIPTIVE METHOD FOR RESIDENTIAL COLD-FORMED STEEL FRAMING

Table 32
Roof Framing Fastening Schedule

Building Element	Number & Type of Fasteners	Spacing of Fasteners
Ceiling Joist to Bearing Wall Top Track[1]	2 # 10 screws	Each joist
Roof Sheathing (oriented strand board or plywood)	# 8 screws	6" on edges 12" at interior supports 6" at gable end truss
Truss to Bearing Wall [1]	2 # 10 screws	Each truss
Gable end truss to endwall top track	# 10 screws	12" oc.

For SI: 1" = 25.4 mm,   1 foot = 304.8 mm,   1 psf = 0.0479 kN/m².

[1]   Screws shall be applied through the flanges of the truss or ceiling joist or a 54 mil clip angle shall be used with 2 # 10 screws in each leg.

PRESCRIPTIVE METHOD FOR RESIDENTIAL COLD-FORMED STEEL FRAMING

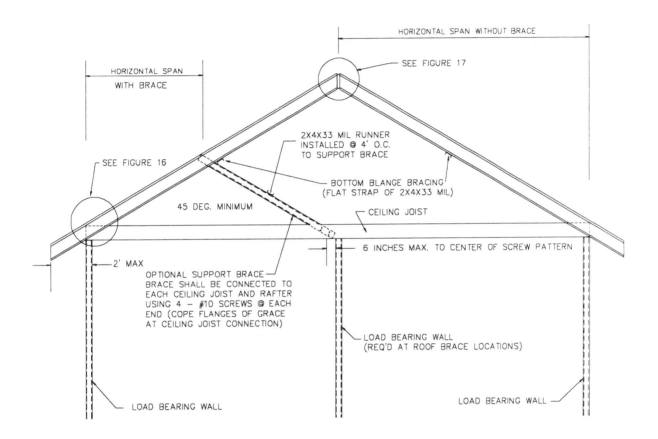

Fig. 15   Roof Assembly

PRESCRIPTIVE METHOD FOR RESIDENTIAL COLD-FORMED STEEL FRAMING

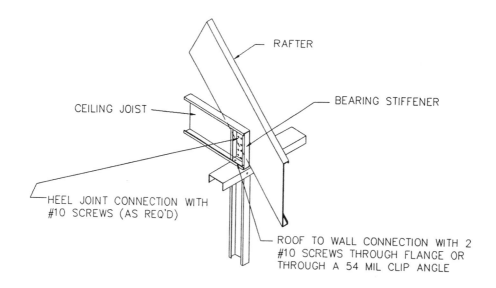

RAFTER

CEILING JOIST

BEARING STIFFENER

HEEL JOINT CONNECTION WITH
#10 SCREWS (AS REQ'D)

ROOF TO WALL CONNECTION WITH 2
#10 SCREWS THROUGH FLANGE OR
THROUGH A 54 MIL CLIP ANGLE

Fig. 16   Heel Connection

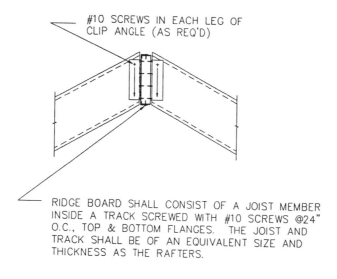

#10 SCREWS IN EACH LEG OF
CLIP ANGLE (AS REQ'D)

RIDGE BOARD SHALL CONSIST OF A JOIST MEMBER
INSIDE A TRACK SCREWED WITH #10 SCREWS @24"
O.C., TOP & BOTTOM FLANGES. THE JOIST AND
TRACK SHALL BE OF AN EQUIVALENT SIZE AND
THICKNESS AS THE RAFTERS.

Fig. 17   Ridge Connection

PRESCRIPTIVE METHOD FOR RESIDENTIAL COLD-FORMED STEEL FRAMING

9.0   MECHANICAL & UTILITIES

9.1   Plumbing

Plumbing shall comply with the applicable plumbing code. Copper and plastic pipes shall be separated from the steel framing by non-conductive grommets or other equivalent means.

9.2   Electrical Systems

Electrical system installation shall comply with the National Electric Code. A grommet bushing, conduit, or equivalent wire protection shall be installed in the service hole or punchout before electrical wiring is pulled through.

9.3   HVAC Systems

HVAC installation shall comply with the applicable mechanical code.

10.0   CONSTRUCTION GUIDELINES

10.1   General Guidelines

a. All structural members shall be aligned vertically (In-Line Framing) to transfer all loads to the foundation.

b. Bearing surfaces for joists, rafters, and trusses shall be uniform and level.

c. All load bearing studs, including king and jack studs, shall be seated in the tracks with a maximum gap of 1/8" between the end of the stud and the web of the track.

d. Temporary bracing shall be provided until permanent bracing has been installed.

e. Track members shall not be used for any load carrying applications without an approved design.

f. Cutting methods which cause significant heating of the steel or damage to the coatings shall only be used when the galvanized coating is repaired.

g. A sill sealer, or equivalent, shall be provided between the underside of the wall when fastened directly to concrete.

h. In addition to the requirements provided in Section 2.10, Table 33 provides some guidance on the screw diameter sizes typically required for various thicknesses of steel.

PRESCRIPTIVE METHOD FOR RESIDENTIAL COLD-FORMED STEEL FRAMING

Table 33
Screw Diameter Size Guidelines Based on
Total Thickness of Steel

Screw size	Nominal Diameter (inches)	Total Thickness of Steel (inches)[1,2]
# 6	0.138	Up to 0.110
# 8	0.164	Up to 0.140
# 10	0.190	Up to 0.175
# 12	0.216	Up to 0.210

For SI: 1" = 25.4 mm.

[1]   Greater thicknesses are possible, consult screw manufacturer.
[2]   Self-drilling tapping screws are required when total thickness of steel exceeds 0.033 inches (0.84 mm). Sharp point screws are acceptable for steel thicknesses of 0.033 inches (0.84 mm) or less.

10.2   Thermal Guidelines

a.   Batt insulation, when required, shall be full width to fill the cavity between studs.

b.   Exterior insulation board, when needed, shall be provided to meet the CABO Model Energy Code or other applicable codes. Refer to the AISI *Thermal Resistance Design Guide* (RG-9405) for details.

C

Commentary on the Prescriptive Method for Residential Cold-Formed Steel Framing

COMMENTARY ON THE PRESCRIPTIVE METHOD FOR RESIDENTIAL COLD-FORMED STEEL FRAMING

First Edition

for

The U.S. Department of Housing and Urban Development
Office of Policy, Development and Research
Washington, D.C.

Co-Sponsored by the American Iron and Steel Institute and the National Association of Home Builders

by

NAHB Research Center
400 Prince George's Boulevard
Upper Marlboro, MD 20774

May 1996

Notice

The U.S. Government does not endorse products or manufacturers. Trade or manufacturer's names appear herein solely because they are considered essential to the object of this report.

The contents of this report are the views of the contractor and do not necessarily reflect the views of the U.S. Department of Housing and Urban Development of the U.S. Government.

Disclaimer

While the material presented in this document has been prepared in accordance with recognized engineering principles and sound judgements, this document should not be used without first securing competent advice with respect to its suitability for any given application. **The use of this document is subject to approval by the local code authority**. Although the information in this document is believed to be accurate, neither the authors, nor reviewers, nor the U.S. Department of Housing and Urban Development of the U.S. Government, nor the NAHB Research Center, nor any of their employees or representatives makes any warranty, guarantee, or representation, expressed or implied, with respect to the accuracy, effectiveness, or usefulness of any information, method, or material in this document, nor assumes any liability for the use of any information, methods, or materials disclosed herein, or for damages arising from such use.

.

Foreword

For centuries home builders in the United States have made wood their material of choice because of its satisfactory performance, abundant supply, and relatively low cost. However, increases and unpredictable fluctuations in prices and problems with quality of framing lumber are causing builders and other providers of affordable housing to seek alternative building products.

Cold-formed steel framing in the residential market has been on the rise for the last few years. The price stability, consistent quality, fire, rot, and termite resistance, similarity to conventional framing, and the successful use of cold-formed steel in the commercial market have attracted the attention of many builders. However, the lack of prescriptive construction requirements has restrained this alternative material from gaining wider acceptance among home builders and code officials.

The *Commentary On The Prescriptive Method For Residential Cold-Formed Steel Framing* provides the background, engineering assumptions and methods, and detailed calculations for the various provisions of the *Prescriptive Method For Residential Cold-Formed Steel Framing*.

As in the *Prescriptive Method For Residential Cold-Formed Steel Framing*, the commentary was developed under a sponsorship of the U.S. Department of Housing and Urban Development (HUD) with co-funding from the National Association of Home Builders (NAHB) and the American Iron and Steel Institute (AISI). The program was conducted by the NAHB Research Center with assistance and input from steering, advisory, and engineering committees. These committees represented steel manufacturers, steel producers, code officials, academia, researchers, professional engineers, and builders who are experienced in residential and cold-formed steel framing.

Michael A. Stegman
Assistant Secretary for Policy
Development and research

Acknowledgments

The *Commentary On The Prescriptive Method For Residential Cold-Formed Steel Framing* was prepared by the NAHB Research Center under sponsorship of the U.S. Department of Housing and Urban Development (HUD). We wish to recognize the National Association of Home Builders (NAHB) and the American Iron and Steel Institute (AISI) whose co-funding made this project possible. Special appreciation is extended to William Freeborne of HUD for guidance throughout the project.

The preparation of this commentary required the talents of many dedicated professionals. The principal authors were Nader R. Elhajj, P.E. and Kevin Bielat with review by Jay Crandell, P.E.

Appreciation is also extended to all the individuals who provided input throughout the development of this document.

COMMENTARY ON THE PRESCRIPTIVE METHOD FOR RESIDENTIAL
COLD-FORMED STEEL FRAMING

———

CONTENTS

COMMENTARY ON THE PRESCRIPTIVE METHOD FOR RESIDENTIAL
COLD-FORMED STEEL FRAMING

COMMENTARY ON THE PRESCRIPTIVE METHOD FOR RESIDENTIAL COLD-FORMED STEEL FRAMING

Executive Summary

This *Commentary On The Prescriptive Method For Residential Cold-Formed Steel Framing* is provided to facilitate the use of and provide background information for the *Prescriptive Method For Residential Cold-Formed Steel Framing* document. The *Prescriptive Method* was developed as an interim guideline for the construction of one- and two-family residential dwellings using cold-formed steel framing. It provides a strictly prescriptive method to stick-frame typical homes with cold-formed steel framing.

The information contained in this commentary represents the design assumptions, engineering methods, industry standards, and construction practices that were utilized in developing the *Prescriptive Method*. This document was put together by the NAHB Research Center with input from an expert steering committee and other knowledgeable individuals to ensure that the decisions were reasonable from a variety of viewpoints and concerns. The steering committee represented several experts in the fields of building code enforcement, engineering, research, and construction. The sections in the *Commentary On The Prescriptive Method* are consistent with those of the *Prescriptive Method*.

COMMENTARY ON THE PRESCRIPTIVE METHOD FOR RESIDENTIAL COLD-FORMED STEEL FRAMING

COMMENTARY ON THE PRESCRIPTIVE METHOD FOR RESIDENTIAL
COLD-FORMED STEEL FRAMING

———

INTRODUCTION

This *Commentary* is provided to facilitate the use of and provide background information for the *Prescriptive Method For Residential Cold-Formed Steel Framing*[1].

The *Prescriptive Method* was developed as an interim guideline for the construction of one-and two-family residential dwellings using cold-formed steel framing.

In this *Commentary*, the individual sections, figures, and tables are presented in the same sequence found in the *Prescriptive Method*. The information contained in this commentary represents the decisions made by the NAHB Research Center with guidance from several experts in the fields of building code enforcement, engineering, research, and construction. The expert input was gathered through a steering committee and other knowledgeable individuals to ensure that the decisions were reasonable from a variety of viewpoints and concerns. Design examples are provided at the end of each division to illustrate the application of the different standards and specifications in the development of the *Prescriptive Method*.

C1.0   GENERAL

C1.1   Purpose

The goal of the *Prescriptive Method* is to present prescriptive criteria (tables, figures, guidelines) for the construction of one-and two-story dwellings framed with cold-formed steel members. Currently, there are no prescriptive standards or building code provisions that can be used by builders to construct simple cold-formed steel houses without the added expense of a design professional. The *Prescriptive Method* is a first step to address that need, provide guidance for building code officials, and reduce the reliance on design professionals for typical residential applications.

The *Prescriptive Method* presents the minimum requirements to provide basic residential construction that is consistent with the safety levels provided in the current U.S. building codes governing residential construction.

The *Prescriptive Method* is not applicable in all areas of the United States and is subject to the applicability limits set forth in Table 1. The applicability limits should be carefully understood as they define important constraints on the use of the *Prescriptive Method*.

C1.2   Approach

The requirements, figures, and tables provided in the *Prescriptive Method* are based on an interpretive application of the AISI *Specification for the Design of Cold-Formed Steel Structural Members* (AISI-86 with 1989 addendum), ASCE *Minimum Design Loads for Buildings and Other Structures* (ASCE 7-93), and the pertinent requirements of the ICBO *Uniform Building Code* (UBC-94), SBCCI *Standard Building Code* (SBC-94), the BOCA *National Building Code* (NBC-93), and the CABO *One- and Two-Family Dwelling Code* (CABO-95).

---

[1]   U.S. Department of Housing and Urban Development, *Prescriptive Method For Residential Cold-Formed Steel Framing*, first edition, prepared by NAHB Research Center, May 1996.

C1.3   Scope

COMMENTARY ON THE PRESCRIPTIVE METHOD FOR RESIDENTIAL COLD-FORMED STEEL FRAMING

———

It is unrealistic to develop an easy to use document that provides prescriptive requirements to cover all types and styles of steel-framed houses. Therefore, the *Prescriptive Method* is limited in its applicability to residential construction. The requirements set forth in the *Prescriptive Method* only apply to the construction of steel-framed houses that meet the limits set forth in Table 1. The applicability limits are necessary to define reasonable boundaries to the conditions that must be analyzed to develop prescriptive construction requirements. The *Prescriptive Method*, however, does not limit the application of engineering designs when the construction does not entirely comply with the constraints of Table 1.

The basic applicability limits were established from industry convention and experience. Detailed applicability limits were documented in the process of developing prescriptive design requirements for various elements of the structure. In some cases, sensitivity analyses were performed to help define appropriate limits.

The applicability limits strike a reasonable balance between engineering theory, available test data, and proven field practices for typical residential construction applications. The applicability limits are intended to prevent misapplication while addressing a reasonably large percentage of new housing conditions. Additional research, design or testing is needed to relax overly-restrictive constraints within the *Prescriptive Method*. Special consideration is directed toward the following items related to the applicability limits.

Building Geometry

The provisions in the *Prescriptive Method* apply to detached one- or two-family dwellings, townhouses, and other attached single-family dwellings not more than two stories in height. Its application to homes with complex architectural configurations is subject to careful interpretation of the user and design support may be required.

The most common house widths (or depths) range from 24 feet to 36 feet, with load bearing wall heights up to ten feet. The house width as used in the *Prescriptive Method* is the dimension measured along the length of the trusses or the joists (floor or ceilings) including cantilevers between the outmost structural walls. A house complying with this document can not be of a length greater than 60 feet. Length is measured in the direction parallel to the roof ridge and width is measured perpendicular to the ridge.

Site Conditions

The snow loads are typically given in a ground snow load map such as provided in ASCE 7-93 or by local experience. The major building codes in the U.S. either adopt the ASCE 7 snow load requirements or have similar map published in the code. The 0 to 70 psf ground snow load used in the *Prescriptive Method* represents approximately 90 percent of the map's regions, which was deemed to include the majority of the houses that are expected to utilize this document. Houses in areas with higher snow loads should not use this document without consulting with a design professional.

All areas of the U.S. fall within the 70 to 110 mph (fastest-mile) range of design wind speeds (ASCE 7-93). The wind exposure category in the *Prescriptive Method* is limited to Exposures B or C. Wind speed and exposure are defined in the *Prescriptive Method*. Wind exposure categories B and C cover the majority of site conditions. Wind exposure is a critical determinant of the wind loads to be expected at a given site, and it should be determined by good judgement on a case-by-case basis. Houses built along the coastline (i.e. beach front property) are classified as Exposure D and therefore, can not use this document without consulting a design professional. Because of additional engineering concerns and complications in high wind conditions (i.e. greater than 90 mph exposure C) engineering design of wall bracing and building anchorage (i.e. uplift straps and hold-down brackets) is required in these regions.

COMMENTARY ON THE PRESCRIPTIVE METHOD FOR RESIDENTIAL
COLD-FORMED STEEL FRAMING

Buildings constructed per the *Prescriptive Method* are limited to regions designated as Seismic Zones 0, 1, 2, 3 and 4. However, because of additional engineering concerns, complications, and limitations in the wall bracing and hold-down requirements, buildings in Seismic Zones 3 and higher will require the services of a design professional to address these particular items.

Loads: Building codes and standards handle loads and load combinations by different methods. Appropriate values were established for design loads in accordance with a review of the major building codes and standards. The results of this load review are embodied in the applicability limits table in the *Prescriptive Method*. Loads and load combinations requiring calculations to analyze the structural components and assemblies of a home are presented in the design examples at the end of each division of this document.

C1.4 Definitions

The definitions in the *Prescriptive Method* are self explanatory. Additional definitions that warrant some technical explanation, are briefly discussed below.

Design Thickness: The minimum uncoated material thickness divided by 0.95. Design thickness is used in developing the requirements and provisions of the *Prescriptive Method* (i.e. design thickness is used in calculating section properties, member capacities, and connections capacities). This adjustment is approved in the current AISI design specification (ref. 11.1) and it conservatively accounts for the normal varieties of material thickness above the minimum required thickness.

Torsional Rigidity: The torsional rigidity of open sections is proportional to cube the thickness. Therefore, relatively thin cold-formed C-sections are susceptible to torsion. Furthermore, the shear center and the centroid of a C-section do not coincide, thus causing rotation of the member due to an applied load in the plane of the web. Due to these facts, torsional-flexural buckling may be a critical factor for compression members when used in a manner inconsistent with the provisions in the *Prescriptive Method* and the loads associated with conventional, repetitive member framing practices.

Local Buckling and Post Buckling Strength: Cold-formed steel C-sections consist of thin elements compared to their overall dimensions, and therefore, may buckle at stresses well below the yield point if they are subjected to high compression, bending, shear, or bearing loads. Because of this, all sections are checked for local buckling. Local buckling may be a critical factor for compression elements of bending members (i.e. studs) or bending members (i.e. headers) when used in a manner inconsistent with the provisions in the *Prescriptive Method* and the conventional practices of repetitive member-framing.

Tests at the NAHB Research Center (unpublished) have shown that cold-formed elements do not fail by local buckling when used in a manner consistent with the *Prescriptive Method*. In fact the elements tend to carry loads above their individual local buckling strength because of the load re-distribution experienced in repetitive member framing. Therefore, the post buckling strength is utilized in the design of cold-formed members for use in the *Prescriptive Method*.

C2.0   MATERIALS AND STANDARD SIZES

C2.1   Types of Cold-Formed Steel

C2.1.1   Structural Members

Standard C-sections are produced by roll forming hot-dipped galvanized steel that conforms to one of the four categories of steel presented in the *Prescriptive Method*, and in accordance with the AISI "Specification for the Design of Cold-Formed Structural Steel" requirements. These four categories are: ASTM A 653, A 792, A 875, and steels conforming to A 653 with the exception of certain elongation and tensile strength to yield

COMMENTARY ON THE PRESCRIPTIVE METHOD FOR RESIDENTIAL
COLD-FORMED STEEL FRAMING

strength ratio requirements. Steel tracks that are used in conjunction with the standard C-section members specified in the *Prescriptive Method* consist of unstiffened channels that are rolled or brake formed from hot-dipped galvanized steel sheets with a specified minimum yield point of 33 ksi before forming

C2.1.2   Non-Structural Members

ASTM C 645 is currently referenced in the three model building codes for non-structural members (non-load bearing) and steel tracks.

C2.2   Physical Dimensions

Because of significant variations in the industry related to dimensions of cold-formed steel members, the dimensions had to undergo industry wide standardization to allow for the development of the *Prescriptive Method* with the consistency demanded for that purpose. Dimensional standardization was focused on the C-shape because of its widespread use in the industry. To achieve the demanded consistency, the web depth, flange width, lip size, uncoated steel thickness, bend radius, and yield strength of the C-section had to be standardized.

Web Depth: The actual web sizes in the *Prescriptive Method* are limited to 3-1/2", 5-1/2", 8", 10", and 12". The 3-1/2" and 5-1/2" web depths were chosen to accommodate current framing practices utilized in the residential building industry (i.e. to accommodate window and door jambs). These sizes can be used directly with conventional home construction practices without having the manufacturers of other home building products alter their sizes. The size of the web, for 8", 10", and 12" members are not of great significance to the installation of different materials in residential construction because they are typically used for horizontal framing members. Therefore, the nominal sizes for 8", 10", and 12" are used.

Flange Width: The *Prescriptive Method* requires the standard C-section to have a minimum of 1-5/8" flange (with a maximum of 2"). A review of current producers of cold formed steel members showed that 1-5/8" flanges are widely used. The range of flange widths, from 1-5/8" to 2", encompasses the majority of the roll forming industry, when combined with the above web depths. An increase in flange size (> 2") may result in decreased capacity for certain members.

Lip Size: The *Prescriptive Method* also provides a minimum size for the stiffening lip of 1/2 inch. This dimension is also common in the industry. Sensitivity analyses performed to optimize the lip dimension showed that the slight increase in capacity resulting from a 5/8" or 3/4" lip size does not justify the increase in weight of the additional steel, although these lip sizes are not prohibited.

The section properties for the standard C-sections are shown in Attachment A. These section properties were used in the engineering calculations to develop the *Prescriptive Method*. Refer to Section C2.11 for an example calculation of the section properties of a standard C-section.

The *Prescriptive Method* requires steel tracks to have a minimum flange dimension of 1-1/4". This dimension is widely used by the steel industry and ensures a sufficient flange width to allow fastening of the track to the framing members and finish materials. Steel track webs are measured from inside to inside of flanges and thus have wider web depths, than the associated standard C-sections, so that the C-sections can properly nest into the track sections. Steel tracks are also available in thicknesses matching those required for the standards C-sections.

C2.3   Uncoated Material Thickness

The steel thickness required by the *Prescriptive Method* is the minimum uncoated material thickness (without the thickness of the galvanization) and is given in mils (1/1000 of an inch). This is a deviation from the

COMMENTARY ON THE PRESCRIPTIVE METHOD FOR RESIDENTIAL COLD-FORMED STEEL FRAMING

current practice which uses gauge for thickness. The gauge represents a range of thicknesses and is therefore, a vague unit of measure. In order to achieve consistency in the thickness of the standardized C-sections, the "mil" designation was adopted with overwhelming industry support. Therefore, the 33 mils, 43 mils, 54 mils, 68 mils, and 97 mils are specified as minimum thicknesses.

The design thickness (shown in the table below) is the value used in the computation of member capacities. The reduction in thickness that occurs at corner bends is ignored, and the design thickness of the flat steel stock, exclusive of coatings, is used in the structural calculations. The design thickness is calculated by dividing the minimum delivered thickness by 0.95 which follows the provisions of the AISI Design Specification. This adjustment reasonably accounts for the normal variation in material thickness above the minimum delivered material thickness required. Refer to Section C2.11 for the section properties calculation example illustrating the use of the design thickness in the computation of section properties.

Minimum Uncoated Material Thickness

Designation (mils)	Minimum Delivered Thickness inches	Design Thickness inches	Reference Gauge Number
18	0.018	0.0188	25
27	0.027	0.0283	22
33	0.033	0.0346	20
43	0.043	0.0451	18
54	0.054	0.0566	16
68	0.068	0.0713	14
97	0.097	0.1017	12

For SI: 1" = 25.4 mm.

C2.4 Bend Radius

The inside bend radius (at bends in cold-formed steel) has an impact on the capacity of structural steel members. As the inside bend radius increases, the effective flat width of the element decreases and thus, the member's capacity decreases. Conversely, strength increases are realized in the regions of bends due to a phenomena known as cold-working. The steering committee established the maximum inside bend radius to be the greater of 3/32" or twice the material thickness. This requirement was established to provide manufacturers with the flexibility of rolling sections at variable bend radii without adversely affecting the strength of the member. The maximum bend radii (shown in the table below) were used in the computations of member capacities for the *Prescriptive Method*. Refer to Section C2.11 for the section properties calculation example illustrating the use of the bend radius in the computation of section properties.

Maximum Inside Bend Radius

Designation (mils)	Minimum Delivered Thickness (t) (inches)	Max. Inside bend Radius	Max. Inside bend Radius (inches)

COMMENTARY ON THE PRESCRIPTIVE METHOD FOR RESIDENTIAL
COLD-FORMED STEEL FRAMING

18	0.018	3/32"	0.09375
27	0.027	3/32"	0.09375
33	0.033	3/32"	0.09375
43	0.043	3/32"	0.09375
54	0.054	2 x t	0.108
68	0.068	2 x t	0.136
97	0.097	2 x t	0.194

For SI: 1" = 25.4 mm.

C2.5   Yield Strength

The *Prescriptive Method* applies to steels with a minimum yield strength of 33 ksi. This minimum yield strength was chosen as an initial step to standardize the steel members. Greater optimization may be achieved by use of steels with a higher yield strength, but an approved design would be required. The AISI Design Specification provides a comprehensive list of ASTM Standards for steels that are acceptable for this document.

Strength increase from cold working is utilized for the standardized C-sections shown in the *Prescriptive Method* concerning bending strength of flexural members, concentrically loaded compression members, and members with combined axial and bending loads. Refer to Section C2.11 for the section properties calculation example illustrating the calculation of the stress increase due to cold-working and the use of the stress increase in calculating the section properties.

C2.6   Corrosion Protection

In residential construction, it is not considered likely for any structural steel framing to be exposed to a severe corrosive environment, except for locations close to coastal areas. The degree of protection is determined by considering the use of the member, its exposure, climate, and other conditions. Galvanizing is required to ensure that the structure will perform its required function during its expected life.

The minimum galvanizing coatings of G-40 for interior non-structural (non-load bearing) members, and G-60 for structural members provide adequate protective coating for steel framed members in normal environments. The user should be cautious in coastal areas and other severe climatic conditions, where a minimum coating of G-90 is recommended. The *Prescriptive Method* provides the minimum coating designation and references the applicable ASTM standards. The requirements for galvanized coatings are given in ASTM A924. Users of cold-formed steel framing are encouraged to locate (store or construct) steel framing members within the building envelope and adequately shield steel materials from direct, long-term contact with moisture from the ground or the outdoor climate, particularly in coastal environments.

The *Prescriptive Method* includes a provision that allows the usage of other metallic coatings that are equivalent to the galvanized coating. However, only coatings that provide sacrificial protection are permissible. Sacrificial protection is necessary to protect bare steel at drilled holes, cuts, and scratches from corrosion.

COMMENTARY ON THE PRESCRIPTIVE METHOD FOR RESIDENTIAL
COLD-FORMED STEEL FRAMING

C2.7 Web Punchouts

The AISI Design Specification does not address punchouts beyond small circular holes. Web punchouts, typically 1-1/2" wide x 4" long are common in the cold-formed steel industry to accommodate plumbing, electrical, and mechanical materials.

The standardized C-sections in the *Prescriptive Method* are all assumed punched with a standard hole of 1-1/2" wide x 4" long (the width of the hole is along the depth of the web and the length of the hole is parallel to the web). The span tables are calculated based on the presence of standard punchouts spaced at 24" oc. with 12 inches minimum distance from center of hole to end of member. The design procedure follows ICBO AC46 "Acceptance Criteria For Steel Studs And Joists" (ref. 11.7).

ICBO AC46 provides a criteria to calculate the allowable axial and bending load of a perforated wall stud or a structural member. The ICBO AC46 procedure, for calculating the effects of punchouts, is briefly described below. The example in Section C2.11 illustrates the use of this procedure in calculating the reduction in the allowable shear capacity due to punchouts.

For perforated members subjected to axial loads, the effective area, A_e, of the member at stress, F_n, can be determined by assuming that the web consists of two unstiffened strips, one on each side of the perforation. Calculation of the major axis bending moment is based on the effective section modulus of the unpunched web (as defined in Section C.1.1(a) of the AISI Design Specification). Calculation of the minor axis bending moment is similar to the determination of the effective area for members subjected to axial loads.

For perforated members subjected to shear alone, ICBO AC46 states that the allowable shear can be determined by applying a reduction factor, q_s, to the design value calculated per Section C.2 of the AISI Design Specification.

$$q_s = 1.0 - 1.1(a'/D)$$ Where $a' =$ Depth of the opening (perforation)

$D =$ overall depth of the web.

For web crippling strength, ICBO AC46 provides a reduction factor for punched members with holes closer than 1.5D from the edge of the bearing. For members with holes having an edge distance greater than 1.5D, no reduction in web crippling strength is required.

The combined bending and web crippling, and bending and shear (as required by the AISI Specification) are checked with the modifications described above.

The ICBO AC46 approach has the following limitations:

Center to center spacing of the perforations shall be at least three times the outside depth of the member, but need not exceed 24 inches.

Perforation width (across the web) shall not exceed half the member depth or 2-1/2 inches. Perforation length shall not exceed 2.67 times the hole width or 4.5 inches.

Minimum distance between the end of the stud and the near edge of the perforation shall be 10 inches.

All web perforations are located along the center line of the web.

COMMENTARY ON THE PRESCRIPTIVE METHOD FOR RESIDENTIAL
COLD-FORMED STEEL FRAMING

C2.8 Cutting, Notching, and Hole Stiffening

As mentioned before, all structural members, in the *Prescriptive Method*, were assumed to be punched when calculating member capacities. The punchout limitations are shown in Fig. 3 of the *Prescriptive Method*. Structural members having holes that do not conform to the standard punchout requirements need to be reinforced to achieve the structural capacity used in the design. The amount and size of reinforcement presented is a minimum. The hole reinforcement detail is applicable only if the punchout (hole) depth does not exceed half the depth of the web, and the punchout length (along the web) does not exceed the depth of the web or 4 inches, whichever is larger.

The hole stiffening detail provided results in a member that has equal or higher capacity than an un-stiffened member with standard holes. Tests performed on floor joists, at the NAHB Research Center (unpublished), showed that a stiffened joist (patched in accordance with these provision) exhibits an ultimate capacity that is equal to or higher than the ultimate capacity of a similar joist with standard holes, subjected to the same loading conditions. Furthermore, the joist section with patched holes showed a slightly higher stiffness than the joist with standard holes.

Ongoing research at the University of Missouri, Rolla is continuing to investigate the sizes and locations of punchouts and their impact on the member's capacity. The hole stiffening detail provided in the *Prescriptive Method* is conservative, and further testing is needed to develop a more economical design.
The *Prescriptive Method* can still be used for members with larger holes, provided that the hole location and size are engineered so that the load carrying capacity of the member as calculated and used in this document is not reduced.

Cutting or notching of structural members are not allowed in the *Prescriptive Method* because cutting or notching of the lips or flanges will greatly reduce the structural capacity of the member. However, a notched or cut member can still be used provided that the member is designed such that its load carrying capacity is equal to or higher than the one used in this document.

C2.9 Bearing Stiffeners

Webs of thin walled cold-formed steel members may cripple (buckle) due to the high local intensity of a concentrated load or reaction. The allowable reactions and concentrated loads for beams having single unreinforced webs depend on web depth, bend radius, web thickness, yield strength, and actual bearing length.

Bearing stiffeners (also called transverse or web stiffeners) are required at all support or bearing point locations for all floor joists and ceiling joists. The specified stiffeners are always required to have a material thickness that is equal to or larger than the thickness of the member they stiffen.

Testing performed at the NAHB Research Center (unpublished) showed that web crippling failure is not a concern for joists and headers with web-depth to thickness ratio (h/t) greater than 200 (but less than 260). Tests performed on 2 x 10 x 43 (h/t = 215) and 2 x 12 x 43 (h/t = 259) members, with concentrated loads at mid-spans and at 1/3 points of the span resulted in ultimate capacities that are much higher than the calculated ones (used in the design). Moreover, none of the members tested showed any signs of web crippling or buckling within the ultimate design load. Joists were found to be controlled by deflection, and headers failed in shear at a point beyond the web stiffener locations.

C2.10 Fasteners

COMMENTARY ON THE PRESCRIPTIVE METHOD FOR RESIDENTIAL COLD-FORMED STEEL FRAMING

This section is not intended to limit the fastening techniques to those listed. Other fastening methods are permitted to be used with this document provided the connection's capacity is shown to exceed that implied in the *Prescriptive Method*. Testing, design, or code approvals may be necessary for alternate fastening techniques.

C2.10.1 Screws

Fastening of cold-formed steel framing members is limited to screws in the *Prescriptive Method*. Self-drilling tapping screws conforming to the requirements of SAE J78 are used in the *Prescriptive Method*. There is a need for additional standards for screws related to applications with cold-formed steel framing. Requirements for sharp point screws, connecting gypsum board and sheathing to steel studs, are found in ASTM C 1002 and ASTM C 954. The edge distance and center-to-center spacing of three screw diameters are well documented in the industry and the AISI guidelines.

The minimum screw size specified in this document is #8 for structural applications, however, larger diameter screws may be required to provide drill capacities through steel thicknesses greater than 100 mils (refer to Table 33 in Section 10.0). In certain applications, # 10 screws are specified in the *Prescriptive Method*. The minimum screw size specified in the *Prescriptive Method* is based on strength requirements and not necessarily optimum constructability for all possible applications. Also, point style will affect the

constructability in certain applications. For example, a sharp point screw may be efficiently used to connect gypsum board and other panel products to steel framing members that are no thicker than 33 mils. For these reasons, screw manufacturer recommendations should be consulted.

Screw capacities given in the tables below are calculated based on the design method given in the CCFSS document (ref. 11.12). This document provides the equations necessary to calculate the shear, pullover, and pullout capacity of a connection based on the thicknesses of the steel sections being fastened together. The equations are conservatively based on tests performed on thousands of screws of many different types and levels of quality.

Minimum Allowable Fastener Capacity for Steel to Steel Connections
[Safety factor = 3.0]

Screw Size	Minimum Shank Diameter (inches)	Minimum Head Diameter (inches)	Minimum Capacity (lbs.)			
			Shear Capacity		Pullout Capacity	
			43 mils[1]	33 mils[1]	43 mils[1]	33 mils[1]
#8	0.164	.322	244	164	94	72
#10	0.190	.384	263	177	109	84

For SI: 1" = 25.4 mm, 1 mil = 1/1000 inch, 1 lb. = 4.5 N

[1] The value represent the minimum thickness of steel being connected.

Minimum Allowable Screw Capacity for Wood to Steel Connections
[Safety factor = 3.0]

Screw Size	Minimum	Steel	Plywood	Minimum Capacity (lbs.)

COMMENTARY ON THE PRESCRIPTIVE METHOD FOR RESIDENTIAL
COLD-FORMED STEEL FRAMING

	Shank Diameter (inches)	Thickness (mils)	Thickness (inches)		
				Shear Capacity	Pullout Capacity
				33 mils	33 mils
#8	0.164	33	1/4	92	33
			≥1/2	180	89
#10	0.190	33	1/4	97	38
			≥1/2	228	120

For SI: 1" = 25.4 mm, 1 mil = 1/1000 inch, 1 lb. = 4.5 N.
EXAMPLE:

The following is an example for calculating one of the entries in the table above, #10 screw capacity connecting two 43 mil steel sections (All referenced equations and sections are to the CCFSS document, ref. 11.12)

Screw diameter = 0.19 inches (# 10 screw)
Ultimate capacity of steel = 45 ksi (tensile)
Steel minimum uncoated thickness = 0.0451 inches (43 mils design thickness)
Factor of safety = 3.0

Calculate Shear capacity: (Section E4.3.1)

The shear force per screw shall not exceed P_{as}
$P_{as} = P_{ns} / 3$

$t_2/t_1 = 1.0 < 1.0$

$P_{ns1} = 4.2(t^3 d)^{1/2} F_{u2}$ [Eq. E4.3.1]
$P_{ns1} = 4.2(.0451^3 \times 019)^{1/2} (45) = 0.7891$ kips

$P_{ns2} = 2.7(t_1 d) F_{u1}$ [Eq. E4.3.2]
$P_{ns2} = 2.7(.0451 \times 019)(45) = 1.0411$ kips

$P_{ns3} = 2.7(t_2 d) F_{u2}$ [Eq. E4.3.3]
$P_{ns1} = 2.7(.0451 \times 019)(45) = 1.0411$ kips

Therefore, $P_{ns} = 0.7891$ kips = 789 lbs.

$P_{as} = P_{ns} / 3 = 263$ lbs.

Pull out in Screw: (Section E4.4.1)

The pullout force shall not exceed P_{not}
where $P_{not} = 0.85 \times t_c \times d \times F_{u2}$ [Eq. E4.4.1.1]

10

COMMENTARY ON THE PRESCRIPTIVE METHOD FOR RESIDENTIAL
COLD-FORMED STEEL FRAMING

P_{not} = 0.85 x 0.0451 x 0.19 x 45 = 0.3278 kips

P_{aot} = 327.8 / 3 = 109.25 lbs.

Pull-over in Screw: (Section E4.4.2)

The pull-over force shall not exceed P_{nov}
where P_{nov} = 1.5 x t_1 x d_w x F_{ul} [Eq. E4.4.2.1]
P_{nov} = 1.5 x 2 x 0.0451 x 0.19 x 45 = 1.1568 kips   (conservatively assuming the washer
                                                                diameter is 2 x screw diameter)

P_{aov} = 1156.8 / 3 = 385.6 lbs.

Therefore, the controlling capacity of the connection is: Shear   = 263 lbs.
                                                                 Pullout = 109 lbs.

C2.10.2   Bolts

Bolts used in cold-formed steel framing are specified to meet or exceed the requirements of ASTM A307. Similar to screws, bolt edge distance and center-to-center spacing are in accordance with industry standards.

C2.11   Section Properties Calculation Example

The following is an illustrative example of calculating the various section properties for a 2 x 8 x 54 mils member. The purpose of this calculation is to show how the section properties are derived and how they relate to each of the values shown in Tables 2 and 3 of the *Prescriptive Method*.

All properties are calculated in accordance with the AISI "*Specification For The Design Of Cold-Formed Steel Structural Members, August 19, 1986 Edition with December 11, 1989 Addendum*". The section properties calculated in this section are based on the requirements of Sections 2.1, 2.2, 2.3, 2.4, and 2.5 of the *Prescriptive Method*.

The following section properties correspond to a 2 x 8 x 54 mil C-section (ref. Attachment A)

Gross Area = 0.667 in^2,   Net Area = 0.582 in^2
I_{xx} = 5.70 in^4 , S_{xx} = 1.425 in^3
I_{yy} = 0.192 in^4 S_{yy} = 0.553 in^3 (gross section modulus)
r_x = 2.93 in, r_y = 0.536 in
S_{yy} = 0.136 in^3 (Effective section modulus)
I_x = 5.699895 in^4 (Effective moment of inertia)
S_x = 1.424974 in^3 (Effective section modulus)
X_o= - 0.935 in,   J = 0.000713 in^4, C_w = 2.503 in^6, r_o= 3.12 in,   β = 0.91.
F_{ya} = 37.2 ks,
M_a = 31,767.17 lb-in. (2,647.5 ft-lb)
V_a = 1,9731.1 lbs. (shear capacity for unpunched section)
V_a = 1,566.3 lbs. (shear capacity for punched section)
P_a = 1029 lbs. (Allowable web crippling capacity, multiple webs)
P_a = 471 lbs. (Allowable web crippling capacity, single web)

(All references to nomenclature, sections, and equation numbers are from the AISI Design Specification):

Design thickness = 0.0566"
Minimum uncoated delivered thickness = 0.0566" x 0.95 = 0.0538"   AISI Section A3.4

COMMENTARY ON THE PRESCRIPTIVE METHOD FOR RESIDENTIAL
COLD-FORMED STEEL FRAMING

Inside bend radius = 2t = 2 x 0.0538" = 0.1076"

Area = t(a + 2b + 4u + 2c)   (AISI Supplementary Information Section 1.2)

Where: a = Web - 2(Radius + thickness) = 8 - 2(0.1076 + 0.0566) = 7.6716"
         b = Flange - 2(Radius + thickness) = 1.625 - 2(0.1076 + 0.0566) = 1.2966"
         c = Lip - (Radius + thickness) = 0.5 - (0.1076 + 0.0566) = 0.335"
         u = 1.57 (Radius + thickness/2) = 1.57(0.1076 + 0.0566/2) = 0.2134"

Gross Area = 0.0566(7.6716 + 2 x 1.2966 + 4 x .2134 + 2 x 0.335) = **0.667 in²**
Net Area = Gross area - area of opening = 0.667 - 1.5 x 0.0566 = **0.582 in²**

Weight per foot = Gross area x density of steel = 0.667 x 490 /144 = **2.27 lb/ft.**

Gross Y_{cg} = 8 / 2 = **4"**

Calculate gross moment of inertia, I_{xx}:

r = Radius + Thickness/2 = .1359

I_{xx}= $2t[0.0417a^3 + b(a/2 + r)^2 + u(a/2 + 0.637r)^2 + 0.149r^3 + 0.0833c^3 + c/4(a - c)^2 + u(a/2 + 0.637r)^2 + 0.149r^3]$ (AISI Supplementary Information Section 1.2)

I_{xx} = **5.70 in⁴**

Gross section modulus, S_{xx}:

S_{xx} = I_{xx} / Y_{cg} = 5.70 / 4 = **1.425 in³**

Radius of gyration, r_x :

r_x = SQRT(I_{xx}/Area) = SQRT(5.70/0.667) = **2.93 in**

Calculate gross moment of inertia, I_{yy}:1   AISI Supplementary Information Section 1.2)

I_{yy}= $2t[b(b/2 + r)^2 + 0.833b^3 + c(b + 2r)^2 + 0.356r^3 - u(b + 1.637r)^2 + 0.149r^3 - Area(x')^2]$

x' = 2t/Area[b(b/2 + r) + u(.363r)(u(b+1.637r) + c(b + 2r))] = 0.3186 in.

I_{yy} = **0.192 in⁴**

Gross X_{cg} = x' + t/2 = **0.347 in.**

Gross section modulus, S_{yy}:

S_{yy} = I_{yy} / X_{cg} = 0.192 / 0.347 = **0.553 in³**

Radius of gyration, r_y :

r_y = SQRT(I_{yy}/Area) = SQRT(0.92/0.667) = **0.536 in**

Torsional Properties:

COMMENTARY ON THE PRESCRIPTIVE METHOD FOR RESIDENTIAL
COLD-FORMED STEEL FRAMING

Calculate distance between centroid and shear center: (AISI Supplementary Information Section 1.2)

$X_o = -(x' + m)$
$m = (b't / 12I_{xx})(6c'a'^2 + 3b'a'^2 - 8c'^3)$

$a'=$	Web - thickness = 8 - 0.0566 = 7.9434"
$b'=$	Flange - thickness = 1.625 - 0.0566 = 1.5684"
$c'=$	Lip - thickness/2 = 0.5 - .0566/2 = 0.4717"
$m=$	0.616
$\mathbf{X_o}=$	$-(0.3186 + 0.616) = \mathbf{-0.935\ in.}$

St. Venant torsional constant, J: (AISI Supplementary Information Section 1.2)

$J = (t^3/3)(a + 2b + 4u + 2c)$
$\mathbf{J = 0.000713\ in^4.}$

Warping constant, C_w:

For calculation of C_w : (refer to AISI Supplementary Information Section 1.2, Section III, page 11)

$C_w = t^2(a')^4(c')^3(4a' + 3c') / 18I_x$
Where a' = Web - [t/2 + αt/2] = 8 - [0.0566/2 + 1(0.0566)/2) = 7.9434 in.
            c' = Lip - (t/2) = 0.5 - .0566/2 = 0.4717 in.

$C_w = (0.0566)^2 (7.9434)^4 (0.4717)^3 (4*7.9434 + 3*0.4717) / 18*5.7$

$\mathbf{C_w = 2.503\ in^6.}$

Polar radii of gyration, r_o:

The polar radii of gyration of a section is calculated about the centroidal principal axis.

$r_o=$	$SQRT(r_x^2 + r_y^2 + X_o^2)$
$r_o =$	$SQRT(2.93^2 + 0.536^2 + 0.935^2)$
$\mathbf{r_o=}$	**3.12 in.**

Torsional flexural constant, β: (AISI Supplementary Information Section 1.2)

$\beta=$	$1 - (X_o / R_o)^2$
$\beta =$	$1 - (-0.935/3.12)^2$
$\mathbf{\beta =}$	**0.91**

Effective Section Properties:

I_x and M_n are first calculated based on initiation of yielding, and then the effective moment and I_x are calculated based on Procedure I for deflection determination at the allowable moment.

Calculate I_{xx}:

Refer to Section B4.2 of the AISI Design Specification:

COMMENTARY ON THE PRESCRIPTIVE METHOD FOR RESIDENTIAL
COLD-FORMED STEEL FRAMING

Flange width/thickness = w/t = 1.2967/0.0566 = 22.91 < 60   OK
Lip width/thickness = 0.33585/0.0566 = 5.93 < 60   OK

$S = 1.28 * SQRT(E/f)$ Equation B4-1
Where E = 29,500,000 psi, and f = 37,200 psi (due to cold forming)

$S = 1.28 * SQRT(29500000/37200) = 36.031$
$S/3 = 12.01$

$36.031 > w/t = 22.91 > 12.01$ Section B4.2 (a)

Therefore, $I_a / t^4 = 399 [(w/t)/S - 0.330)]^3$ Equation B4.2-6
$I_a / t^4 = 11.415$ $I_a = 0.000117$ in^4

$n = 0.5$
The moment of inertia of the full edge stiffener (lip), I_s:
$d = D - (R + t) = 0.5 - (0.10754 + 0.0566) = 0.33586$ in.
$d/t = 0.33586/0.0566 = 5.93 < 14$ (max. d/t)
$I_s = td^3/12 = (0.33585)^3(0.0566) / 12 = 0.000179$ in^4

$I_s/ I_a = 0.000179 / 0.000117 = 1.525 > 1$ Use 1.0
$C_2 = 1.0$ Equation B4.2-7
$C_1 = 2 - 1 = 1$ Equation B4.2-8

$D/w = 0.5/1.2967 = 0.3856$

$0.8 > D/w = 0.3856 > 0.25$

$k = 4.82 - 5D/w(I_s/ I_a)^n + 0.43 =< 5.25 -5D/w$ Equation B4.2-9
$k = 3.322$

Use k = 3.322 to calculate λ the plate slenderness factor for the compression flange per section B2.1:

$\lambda = [1.052/(k)^{1/2}](w / t)(f / E)^{1/2} = 0.469759$ Equation B2.1-4

$\lambda < 0.673$, Therefore, **b = w = 1.29672"**   Equation 2.1-1

[Compression flange is fully effective]

The effective width of the edge stiffener can be computed in accordance with section B3.2 of the AISI Design Specification. Using k = 0.43, d/t = 5.93 and a conservative $f = F_y$, the slenderness factor is

$\lambda = [1.052/(0.43)^{1/2}](5.93)(f / E)^{1/2} = 0.338187$ Equation B2.1-4

$\lambda < 0.673$, Therefore, $d'_s = d_s$ Equation 2.1-1
Where d'_s is the effective width of the edge stiffener.

$d_s = d's(I_s/I_a) = $ **0.33586 in.**
Compression stiffener (lip) is fully effective.

COMMENTARY ON THE PRESCIPTIVE METHOD FOR RESIDENTIAL
COLD-FORMED STEEL FRAMING

Check if web is fully effective:

Locate neutral axis based on a full element. Assume the web is fully effective, and top fiber stress is F_{ya}:
(F_{ya} is calculated in a later section = 37.229 ksi)

Element	Effective Length (L) (in.)	Dist. From Top Fiber, y (in.)	L_y (in^2)	L_y^2 (in^3)	I_1' About Own Axis (in^4)
Web	7.6717	4.00	30.6869	122.7475	37.6268
Tension Flange	1.2967	7.9717	10.3371	82.4040	
Compression Flange	1.2967	.0283	.0367	.0010	
Upper Corners	.4265	.0776	.0331	.0026	
Lower Corners	.4265	7.9224	3.3792	26.7713	
Upper Stiffener	.3359	.3321	.1115	.0370	0.0032
Lower Stiffener	.3359	7.6679	2.5754	19.7476	0.0032
Sum	11.78996	28	47.15982	251.7111	37.63309

$y_{cg} = \Sigma (L_y) / \Sigma L = 47.15982 / 11.78996 = 4.00$ inches. (Distance from top fiber)

Since distance from of compression fiber from neutral axis is equal one half the beam depth, a compression stress of F_{ya} will govern as assumed (i.e. initial yield is in compression)

Use section B2.3 of the AISI Design Specification to check the effectiveness of the web element.

$f_1 = F_{ya} (y_{cg} - \text{thickness} - \text{radius}) / y_{cg} = 37,229(4 - 0.0566 - 0.1076) / 4 = 35,701.86$ psi (Compression)
$f_2 = - F_{ya} (\text{Web} - y_{cg} - \text{thickness} - \text{radius})/y_{cg} = -37,229 (8 - 4 - 0.0566 - 0.1076) / 4$
$f_2 = - 35,701.86$ psi (Tension)

$\psi = f_2/f_1 = -1$

$\psi < - 0.236$ Equation B2.3-1

$k = 4 + 2(1 - \psi)^3 + 2(1 - \psi) = 24$ Equation B2.3-4

$b_1 = b_e / (3 - \psi)$ Equation B2.3-1

$b_2 = b_e / 2$ Equation B2.3-2

Where b_e is calculated per section B2.1 of the AISI Design Specification with f_1 substituted for f and k as determined above:

$\lambda = [1.052/(24)^{1/2}](7.67172 / 0.0566)(35701.86 /29500000)^{1/2} = 1.01256$ Equation B2.1-4

$\lambda > 0.673$, Therefore, $b_e = p$ (web)   Equation B2.1-2

$p = (1 - 0.22/\lambda) / \lambda = (1 - 0.22/1.01256) / 1.01256 = 0.77302$

$b_e = 0.77302(7.67172) = 5.93039$ inches

$b_2 = 5.93039 / 2 = 2.9652$ inches.

$b_1 = 5.93039 / (3 + 1) = 1.4826"$

$b_1 + b_2 = 2.9652 + 1.4826 = 4.4478 "$

Compression portion of the web calculated on the basis of the effective section =
$\quad y_{cg} - (R + t) = 4.00 - (0.1076 + 0.0566) = 3.83586$ in.

Since $b_1 + b_2 = 4.4478" > 3.83586"$, the web element is fully effective as assumed.
$b_1 + b_2$ shall be taken as 3.83586"

$y_{cg} = 4.0"$

$I'_{(x)} = L_y^2 + I'_1 - Ly_{cg}^2 = 251.7111 + 37.63309 - 11.78996(4)^2 = 100.7049$ in^3

$I_x = I'_{(x)} * t = 100.7409 (0.0566) = 5.699895$ in^4

$S_x = I_x / y_{cg} = 5.6990895/4 = 1.424974$ in^3   (Effective section modulus)

$M_n = S_e F_y = 1.424974 (37229) = 53,051.17$ lb-in.

Safety factor = 1.67

$M_a = M_n / S.F. = 53051.17 / 1.67 = 31,767.17$ lb-in. (2,647.5 ft-lb)

<u>Procedure I for Deflection Determination at the Allowable Moment:</u>

Refer to section B2 of the AISI Design Specification.

This procedure is iterative: one assumes the actual compressive stress f under this allowable moment. Knowing f, one proceeds as usual to obtain S_e and checks to see if (f x S_e) is equal to M_a as it should. If not, reiterate until the desirable level of accuracy is obtained.

Since it has been determined that the compression flange is fully effective, the top compression stress should be equal to the one calculated before. For the first iteration, assume f = 35,702 psi, and the web is fully effective.

$S = 1.28 \text{SQRT}(29500000/35702) = 36.7938$
$S/3 = 12.2646 < w/t = 1.29672/0.0566 = 22.91 < S = 36.7938$ Section B4.2
$I_a/t^4 = 399 [(w/t)/S - 0.330)]^3$ Equation B4.2-6
$I_a = 399(0.0566)^4[(22.91/36.7938) - 0.33]^3$
$I_a = 0.00010$ in^4
$I_s = td^3/12 = (0.33585)^3(0.0566)/12 = 0.000179$ in^4

COMMENTARY ON THE PRESCRIPTIVE METHOD FOR RESIDENTIAL
COLD-FORMED STEEL FRAMING

$I_s/I_a = 0.000179/0.00010 = 1.79$

$D/w = 0.5/1.2967 = 0.3856$

$k = [4.82 - 5(0.3856)](1.79)^{.5} + 0.43 =< 5.25 - 5(0.3856)$ Equation B4.2-9

$k = 3.322$

Use $k = 3.322$ to calculate λ the plate slenderness factor for the compression flange per section B2.1:

$\lambda = [1.052/(3.322)^{1/2}](22.91)(35702/29500000E)^{1/2} = 0.46002$ Equation B2.1-4

$\lambda < 0.673$, Therefore, **b = w = 1.29672"**   Equation 2.1-1
Compression flange is fully effective

Again assume $f = 35,702$ psi as in top compression fiber. Calculate effective width of edge stiffener:

$\lambda = [1.052/(0.43)^{1/2}](5.93)(35,702/29500000E)^{1/2} = 0.33096$ Equation B2.1-4

$\lambda < 0.673$, Therefore, **$d'_s = 0.33586"$**   Equation 2.1-1

Since $I_s/I_a = 1.79 > 1.0$, it follows that $d_s = d'_s = 0.33586"$ (i.e. compression stiffener is fully effective)

Therefore the section is fully effective.

Using the full width of the compression flange and assuming the web is fully effective, the neutral axis is located at mid-depth (i.e. $y_{cg} - 4.0"$). Prior to computing the moment of inertia, check the web for effectiveness as follows:

$f_1 = [(4.0 - 0.10754 - 0.0566)/4.0](35,702) = 34,237$ psi (compression)
$f_2 = -[(4.0 - 0.10754 - 0.0566)/4.0](35,702) = -34,237$ psi (tension)

$\psi = f_2/f_1 = -1$

$\psi < -0.236$ Equation B2.3-1

$k = 4 + 2(1 - \psi)^3 + 2(1 - \psi) = 24$ Equation B2.3-4

$b_1 = b_e/(3 - \psi)$ Equation B2.3-1
$b_2 = b_e/2$ Equation B2.3-2

be is calculated per section B2.1 of the AISI Design Specification with f_1 is substituted for f and k as determined above:

$\lambda = [1.052/(24)^{1/2}](7.67172/0.0566)(34237/29500000)^{1/2} = 0.99157$ Equation B2.1-4

$\lambda > 0.673$, Therefore, be = p(web)   Equation B2.1-2

$p = (1 - 0.22/\lambda)/\lambda = (1 - 0.22/.99157)/.99157 = 0.78475$

$b_e = 0.78475(7.67172) = 6.02034$ inches

COMMENTARY ON THE PRESCRIPTIVE METHOD FOR RESIDENTIAL
COLD-FORMED STEEL FRAMING

$b_2 = 6.02034 / 2 = 3.01017$ inches.

$b_1 = 6.02034 / (3 + 1) = 1.50509''$

$b_1 + b_2 = 4.51526''$

Compression portion of the web calculated on the basis of the effective section =
$y_{cg} - (R + t) = 4.00 - (0.1076 + 0.0566) = 3.83586$ in.

Since $b_1 + b_2 = 4.51526'' > 3.83586''$, the web element is fully effective as assumed.

$b_1 + b_2$ shall be taken as 3.83586''

$y_{cg} = 4.0''$

Full section properties about X-axis

$y_{cg} = 8/2 = 4.0''$

Element	L (in.)	y Distance from centerline of section (in.)	L_y^2 (in³)	I'_1 About Own Axis (in³)
Web	7.67172	-----	-----	37.62677
Stiffeners	2x 0.33586=0.6717	3.67172	9.05554	
Corners	4x.213269=.85308	3.91793	13.09493	
Flanges	2x1.29672=2.59344	3.9717	40.90996	
Sum	11.78994		63.06043	37.62677

$I'_{(x)} = L_y^2 + I'_1 = 63.06043 + 37.62677 = 100.6872$ in³

$I_x = I'_{(x)} * t = 100.6872 (0.0566) = $ **5.6989 in⁴** (Effective moment of inertia)

$S_e = I_x / y_{cg} = 5.6989/4 = $ **1.4247 in³** (Effective section modulus)

$M_n = S_e f = 1.42472 (34,237) = 48,778.27$ lb-in.

Safety factor = 1.67
$M_a = M_n / S.F. = 48778.271 / 1.67 = $ 29,208.54 lb-in. (2,434.05 ft-lb) $< M_a = $ 2,647.26 ft-lb.

Need to do another iteration and also to increase f.

The final iteration will yield the following results:

$I_x = 5.699895$ in⁴

COMMENTARY ON THE PRESCRIPTIVE METHOD FOR RESIDENTIAL
COLD-FORMED STEEL FRAMING

S_e = 1.424974 in^3
M_n = 2,647.5 ft-lb.

Calculate Allowable Moment, M_y

M_y = $(F_y)(S_y)$ / 1.67 (where S_y is the effective section modulus)

$\mathbf{M_y}$ = 33,000 (0.136) / 1.67 = 2687.43 lb-in. = **223.95 ft-lb.**

Calculate Allowable Shear:

Refer to section C.2 of the AISI Design Specification, for unpunched web.

Compute the depth of the flat portion of the web (h)

h = h - 2(R + t) = 8 - 2(0.1076 + 0.0566) = 7.67172"
h/t = 7.67172/0.0566 = 135.5427 < 200   OK

Calculate 1.38 SQRT(E_v/F_y) = 1.38 SQRT(29500,000 x 5.34 / 37229) = 95.34628

h/t > 95.34628, Therefore, V_a = 0.53 $E_v(t)^3$ / h   Equation C.2-2

$\mathbf{V_a}$ = 0.53(29500000)(5.43)(0.0566)^3 / 8 = **1973.1 lbs.** (unpunched)

This value matches the one tabulated at the begining of this section.

Calculate Allowable Web Crippling Strength:

Refer to section C.4 of the AISI Design Specification, for unpunched web.

The equations given in Table C.4-1 of the AISI Design Specification are applicable to sections with   h/t < 200 and R/t <= 6. For the 2 x 8 x 54 member, both provisions are met.

The length of bearing (N) is assumed to be a minimum of 1.5".

Use equation C.4-1 for single webs, stiffened flanges:

P_a = $t^2 K C_3 C_4 C_\theta$ [179 - 0.33(h/t)][1 + 0.01(N/t)]   Equation C.4-1

K = F_y/33 = 33/33 = 1.00   (Note F_{ya} can not be used for this section)   Equation C.4-21
C_3 = 1.33 - .33(1.00) = 1.00   Equation C.4-12
C_4 = 1.15 - 0.15R/t = 0.8648   Equation C3.4-13
C_6 = 1 + (h/t) / 750 when h/t <= 150 = 1.180724   Equation C.4-13
C_θ = 0.7 + 0.3$(\theta/90)^2$ = 1.0   Equation C.4-20

P_a = (0.0566)^2 (1.00)(1.00)(0.8648)(1.0)[179 - 0.33(95.34628)][1 + 0.01(1.5/0.0566)]

$\mathbf{P_a}$ = **0.471 kips = 471 lbs.**

Calculate the web crippling capacity for a multiple web, stiffened flanges section using equation C.4-3:

P_a = (0.0566)^2 (33,000)(1.180724)[5.0 + 0.63(1.5/0.0566)$^{0.5}$]

COMMENTARY ON THE PRESCRIPTIVE METHOD FOR RESIDENTIAL
COLD-FORMED STEEL FRAMING

$P_a = 1029$ lbs.

Calculate the yield strength due to cold working, F_{ya}:

In order to use equation A5.2.2-3 of the AISI Design Specification, the channel must have a compact compression flange, that is $p = 1$. For this section, assume the reduction factor p is 1.

$$F_{ya} = CF_{yc} + (1 - C)F_{yf} \qquad\qquad \text{Equation A5.2.2-1}$$

Limitation of this equation: $F_u/F_y \geq\geq 1.2$
                                $R/t \quad \leq 7$
                                $\theta \qquad \leq 120$

$F_y = 33$ ksi, $F_u = 45$ ksi, $t = .0566"$, $R = 0.108"$, $\theta = 90$ degrees,
$F_{yf} = 33$ ksi (Virgin yield point)

$F_u/F_y = 45/33 = 1.3636 \geq 1.2$ OK
$R/t = 0.108/0.0566 = 1.908 \leq 7$ OK
$\theta = 90 \leq 120$

$F_{yc} = [(B_c)(F_y)/(R/t)^m]$ Equation A5.2.2-2
$m = 0.192(F_u/F_y)$ Equation A5.2.2-4

$m = 0.192(1.3636) = 0.2618$

$B_c = 3.69(F_u/F_y) - 0.819((F_u/F_y)^2 - 1.79$ Equation A5.2.2-3
$B_c = 3.69(1.3636) - 0.819(1.3636)^2 - 1.79 = 1.7188$

$C = $ (total cross sectional area of two corners) / (Full cross sectional area of flange)

Flange width $= 1.625 - 2(R + t) = 1.625 - 2(0.108 + 0.0566) = 1.2958"$

Arc length of radius $= \qquad\qquad 1.67 R'$ (Where $R' = R + t/2$)
       $1.67(0.108 + .0566/2) = 0.2276$

$C = (2 \times 0.2276) / (2 \times 0.2276 + 1.2958) = 0.26$

$F_{yc} = [(1.7188)(33)/(1.908)^{.2618} = 47.89$ ksi

$F_{ya} = 0.26(47.89) + (1 - .26)(33) = 37.2$ ksi

Calculate the allowable shear value for the 2 x 8 x 54 mil member with punched web, per the ICBO AC46 method. Punchouts are standard 1.5" wide x 4" long, located along the centerline of the web.

The allowable shear capacity of a 2 x 8 x 54 member was calculated above as 1973.1 lbs. for a section with no punchouts.

Calculate reduction factor: q_s
$q_s = 1 - 1.1 (a/d) = 1 - 1.1(1.5 / 8) = 0.79375$
       where $a = $ web size $= 3.5"$ and $d = $ width of punchout $= 1.5"$

COMMENTARY ON THE PRESCRIPTIVE METHOD FOR RESIDENTIAL
COLD-FORMED STEEL FRAMING

Allowable shear V_a $= 1973.1 * q_s = 1973.1 (0.79375) =$ **1566.3 lbs**.   (Punched)
This value matches the one tabulated at the beginning of this section.

C3.0 FIELD IDENTIFICATION SYSTEM

The identification system specified in the *Prescriptive Method* is a minimum requirement. Additional information may be added to the stamp, logo, stencil, or embossment. The requirements in this section closely follow what is currently proposed by the industry. The identification requirements are necessary for building code enforcement, construction coordination, and quality control.

C4.0 FOUNDATION

The *Prescriptive Method* does not provide prescriptive requirements for foundation construction. It references the model building codes for the design and installation of the foundation. The designer needs to ensure that the foundation is designed to support all superimposed loads, and complies with applicable model building codes. The *Prescriptive Method* provides prescriptive requirements for the connection for steel framing to the foundation. These requirements are provided in Sections 5 and 6.

Anchor bolts are installed in concrete in accordance with current building code requirements.

C5.0 FLOORS

C5.1 Floor Construction

The *Prescriptive Method* shows a typical floor construction layout with all the different elements of the floor identified. Floor construction in the *Prescriptive Method* is limited to design wind speeds less than or equal to 110 mph (fastest-mile) and to Seismic Zones 0, 1, 2, 3, and 4. However, as mentioned in the applicability limit table of the *Prescriptive Method*, an approved design will be required for hold-down requirements, and/or uplift straps for steel floors in buildings subjected to wind speeds greater than 90 mph, Exposure C, and buildings located in Seismic Zones 3 or higher. This limitation is a result of the additional design required for connections of the floor to the foundation or bearing walls.

C5.2 Floor to Foundation or Bearing Wall Connection:

The *Prescriptive Method* provides two different details to connect floor joists to foundation or bearing walls. The details are self explanatory and reflect a selection of what is currently being used in residential steel framing.

C5.3 Allowable Joist Spans

The *Prescriptive Method* provides tables that show the maximum allowable floor joist spans for two live load conditions: 30 psf and 40 psf. The spans shown in the *Prescriptive Method* assume bearing stiffeners are installed at each bearing point. Stiffener requirements are also provided in the *Prescriptive Method*. The floor joist spans are calculated based on the loading conditions, deflection criteria, and engineering method described below.

Loading conditions:

COMMENTARY ON THE PRESCRIPTIVE METHOD FOR RESIDENTIAL COLD-FORMED STEEL FRAMING

In the design of floor joists, any one of several engineering criteria may control the prescriptive requirements depending on the configuration of the section, thickness of material, and member length. The analysis must include checks for:

* Yielding
« Flexural buckling
« Web crippling
* Shear
* Deflection
           * Combined bending and shear (for multiple spans)

The engineering approach used in the *Prescriptive Method* carefully considers floor joists dead and live load combinations as they apply to the design of joists. The objective was to maintain a simple table format without sacrificing the economies possible through case-by-case design.

Floor joists were designed using conservative engineering assumptions which neglect the benefits of composite action of floor assemblies. The composite strength of typical residential floor assemblies may provide higher resistance to bending and deflection. Web crippling check, and combined bending and web crippling check are not required because web stiffeners are specified for all bearing points.

The following applied loads were used in developing the joist tables:

Dead Load:	Floor Dead Load	= 10 psf
Live Loads:	Sleeping Quarters	= 30 psf
	Other Rooms	= 40 psf

These loads are widely accepted by the engineering community (i.e. ASCE 7-93), and are specified in the major building codes: CABO, SBCCI, UBC, and BOCA (UBC does not permit the 30 psf live load).
The section properties, moment and shear capacities were determined in accordance with the AISI Design Specification requirement, as shown previously.

The allowable joist spans were based on the minimum spans resulting from the following equations based on a simply supported span:

Bending: $S_b = [(F_b \times S_x \times 8000)/(w \times spacing)]^{1/2}$

Deflection $S_d = [(188800000 \times I_x)/(w \times spacing \times Deflection\ limit)]^{1/3}$

Web Crippling $S_w = (P_{allow} \times 24)/(w \times spacing)$

Shear   Does not control

Where: S = Single span, feet
                F_b = Allowable bending (compressive) stress = 33/1.67 = 19.76 ksi
                I_x = Moment of inertia, inches^4.
w = Load per square foot, psf
S_x = Section modulus, inches^3.
Spacing = Spacing of joists, inches.
Deflection limit = Allowable deflection limit (S/480 for live loads and S/240 for total load)
P_{allow} = Allowable end reaction for 1.5" bearing length, lbs.

COMMENTARY ON THE PRESCRIPTIVE METHOD FOR RESIDENTIAL
COLD-FORMED STEEL FRAMING

All joists are considered punched with 1-1/2" wide x 4" long hole. The compression flange of the floor joists are also assumed to be continually braced by the subflooring, thus providing lateral restraint for the top flanges.

The provisions for the combined bending and shear were taken directly from the AISI Design Specification.

The joist span tables are calculated based on deflection limit of L/480 for live load and L/240 for total loads, where L is the span length. Building codes vary in requirements for deflection criteria. Also, deflection requirements which have worked well in the past for one type of material or method have created problems when applied to different members. One particular serviceability problem is related to floor vibrations and many practitioners and standards use more stringent deflection criteria than the L/360 typically required for residential floors.

Deflection limits are primarily established with regard to serviceability concerns. The intent is to prevent excessive deflection which might result in cracking of finishes. The defection criteria also affects the "feel" of the building in terms of rigidity and vibratory response to normal occupant loads. For materials like steel framing which have high material strength, longer spans are possible with members of lower apparent stiffness (i.e. E x I). In such cases, typical deflection criteria may not be appropriate. For example, industry experience indicates that an L/360 deflection limit often results in these floors being perceived to be "bouncy" by occupants. To the normal home-owner this condition may be misconstrued as a sign of weakness.

While a deflection-to-span- ratio of L/360 may be adequate under static loading, it is suggested that a significantly tighter deflection-to-span ratio under the full design live load only may be appropriate to ensure adequate performance. The deflection-to-span ratio has always been a controversial issue when it is used to assess human comfort. A higher deflection limit is usually recommended to overcome the human concern about occupant induced floor vibrations. To ensure dynamic performance, the Australian code (AISC) for example, considers a deflection-to-span ratio of L/750 (under full live loading) to be appropriate in the absence of full dynamic analysis. Furthermore, the Australian (AISC) "Floor Vibration in Buildings, Design Methods" (ref. 11.13) provides a criterion to determine what is an acceptable "house" system, based on critical damping. This method is not generic and requires the calculation of the first natural frequency and percent damping of the floor which depend on the physical dimensions, stiffness, and attachments of the house. Due to this fact, a more simplistic approach would be to tighten the deflection criteria. A floor deflection-to-span ratio of L/480 typically results in an increase in the percent of critical damping as suggested by the AISC, and thus ensures that vibration does not exceed a tolerable level. Many engineers apply a L/480 deflection criteria when designing steel floor joists. Additional research is planned to develop a reasonable design criteria to more efficiently address vibration concerns.

Multiple Spans:

Multiple spans are commonly used in the residential steel industry. A few concerns are encountered when using multiple spans. The first concern is the magnitude of the reaction at the middle support which is greater than the end reactions. This may create a web crippling failure at the middle support. This is resolved by requiring web stiffeners at all bearing points. The second concern is the presence of negative moments at the middle reaction region, causing the compression flanges to be at the bottom rather than the top. If left unbraced, this would cause lateral instability and may cause premature failure of the joists under extreme loading conditions. Furthermore, shear and bending interaction are checked for multiple spans due to the presence of high shear and bending stresses at the middle reactions, creating greater susceptibility of web buckling.

COMMENTARY ON THE PRESCRIPTIVE METHOD FOR RESIDENTIAL
COLD-FORMED STEEL FRAMING

Bottom flange bracing requirements at interior supports of multiple spans consider the inherent lateral support provided by ceiling finishes (when present) and connection to the bearing element at the relatively small region of negative moment. Benefits from composite action with the floor diaphragm and connection to the bearing support are also important considerations that were not utilized in the development of the *Prescriptive Method*. The objective is to minimize unnecessary redundancy in bracing requirements.

C5.4 Joist Bracing

Steel floors have long been designed by considering the joists to act as simple beams acting independently without consideration of composite action from other materials composing the floor. For typical residential floors, it has been assumed that the floor sheathing's only function is to transfer the loads to the joists, and to provide continuous lateral bracing to the compression flanges. These assumptions neglect many factors that affect the strength and stiffness of a floor. Testing has indicated that the cold formed steel joist floors with nailed plywood decks should be considered as one-way noncomposite slabs (ref. 11.8). Therefore, using a single joist for strength calculation agrees with the tested behavior.

Bracing of the bottom flanges of joists as specified in the *Prescriptive Method* is based on industry practice and engineering judgement. Steel strapping as well as finished ceilings are considered to be adequate bracing for the tension flanges. It is necessary, however, for steel strapping to have blocking (or bridging) installed at a maximum spacing of 12 feet and at the termination of all straps. Moreover, the ends of steel straps are required to be fastened to a stable component of the building or blocking.

C5.5 Floor Cantilevers

Cantilevered floor joists are often used in the construction of houses. In many cases, cantilevers are supporting load bearing walls which create special loading conditions that complicate engineering analysis. Floor cantilevers are limited to a maximum of 24" for floors supporting one wall and roof only (one story). This limitation is imposed to minimize the impact of the added load on the floor joists. To fully utilize the section, punchouts are not permitted in cantilevered sections.

C5.6 Splicing

Splicing of structural members is not a typical application, but some situations may arise where splicing requirements would be useful. Applications may include repair of damaged joists, simplified details for dropped floors, and others.

Splices, generally, are required to transfer shear, bending moments, and axial loads. Some splices may occur over points of bearing, and splicing may only be required to transmit nominal axial loads. The Floor joist spans provided in the *Prescriptive Method* are based on the assumption that the joists are continuous, with no splices (except for punchouts and hole stiffeners). Therefore, splicing of joist members requires an approved design except lapped joists at bearing points. A detail is shown in the *Prescriptive Method* illustrating the splicing of non-structural members such as joist tracks.

C5.7 Framing of Openings

Openings in floors are needed for several reasons, such as to install stairs. The *Prescriptive Method* limits the width of the floor opening to 8 feet and provides a provision for reinforcing the members around floor openings. All members around floor openings (i.e. header and trimmer joists) are required to be box-type members made by nesting a joist into a joist track and fastening them together along the flanges. These members are required to be the same size and thickness (or thicker) than the floor joists. Each header joist is required to be connected to the trimmer joist with a clip angle on either side of each connection. The clip angle is of a thickness equivalent to the floor joists.

C5.8 Floor Trusses

This section is included so that engineered trusses can be used in conjunction with this document.

COMMENTARY ON THE PRESCRIPTIVE METHOD FOR RESIDENTIAL
COLD-FORMED STEEL FRAMING

C5.9 Design Example

The following is an example for calculating the maximum allowable single span for a 2 x 8 x 54 mils floor joist (punched C-section):

The section properties, and allowable moments are taken from previous calculations.

F_y = 33 ksi (F_b = 33/1.67 = 19.76 ksi)
S_x = 1.425 in^3 Spacing= 24" oc
I_x = 5.7 in^4 Dead Load = 10 psf
M_a= 2,647.5 ft-lb.   Live Load = 30 psf
Deflection criteria: L/480 for live load (L/240 for total load)
Allowable web crippling load = P_a = 471 lbs.
Allowable web crippling load, with stiffener = Pa = 1,029 x 2 = 2,058 lbs
Allowable shear for punches section = V_a = 1566.3 lbs.

Maximum allowable span due to bending = S_b = $[(19.76 \times 1.425 \times 8000)/(40 \times 24)]^{1/2}$ = 15'-4"

Maximum allowable span due to Live load deflection = S_{d1} = $[(188800000 \times 5.7)/(30 \times 24 \times 480)]^{1/3}$
= 14'-7"

Maximum allowable span due to Dead load deflection = S_{d2} = $[(188800000 \times 5.7)/(40 \times 24 \times 240)]^{1/3}$
= 16'-9"

Maximum allowable span due to web crippling load = S_w = (471 x 24)/(40 x 24) = 11'-9"

Maximum allowable span due to shear = S_s = (1566.3 x 24)/(40 x 24) = 39'-2"

Based on the above calculation, the span due to web crippling load (11'-9") controls for unstiffened webs. For stiffened webs, the span due to live load deflection controls (14'-7").

Check the combined bending and web crippling interaction equation per AISI Specification:

$1.2(P/P_a) + (M/M_{axo}) < 1.5$ Equation C3.5-1

Calculate P and M from the maximum allowable span above (14'-7")

P = (40 x 2)(14.58)/2 = 583.2 lbs.

M = $wl^2/8$ = (40 x 2)(14.58)^2/2 = 2125.76 ft-lbs.

1.2(583/2058) + (2125.76/2647.5) < 1.14 < 1.5   OK

COMMENTARY ON THE PRESCRIPTIVE METHOD FOR RESIDENTIAL
COLD-FORMED STEEL FRAMING

C6.0 STRUCTURAL WALLS

C6.1 Wall Construction

The *Prescriptive Method* is applicable to wall studs in buildings within the applicability limits of Table 1.0 of the *Prescriptive Method*. The maximum design conditions are Seismic Zone 4, ground snow load of 70 psf, and design wind speed of 110 mph fastest-mile, Exposure C. However, as mentioned in the applicability limit table of the *Prescriptive Method*, approved design will be required for wall bracing (shearwall), hold-down requirements, and/or uplift straps for steel walls in buildings subjected to wind speeds greater than 90 mph, Exposure C, and buildings located in Seismic Zones 3 or higher.

C6.2 Wall to Foundation or Floor Connection

The *Prescriptive Method* provides a commonly used detail to connect wall studs to foundation or bearing walls. The detail is self explanatory and reflects what is currently being used in residential steel framing.

C6.3 Load Bearing Walls

The *Prescriptive Method* provides the minimum required thicknesses of steel studs for different wind speeds, exposure categories, wall heights, building widths, and ground snow loads. Stud selection tables are limited to one- and two-story buildings with load bearing wall heights up to 10 feet.

The 8-foot walls are widely used in residential construction, however, steel framed homes often take advantage of higher ceilings such as 9- and 10-foot walls.

The wall studs are grouped in two categories:

- Studs for one story or second floor of two story building (supporting roof only)
- Studs for first story of a two story building (supporting roof + one floor)

All studs in exterior walls are treated as load bearing members.

Loading:

In the design of compression members, such as wall studs, any one of several engineering criteria may control the prescriptive requirements depending on the configuration of the section, thickness of material, and column length. The analysis must include checks for:

- Yielding,
- Overall column buckling,
- Flexural buckling,
- Torsional buckling,
- Torsional-flexural buckling, and
- Local buckling of an individual member.

The engineering approach used in the *Prescriptive Method* considers wind, snow, and live load combinations as they apply to the design of studs. The objective was to maintain a simple table format without sacrificing the economies possible through case-by-case design.

COMMENTARY ON THE PRESCRIPTIVE METHOD FOR RESIDENTIAL COLD-FORMED STEEL FRAMING

Wall studs were designed using conservative engineering assumptions which neglect the benefits of system effects of wall assemblies. The composite strength of typical residential wall assemblies provides higher resistance to out-of-plane bending and axial load conditions.

The following applied loads were used in developing the wall stud tables:

Dead Loads:	Floor Dead Load = 10 psf
	Wall Dead Load = 10 psf (of surface area)
	Ceiling Dead Load = 5 psf
	Roof Dead Load = 7 psf

Live Loads:	Second Floor Live Load = 30 psf
	First Floor Live Load = 40 psf
	Roof Live Load = Greater of 16 psf or Snow Load.

The ASCE 7-93 ground snow map was divided into four regions as follows:

Region	Ground snow load
1	0 to 20 psf
2	21 to 30 psf
3	31 to 50 psf
4	51 to 70 psf

Applied roof snow loads are calculated using the ASCE 7-93 criteria by multiplying the ground snow load by 0.7 (no further reductions were made for special cases). All roofs are assumed to have slopes from 3:12 to 12:12.

Ground Snow Load (psf)	20	30	50	70
Applied Roof Snow Load (psf)	16	21	35	49

The sloped roof snow load, $P_s = C_s \times P_f$, where P_f is the flat roof snow load.

$$P_f = 0.7\, C_e\, C_t\, I\, P_g$$

I is the importance factor depending on building classification. Houses are typically class I structures, with an importance factor of 1.0.

P_g = Ground snow load from the ASCE 7-93 estimated ground snow map (psf). This map is also included in all major building codes.

C_e is the exposure factor depending on the location of the house. C_e varies from 0.8 for windy, unsheltered areas, to 1.2 for heavily sheltered areas. A factor of 1.0 is deemed reasonable for houses.

C_t is a thermal factor that varies from 1.0 for heated structures to 1.2 for unheated structures. The thermal factor should be used based on the thermal conditions which is likely to exist during the life of the structure. Houses are typically considered heated structures with $C_t = 1.0$. Although it is possible that a brief interruption of power will cause temporary cooling of a heated house, the joint probability of this with a peak snow load is highly unlikely. Houses that are unoccupied during cold seasons, may experience a higher thermal factor, however, for unoccupied buildings the importance factor drops to 0.8, thus reducing the design loads by 20%, which offsets the 20% increase in the thermal factor.

COMMENTARY ON THE PRESCRIPTIVE METHOD FOR RESIDENTIAL
COLD-FORMED STEEL FRAMING

C_s is the roof slope factor ranging from approximately 0.1 to 1.0. For warm roofs (i.e. house roofs) the C_s curve is slightly smoother than that for cold roofs. Roofs with slopes up to 6:12 have a slope factor of 1.0, while roofs with slopes greater than 7:12 have a slope factor between 0.4 to 1.0. A slope factor of 1.0 is judged to be conservative for houses with roof slopes from 3:12 to 12:12.

Unbalanced snow loads, sliding snow loads, and snow drifts on lower roofs were not considered due to the lack of evidence for damage from unbalanced loads on homes and the lack of data to typify the statistical uncertainties associated with this load pattern on residential structures. Rain-on-snow surcharge load was also not considered in the calculations. Roof slopes in the *Prescriptive Method* exceed the 1/2" per foot requirement by ASCE 7-93 for the added load to be considered.

Therefore, the snow load can be computed as: $1.0 * 0.7 * 1.0 * 1.0 * P_g = 0.7\ P_g$

Wind loads are also calculated using the ASCE 7-93 criteria. The stud tables are based on variable wind speeds ranging from 70 mph to 110 mph, for Exposures B and C. The studs were designed using the component and cladding coefficients given in ASCE 7-93. Wind loads are calculated as follows:

$q = 0.00256\ K_z\ (GC_p + GC_{pi})\ (VI)^2$
  Where:

K_z	= 0.87 for exposure C at 20 ft.
GC_p	= ± 1.5 (Component and cladding, wall interior, Zone 4)
GC_{pi}	= ± 0.25 (Internal pressure, enclosed building)
I	= 1.0 (Residential building in 90 mph wind speed or less)
	= 1.05 (Residential buildings in 100 mph wind speed or greater)

Exposure B values = Exposure C values x .85 (per ASCE 7-95)

Applying the above equation, the design wind loads used in generating the stud tables are as follows:

Basic Wind Speed (mph)	70	80	90	100	110
Wind Load (psf) Exposure C	20	25	32	43	52
Wind Load (psf) Exposure B	15	20	25	32	43

Several load combinations were examined in the process of developing the stud tables. The load combination requirements of ASCE 7-93, BOCA, SBCCI, and the UBC were compared and the UBC load combination was chosen to be the most applicable to one-and two-family dwellings.

For single-story homes or the upper story of two-story homes, the design load is the higher of:

i   Dead load of roof plus uniform live load on roof

ii   Dead load of roof plus Wind load on roof and wall

iii   (Dead load of roof + Roof live load + 1/2 Wind load)

iv   (Dead load of roof + 1/2 Roof live load + Wind load)

For the lower story of two-story homes, the design load is the higher of:

i   Dead load of roof, upper story walls, upper story floor plus roof and upper story floor live load

COMMENTARY ON THE PRESCRIPTIVE METHOD FOR RESIDENTIAL COLD-FORMED STEEL FRAMING

---

ii   Dead load of roof, upper story walls, upper story floors plus wind load on roof and wall

iii   [Roof & upper story floor and wall dead loads + Roof, upper story floor, and wall live loads +1/2 Wind load]

iv   [Roof & upper story floor and wall dead loads + Upper story floor and wall live loadsWind load + 1/2 Snow load]

The following loading conditions were found to control the design:

One Story house or the second floor of a two story house, the maximum of the following two conditions:

Total Load = (RDL + SL + 1/2 WL) or
Total Load = (RDL + 1/2 SL + WL)

First floor of a two story house, the maximum of the following two conditions:

Total Load = TL = (RDL + SL + 1/2 WL + DLW + DLF + LLF) or
Total Load = TL = (RDL + 1/2 SL + WL + DLW + DLF + LLF)

Where: RDL = Roof + ceiling dead load
   SL   = Snow load
   WL   = Wind load
   TL   = Total load
   DLF = Dead load of the second floor
   LLF = Live load of the second floor
   DLW = Dead load of the second floor walls

<u>Design Assumptions</u>:

+   Simply supported beam.

+   Bracing of the interior exterior flanges of the studs (Sheathing or bracing).

+   Maximum of 24" for roof overhangs.

+   Roof slopes 3:12 to 12:12

+   Deflection Criteria: L/240 (total loads)

+   Allowable stress increase: 1.33

Deflection limits are primarily established with regard to serviceability concerns. The intent is to prevent excessive deflection which might result in cracking of finishes. For walls, most codes generally agree that L/240 represents an acceptable serviceability limit for deflection. For walls with flexible finishes, L/180 is often used. SBCCI allows a more liberal deflection limit for steel siding and roofing construction.

C6.4   Stud Bracing (Buckling Resistance)

Studs in load bearing walls are laterally braced on each flange by either a continuous 1-1/2" x 33 mil (min.) strap at mid-height or by direct attachment of structural sheathing or rigid wall finishes (i.e. APA rated plywood or OSB, gypsum board) according to the requirements of the *Prescriptive Method*. Therefore, all studs were assumed to be braced at mid-height for engineering analysis of the stud tables. The benefit of structurally sheathed walls on the required stud thickness and the composite wall strength are recognized in the footnotes to the stud tables.

Temporary bracing is necessary to facilitate safe construction practices and to ensure that the structural integrity of the wall assembly is maintained. Prior to the installation of cladding, a wall stud is free to twist, thus making the stud subject to premature failure under construction loads (i.e. stack of gypsum wallboard or roof shingles). In this case, all buckling modes must be considered in design or adequate temporary bracing needs to be applied.

C6.5   Splicing

Splicing of structural members is not a typical application, but some situations may arise where splicing requirements would be useful. Applications may include repair of damaged member or simplified details.

Splices, generally, are required to transfer shear, bending moments, and axial loads. The stud tables provided in the *Prescriptive Method* are based on the assumption that the studs are continuous, with no splices. Therefore, load bearing steel studs shall not be spliced without an approved design. Non-structural members such as wall tracks are spliced according to the requirements and details in the *Prescriptive Method*.

C6.6   Corner Framing

The *Prescriptive Method* utilizes a typical industry detail for framing corners.

C6.7   Headers

Headers are members used to transfer loads around openings in load bearing walls. Headers specified in the *Prescriptive Method* are allowed only on top of the opening adjacent to the wall top track (high headers). Headers are formed from two, equal sized C-sections in a back-to-back or box type configuration. Cripple studs, installed below headers or window sills are required to be of the same size and thickness as the adjacent full height wall studs. Tracks used to frame around openings are also required to be of an equivalent thickness as the wall studs.

Headers were divided into two categories: headers supporting roof and ceiling only (i.e. one story house or upper story of two story house); and headers supporting one story, roof and ceiling (i.e. bottom floor of a two story house).

Header tables were generated for four different ground snow load conditions, similar to the wall studs: 70 psf, 50 psf, 30 psf, and 20 psf. The ground snow load is transformed into a roof load in accordance with the ASCE 7-93 method as previously described in the wall stud section. The following roof snow loads are, thus, used:

Ground Snow Load (psf)	20	30	50	70
Applied Snow Load (psf)	16	21	35	49

The following loads, loading conditions, and assumptions were used in generating the header tables:

Headers supporting roof and ceiling only:

Roof dead load   = 12 psf

COMMENTARY ON THE PRESCRIPTIVE METHOD FOR RESIDENTIAL
COLD-FORMED STEEL FRAMING

Roof live load = Max. of snow load or roof live load (16 psf min.)
Attic live load = 10 psf
Ceiling dead load = 5 psf

Vertical load acting on a header supporting one floor only = Roof dead load + roof live load + ceiling dead load + attic live load

Headers supporting one floor, roof and ceiling:

Roof dead load = 12 psf
Roof live load = Max. of snow load or roof live load (16 psf min.)
Attic live load = 10 psf
Floor dead load = 10 psf
Floor live load = 30 psf
Wall dead load = 10 psf
Ceiling dead load = 5 psf

Vertical load acting on a header supporting one floor and a roof = Roof dead load + roof live load + ceiling dead load + attic live load + second floor live and dead load+ second floor wall dead load.

The number of king and jack studs (located adjacent to a wall opening) are calculated as follows:

a. Calculate the actual number of full height wall studs for a wall with no opening.

b. Provide an equivalent number of studs, to the one calculated above, at each end of the header.

Special detailing to accommodate lateral wind loads on window openings is provided in the framing details of the *Prescriptive Method.*

6.8 Wall Bracing

The wall bracing requirements in the *Prescriptive Method* are limited to wind loads less than 90 C (or 100 B) and seismic Zones 0, 1, and 2 (excluding Zones 3 and 4). These limits were established because of the additional concerns for building geometry constraints and design issues such as hold-down requirements for shearwalls. Homes built in conditions that exceed these loading limits must have the wall bracing engineered by a design professional.

The development of wall bracing requirements for the *Prescriptive Method* involved several activities including:

+ a review of applicable wall bracing tests and data;

+ development of the engineering approach to apply loads and determine lateral resistance;

+ selection of wall bracing methods and materials for the *Prescriptive Method*;

+ determination of appropriate safety factors and allowable capacities;

COMMENTARY ON THE PRESCRIPTIVE METHOD FOR RESIDENTIAL
COLD-FORMED STEEL FRAMING

+   sensitivity studies of the effects of lateral loads (wind and seismic) relative to building geometry (i.e. roof slope, width, length, and wall height);

+   comparison of earthquake induced lateral loads to those from wind based on the applicability limits for the *Prescriptive Method* (i.e. dead loads, geometry, etc.);

+   sensitivity (parametric) studies to guide simplification of the prescriptive requirements; and,

+   testing of wall bracing requirements proposed for the *Prescriptive Method*.

Wall bracing in cold-formed steel structures has typically relied on case-by-case design of diagonal steel strapping, its anchorage, and associated details. Limited shearwall testing of steel-framed walls using monotonic and cyclic loads has recently provided data for applications using structural sheathing (15/32" plywood or 7/16" OSB) as the bracing method.

In wood-frame homes, structural sheathing (eg. plywood or OSB) is a common bracing method, particularly in conditions of high lateral loads. However, many conventionally built homes in normal load conditions use wood let-in braces (diagonal 1x4s) or steel straps for racking resistance. Alternatively, 4'x8' bracing panels of structural sheathing may be substituted for the let-in braces forming a "partially" or "intermittently" sheathed wall. While these methods are non-engineered and have been established by experience, they have performed well in regions not subject to high winds. For one- and two-story detached homes with simple geometries and limited wall openings, these conventional bracing practices have also performed satisfactorily in past seismic events. The trend in high seismic and high wind regions is toward engineered designs using structural sheathing and numerous hold-down brackets.

The wall bracing requirements in the *Prescriptive Method* are based on an approach that considers both the available technical knowledge and a conservative understanding of experience with similar wall bracing practices currently utilized in the building codes for conventionally-built, one- and two-family dwellings.

Two wall bracing methods were investigated based on engineering test data or rational analysis to provide wall bracing requirements for the *Prescriptive Method*. These methods are diagonal steel straps and structural sheathing. Verification tests were performed on proposed diagonal steel strap requirements. These tests indicated that additional testing or design work was needed to develop simple diagonal strap bracing requirements that may be used efficiently in the *Prescriptive Method*. This concern relates to the limited capability of the "non-structural" top and bottom wall tracks to adequately resist compressive or bending loads that result when strap bracing or non-continuous sheathing ("partially sheathed" walls) are used for lateral support. For this reason, the wall bracing in the *Prescriptive Method* was conservatively limited to the use of continuously sheathed walls with the previously described limitations on loading conditions and building geometry. Even so, judgements in the application of the shearwall requirements must be exercised for specific cases, as is currently the situation in conventionally-built homes.

C6.8.1   Strap Bracing

Diagonal steel strap braces or 'X-braces' must be designed in accordance with approved engineering practices for the reasons mentioned previously. Simple and efficient prescriptive strap bracing requirements can be developed through testing and engineering analysis.

C6.8.2   Structural Sheathing

The allowable shear capacities for plywood sheathing and oriented strand board sheathing are based on recent racking tests performed at the University of Santa Clara for the American Iron and Steel Institute. The test set-up was similar to that required by the ASTM E564 standard for static load tests of framed shearwalls. A cyclic

COMMENTARY ON THE PRESCRIPTIVE METHOD FOR RESIDENTIAL COLD-FORMED STEEL FRAMING

loading protocol was also performed on the test specimens. The monotonic load test results relevant to this document are summarized as follows:

Description of Wall	Avg. Ultimate Capacity (plf)	Avg. Load at 1/2" Defl. (plf)	Allowable Load Capacity (plf)
15/32" Plywood APA rated sheathing w/ panels on one side	1063	508	425
7/16" OSB APA rated sheathing w/ panels on one side	911	593	364

Notes:
1. Framing at 24"oc
2. Studs and track are 3-1/2 in. 33 mils. -- 33ksi steel
3. Studs flange: 1-5/8 in.; stud lip: 3/8 in.; track flange: 1-1/4 in.
4. Framing screws -- No. 8 x 5/8 in. wafer head self-drilling tapping screws;
   Sheathing screws -- No. 8 x 1 in. flat head (coarse thread)
   @ spacing of 6" edge / 12" field of panels
5. A safety factor of 2.5 is used to determine allowable load capacity			

An allowable shear capacity of 364 plf was used for the determination of wall bracing requirements. The allowable capacity is less than the load recorded at 1/2" deflection of the wall test specimens.

For the determination of lateral forces, wind loads were calculated for the various building surfaces using the provisions of ASCE 7-93. A load combination of D + 0.8W was used. The 0.8 factor was used to account for uncertainties in wind loads associated with wind direction, building siting, and correlation of windward and leeward maximum surface pressure coefficients in ASCE 7-93. Next, standard tributary areas comprised of the leeward and windward building surfaces, were assigned to each exterior shearwall (i.e. sidewalls and endwalls) to determine the lateral, in-plane shear loads to be resisted by these walls, depending on the two worst-case orthogonal wind directions. No interior walls or alternate shear pathways were considered in the building. A computer spreadsheet was designed to perform these calculations so that a wide range of building geometries and loading conditions could be investigated.

Based on the loads calculated as described above, the amount of full-height structural sheathing required was determined using the allowable sheathing capacity of 364 plf as determined above. The length of full-height sheathing required was then tabulated as a percentage of wall length for sidewalls and endwalls over the range of building geometries defined in the *Prescriptive Method's* applicability limits. The length of wall with full-height sheathing is defined as the sum of segments of walls that have sheathing extending from the base of the wall to the top, without interruption of openings (i.e., the total of lengths of wall between openings). Further, the individual segments of wall must be 48" in length or greater to contribute to the required length of full-height sheathing for a given wall line.

As a final step necessary for a basic prescriptive approach, the requirements were conservatively reduced to the minimum percent lengths of full-height sheathed wall shown in the wall bracing table of the *Prescriptive Method*. The only building geometry parameter retained was roof slope because of its large impact on the wind loads transferred to the wall bracing. Footnotes to the wall bracing table provide additional information related to the proper use of the requirements.

COMMENTARY ON THE PRESCRIPTIVE METHOD FOR RESIDENTIAL
COLD-FORMED STEEL FRAMING

6.9   Hold-down Requirements in High Wind or Seismic Conditions

In wind conditions greater than 90 mph exposure C and in Seismic Zones 3 and 4, hold-down brackets to stabilize shearwalls must be designed by a design professional.

6.10   Uplift Straps

In wind conditions greater than 90 mph exposure C and in Seismic Zones 3 and 4, uplift straps to offset potential wind uplift loads must be designed by a design professional.

COMMENTARY ON THE PRESCRIPTIVE METHOD FOR RESIDENTIAL COLD-FORMED STEEL FRAMING

6.11   Design Example

The following is an example for calculating the applicable loads on wall studs and designing the wall studs to resist such loads. A design for an 8-foot header, with its associated jack and king studs, is also included in this example.

A hypothetical two story steel building is used to demonstrate the engineering calculations used in the *Prescriptive Method*. The building has the following physical dimensions:

Building type:	Two story house with basement.
Building width:	28 feet (with 2 feet roof overhang)
Building length:	40 feet.
First story wall height:	8 feet.
Second story wall height:	8 feet.
Wall stud spacing:	24" oc.
Wind speed and Exposure:	70 mph Exposure C (or 80 mph Exposure B)
Ground snow load:	30 psf
Opening size:	8 feet.

Calculate Applied Loads:

The applied loads on the two story building are calculated as follows:

Roof live load = maximum of snow load or roof live load (16 psf minimum).

Snow load = 0.7 * 30 = 21 psf   (calculated per ASCE 7-93 criteria, no other reduction is used)

Wind loads are calculated in accordance with ASCE 7-93.

Wind load = WL = $0.00256K_z(GC_p + GC_{pi})(VI)^2$
$K_z = 0.87$ (at 20 feet)
$GC_p = \pm 1.5$ (for components and cladding, zone 4)
$GC_{pi} = \pm 0.25$ (for interior pressure, enclosed buildings, components and cladding)
$I = 1$ (for residential buildings in areas with wind speed < 100 mph)
$V = 70$ mph
WL = $0.00256(0.87)(1.5 + 0.25)(70 * 1)^2 = 19.10$ psf round up to 20 psf.

Summary of applied gravity loads:

Snow load =	21 psf * (28' + 2' + 2')/2 = 336 plf
Roof dead load =	7 psf * (28' + 2' + 2')/2 = 112 plf
Wall dead load = 10 psf * 8' = 80 plf	
Ceiling dead load =	5 psf * (28')/2 = 70 plf
Second floor dead load =	10 psf * (28' + 2' + 2')/2 = 140 plf
Second floor live load =	30 psf * (28' + 2' + 2')/2 = 420 plf

Case 1:   Dead load + live load + half wind + full snow

a.   One story building or the top floor of a two story building:

COMMENTARY ON THE PRESCRIPTIVE METHOD FOR RESIDENTIAL
COLD-FORMED STEEL FRAMING

Applied axial load = (roof dead load + full snow load + ceiling dead load) x 24/12

Applied axial load = (336 + 112 + 70)(2) = **1036 lbs.**

Applied lateral load = Half wind load = 20 / 2 = **10 psf**

b.   Bottom floor of a two story building:

Applied axial load = (roof dead load + full snow load + wall load + ceiling dead load + second floor dead load + second floor live load.) x 24/12

Applied axial load = (336 + 112 + 80 + 70 + 140 + 420)(2) = **2316 lbs.**

Applied lateral load = Half wind load = 20 / 2 = **10 psf**

Case 2:   Dead load + live load + half snow + full wind

a.   One story building or the top floor of a two story building:

Applied axial load = (roof dead load + half snow load + ceiling dead load) x 24/12

Applied axial load = (336/2 + 112 + 70)(2) = **700 lbs.**

Applied lateral load = Full wind load = **20 psf**

b.   Bottom floor of a two story building:

Applied axial load = (roof dead load + half snow load + wall load + ceiling dead load + second floor dead load + second floor live load.) x 24/12

Applied axial load = (336/2 + 112 + 80 + 70 + 140 + 420)(2) = **1980 lbs.**

Applied lateral load = Full wind load = **20 psf**

Therefore, the load combinations to be checked are:

a)   One story building: 1036 lbs. axial with 10 psf lateral
                                        700 lbs. axial with 20 psf lateral

b)   Bottom story: 2316 lbs. axial with 10 psf lateral
                                        1980 lbs. axial with 20 psf lateral

Design Wall Studs

Select a stud size and check its capacity to see if it can resist the applied loads. A 2 x 4 x 43 stud is initially selected. Calculate the capacity of a 2 x 4 x 43:

i.   *Allowable Moment*

COMMENTARY ON THE PRESCRIPTIVE METHOD FOR RESIDENTIAL COLD-FORMED STEEL FRAMING

Calculate the allowable moment per Section C of the AISI Design Specification:

The allowable moment is calculated as the smallest nominal moment calculated per AISI Design Specification Sections C.1.1, C.1.2, and C.1.3 divided by a factor of safety of 1.67.

<u>Nominal section strength per AISI Design Specification Section C.1.1:</u>

Use procedure (a) based on initiation of yielding.

$$M_n = S_e F_y$$ Eq. C.1.1-1

Where:
F_y = Design yield strength.
S_e = Elastic section modulus of the effective section calculated with the extreme compression or tension fiber at F_y.

Calculate strength increase from cold working as per Section A5.2.2 of the AISI Design Specification:

$$F_{ya} = CF_{yc} + (1 - C)F_{yf}$$ Eq. A5.2.2-1

F_{yf} = 33,000 psi

C = corner cross sectional area of flange/full cross sectional area of flange
u = 1.57 (0.09375 + 0.0451/2) = 0.182591 in.
Area of corners = 2(0.0451)(0.182591) = 0.01647 in.^{2}
Area of flanges = 2(0.0451)(0.18259) + 1.3473(0.0451) = 0.07723 in.^{2}

C = 0.01647 / 0.07723 = 0.213

$$F_{yc} = B_c F_{yv} / (R/t)^m$$ Eq. A5.2.2-2

F_{uv} / F_{yv} = 45000/33000 = 1.36 > 1.2   OK
R/t = 0.09375/0.0451 = 2.08 < 7   OK
Angle < 120 degrees   OK

$$B_c = 3.69(F_{uv}/ F_{yv}) - 0.819 (F_{uv} / F_{yv})^2 - 1.79$$ Eq. A5.2.2-3

$$B_c = 3.69(1.36) - 0.819 (1.36)^2 - 1.79 = 1.7136$$

$$m = 0.192(F_{uv}/ F_{yv}) - 0.068$$ Eq. A5.2.2-4
$$m = 0.192(1.36) - 0.068 = 0.1938$$

$$F_{yc} = 1.7136(33000) / (0.09375/0.0451)^{0.1938} = 49222.79 \text{ psi}$$

$$\mathbf{F_{ya}} = 0.213(49222.79)+ (1 - 0.213)33000 = \mathbf{36,455.5 \ psi}$$

Calculate the elastic section modulus of the effective section calculated with the extreme compression or tension fiber at F_y = 36,455.5 psi

w = Flange width - 2(radius + Thickness) = 1.3473"
w/t = 1.3473/.0451 = 29.87

COMMENTARY ON THE PRESCRIPTIVE METHOD FOR RESIDENTIAL
COLD-FORMED STEEL FRAMING

---

d = Lip - radius - thickness = 0.5 - .09375 - .0451 = 0.36115 in.

$S = 1.28\int(E/f) = 1.28\sqrt{(29500000/36455.5)} = 36.41$ Eq. B4-1
$S/3 = 12.134 < w/t = 29.87 < S = 36.41$

$I_a = 399t^4\{[(w/t)/S] - 0.33\}^3$ Eq. B4.2-6
$I_a = 399(.0451)^4\{(29.87/36.41) - 0.33\}^3 = 0.000195$ in.^{4}
$n = 1/2$

$D/w = 0.5 / 1.3473 = 0.3711$, $0.25 < D/w = 0.3711 < 0.8$

$k = [4.82 - 5(D/w)](Is/Ia)^n + 0.43 \leq 5.25 - 5(D/w)$ Eq. B4.2-9

$I_s = d^3t/12 = (0.3612)^3(0.0451)/12 = 0.0001777$ in.^{4}

$k = [4.82 - 5(0.3711)](0.0001777 / 0.000195)^{0.5} + 0.43 \leq 5.25 - 5(0.3711)$
$k = 3.3944$

Effective width of the flange shall be calculated according to Section Design Specification B2.1 using k as calculated above, and $f = M_c / S_f$.

$w/t = 29.87 < 90$ OK   Section B1.1-(a)-(3)

$\lambda = [1.052 / \sqrt{k}](w/t)\int f/E$ Eq. B2.1-4
Where $f = M_c / S_f = 36,455.5$ psi

$\lambda = [1.052/\int3.3944](29.81)\sqrt{36455.5/29500000} = 0.59837$
$\lambda \leq 0.673 \Rightarrow b = w$ Eq. B2.1-1

Effective width **b = 1.3473 inches**   (i.e. compression flange is fully effective)

Compression Stiffener:

Determine the effective width of the compression stiffener (lip):

D = 0.5 inches
d = Lip size - radius - thickness = 0.5 - 0.09375 - 0.0451 = 0.3612 inches.
d/t = 0.3612/.0451 = 8.01 < 60 OK   Section B1.1-(a)-(3)

D/w = 0.5 / 1.3473 = 0.3711, $0.25 < D/w = 0.3711 < 0.8$

Calculate the effective width for the lip per AISI Design Specification Section B2.1 :

$\lambda = [1.052/\int k](w/t)\sqrt{f/E}$ Eq. B2.1-4

Where $f = M_c / S_f = 36455.5$ psi and $k = 0.43$ per AISI Design Specification Section B3.1

$\lambda = [1.052 / \sqrt{0.43}](8.01)\sqrt{36455.5/29500000} = 0.451634$
$\lambda \leq 0.673 \Rightarrow d'_s = d = 0.36115"$ Eq. B2.1-1

COMMENTARY ON THE PRESCRIPTIVE METHOD FOR RESIDENTIAL
COLD-FORMED STEEL FRAMING

$d_s = d_s' (I_s/I_a) \leq d_s'$ Eq. B4.2-11

$d_s = 0.36115 (0.0001777/0.000195) \leq 0.36115$

$d_s = 0.328234$ in.

Effective width of lip $d_s = \mathbf{0.328234}$ **inches** (i.e. compression stiffener is not fully effective)

Assume the web is fully effective, and top fiber stress is 31,900.27 psi:

Element	Effective length (in)	Distance from top fiber, y (in.)	Ly in.^{2}	Ly^2 in.^{3}	I$_1'$ About own axis in.^{3}
Web	3.2223	1.75	5.6390	9.8683	2.7882
Tension flange	1.3473	3.4775	4.6852	16.2924	-
Comp. Flange	1.3473	0.0226	0.0304	0.0007	-
Upper corners	0.3652^1	0.0648	0.0237	0.0015	-
Lower corners	0.3652^1	3.4352	1.2545	4.3094	-
Upper stiffener	0.3282	0.3030	0.0994	0.0301	0.0029
Lower stiffener	0.3612	3.1806	1.1487	3.6534	0.0039
Sum	7.33665	12.2335	12.8808	34.1559	2.795026

^{1} Where corner length = Length of arc = 1.57(R + t/2) = 0.18259. Two corners = 0.3652"

Distance from top fiber to x-axis is Y_{cg} = 12.8808 / 7.33665 = 1.755682 in.

Since distance of top compression fiber from neutral axis is greater than one half the beam depth, a compression stress of 36,455.5 psi will govern as assumed (i.e. initial yielding is in compression).

Check if web is fully effective:

$f_1 = +f (Y_{cg} - t - R)/Y_{cg} = 36455.5(1.755682 - 0.0451 - 0.09375)/1.755682 = 33,576.03$ psi

$f_2 = -f (\text{Web size} - Y_{cg} - t - R)/Y_{cg} = 36,455.5(3.5 - 1.755682 - 0.0451 - 0.09375)/1.755682$
$f_2 = -33,340.03$ psi (tension)

The effective widths are calculated per AISI Design Specification Section B2.3 (a) for load capacity determination.

$\Psi = f_2 / f_1 = -0.99297 \leq -0.236$ Section B2.3
$b_1 = b_e / (3-\Psi)$ Eq. B2.3-1
$b_2 = b_e / 2$ Eq. B2.3-2
$b_1 + b_2$ shall not exceed the compression portion of the web calculated on the basis of effective section.

b_e = Effective width b determined in accordance with AISI Design Specification Section B2.1 with f_1 substituted for f and k determined as follows:
$k = 4 + 2(1 - \Psi)^3 + 2(1 - \Psi)$ Eq. B2.3-4

$k = 4 + 2(1 + 0.99297)^3 + 2(1 + 0.99297) = 23.81785$

COMMENTARY ON THE PRESCRIPTIVE METHOD FOR RESIDENTIAL COLD-FORMED STEEL FRAMING

$h = w = $ Web - 2(Radius + thickness) = 3.5 - 2(0.09375 + 0.0451) = 3.222 inches.

$h/t = 3.222 / 0.0451 = 71.45 < 200$ OK   Section B1.2-(a)

$\lambda_{(web)} = [1.052 / \sqrt{23.81785}](71.45)\sqrt{33,576.03/29500000} = 0.519586 < 0.67$ Eq. B2.1-4

$b_e = w = 3.222$ "   Eq. B2.1-1
$b_2 = 3.222 / 2 = 1.611$ inches
$b_1 = 3.222 / (3 + 0.99297) = 0.8069$ inches.
$b_1 + b_2 = 1.611 + 0.8069 = 2.4181$ inches

Compression portion of the web = y_{cg} - (Radius + thickness)
                = 1.755682 - (0.09375 + 0.0451) = 1.616832 inches.

Since $b_1 + b_2 = 2.4181" > 1.616832"$, $b_1 + b_2$ shall be taken as 1.616832 in.
This verifies the assumption that the web is fully effective.

$I'_x = Ly^2 + I'_1 - Ly^2_{cg}$
$I'_x = 34.1559 + 2.795026 - 7.33665(1.755682)^2 = 14.33634$ in.^{3}

Actual $I_x = I'_x t = (14.33634)(0.0451) = 0.646569$ in.^{4}

$S_e = I_x / y_{cg} = 0.646569 / 1.755682 = \textbf{0.368272 in.}^3$

$M_n = S_e F_y$ Eq. C.1.1-1
$\textbf{M}_n = 0.368272 * 36,455.5 = \textbf{13427.01 lb.-in.} = \textbf{1118.9 ft.-lb.}$

<u>Nominal section strength per AISI Design Specification Section C.1.2:</u>

$M_n = S_c (M_c / S_f)$ Eq. C.1.2-1
$S_f = I_x / Y_{cg} = 0.649 / 1.75 = 0.37086$ in.^{4}
Where I_x is taken from Attachment A.

$M_c = M_y (1 - M_y / 4M_e)$ Eq. C.1.2-2
$M_y = S_f F_y$ Eq. C.1.2-4

$M_y = 0.37086 * 36,500 = 13,536$ lb.-in.   (F_y is rounded up to 36,500 psi)
$C_b = 1$ (Combined axial and bending, Section C.1.2)

$M_e = C_b r_o A \sqrt{\sigma_{ey}\sigma_t}$ Eq. C.1.2-5
$\sigma_{ey} = (\pi^2 E)/(K_y L_y/r_y)^2$ Eq. C.1.2-8
$\sigma_t = (1/Ar_o^2)/[GJ + \pi^2 EC_w/(K_t L_t)^2] \ \sigma$ Eq. C.1.2-9

Section properties are taken from Attachment A.
$\sigma_{ey} = (\pi^2 * 29500000)/(1 * 48 / 0.6107)^2 = 47129.67$ psi

$\sigma_t = [1 / 0.3324 * (2.033)^2] / [11,300,000 * 0.000225 + \pi^2 * 29,500,000 * 0.3441 / (1 * 48)^2]$
$\sigma_t = 33,504.95$ psi

$M_e = 1* 0.3324 * 2.033 \sqrt{(47129.67)(33504.95)} = 26,853.45$ lb.-in. $> 0.5 M_y = 6,768$ lb.-in.

$M_c = 13,536 (1 - 13,536 / 4 * 26853.45) = 11,830.44$ lb.-in.

COMMENTARY ON THE PRESCRIPTIVE METHOD FOR RESIDENTIAL COLD-FORMED STEEL FRAMING

Determine S_c, the elastic section modulus of the effective section calculated at a stress of M_c / S_f in the extreme compression fiber. $M_c / S_f = 11,830.44 / 0.37086 = 31,900.27$ psi

Compression Flange:

Determine the effective width of the compression flange:

w = Flange width - 2(radius + Thickness) = 1.3473"
$w/t = 1.3473/.0451 = 29.87$

$S = 1.28\sqrt{(E/f)} = 1.28\sqrt{(29500000/31900.27)} = 38.9246$ Eq. B4-1

$S/3 = 12.9747 < w/t = 29.87 < S = 38.9246$

$I_a = 399t^4\{[(w/t)/S] - 0.33\}^3$ Eq. B4.2-6
$I_a = 399(.0451)^4\{[(29.87)/38.9246] - 0.33\}^3 = 0.000138$ in.^{4}
$n = 1/2$

$D/w = 0.5 / 1.3473 = 0.3711$, $0.25 < D/w = 0.3711 < 0.8$

$k = [4.82 - 5(D/w)](Is/Ia)^n + 0.43 \leq 5.25 - 5(D/w)$ Eq. B4.2-9

$I_s = d^3t/12 = (0.3612)^3 (0.0451)/12 = 0.0001777$ in.^{4}

$k = [4.82 - 5(0.3711)](0.0001777 / 0.000138)^{0.5} + 0.43 \leq 5.25 - 5(0.3711)$
$k = 3.3944$

Effective width of the flange shall be calculated according to AISI Design Specification Section B2.1 using k as calculated above, and $f = M_c / S_f$.

Since Is > Ia and D/w < 0.8, the stiffener is not considered as a simple lip.

$w/t = 29.87 < 90$ OK   Section B1.1-(a)-(3)
$\lambda = [1.052/\sqrt{k}](w/t)\sqrt{f/E}$ Eq. B2.1-4
Where $f = M_c / S_f = 31,900.27$ psi
$\lambda = [1.052/\sqrt{3.3944}](29.81)\sqrt{31900.27/29500000} = 0.5609 \leq 0.673 \Rightarrow b = w$ Eq. B2.1-1

Effective width   **b = 1.3473 inches**   (i.e. compression flange is fully effective)

Compression Stiffener:

Determine the effective width of the compression stiffener (lip):
$D = 0.5$ inches
d = Lip size - radius - thickness = 0.5 - 0.09375 - 0.0451 = 0.3612 inches.
$d/t = 0.3612/.0451 = 8.01 < 60$ OK   Sec. B1.1-(a)-(3)

$D/w = 0.5 / 1.3473 = 0.3711$, $0.25 < D/w = 0.3711 < 0.8$

Calculate the effective width for the lip per AISI Section B2.1 :

COMMENTARY ON THE PRESCRIPTIVE METHOD FOR RESIDENTIAL
COLD-FORMED STEEL FRAMING

$\lambda = [1.052/\sqrt{k}](w/t)\sqrt{f/E}$

Where $f = M_c / S_f = 31,900.27$ psi and $k = 0.43$ per AISI Design Specification Section B3.1   Eq. B2.1-4

$\lambda = [1.052/\sqrt{0.43}](8.01)\sqrt{31900.27/29500000} = 0.42257$

$\lambda \leq 0.673 \Rightarrow d_s' = d = 0.3612''$ Eq. B2.1-1

$d_s = d_s' \, (I_s/I_a) \leq d_s'$ Eq. B4.2-11

Effective width of lip **$d_s = 0.3612$ inches** (i.e. compression stiffener is fully effective)

Thus, one concludes that the section is fully effective. $Y_{cg} = 3.5 / 2 = 1.75$ in. (from symmetry)

Web:

Check if web is fully effective: (AISI Design Specification Section B2.3)
Assume the web is fully effective, and top fiber stress is 31,900.27 psi:

Element	Effective length (in)	Distance from top fiber, y (in.)	Ly (in.^{2})	Ly^2 (in.^{3})	I$_1'$ About own axis (in.^{3})
Web	3.2223	1.75	5.6390	9.8683	2.7882
Tension flange	1.3473	3.4775	4.6852	16.2924	-
Comp. Flange	1.3473	0.0226	0.0304	0.0007	-
Upper corners	0.3652[1]	0.0648	0.0237	0.0015	-
Lower corners	0.3652[1]	3.4352	1.2545	4.3094	-
Upper stiffener	0.3612	0.3194	0.1154	0.0368	0.0039
Lower stiffener	0.3612	3.1806	1.1487	3.6534	0.0039
Sum	7.3696	12.25	12.8967	34.1627	2.796004

[1] Where corner length = Length of arc = $1.57(R + t/2) = 0.18259$. Two corners = 0.3652''

Distance from top fiber to x-axis is $Y_{cg} = 12.8967 / 7.3696 = 1.75$ in

$f_1 = +f \, (Y_{cg} - t - R)/Y_{cg} = 31,900.27(1.75 - 0.0451 - 0.09375)/1.75 = 29,369.21$ psi

$f_2 = -f \, (\text{Web size} - Y_{cg} - t - R)/Y_{cg} = 31,900.27(3.5 - 1.75 - 0.0451 - 0.09375)/1.75$
$f_2 = -29,369.21$ psi

The effective widths are calculated per AISI Design Specification Section B2.3 (a) for load capacity determination.

$\psi = f_2 / f_1 = -1 \leq -0.236$ Section B2.3

$b_1 = b_e / (3-\psi)$ Eq. B2.3-1

$b_2 = b_e / 2$ Eq. B2.3-2

$b_1 + b_2$ shall not exceed the compression portion of the web calculated on the basis of effective section.

b_e = Effective width b determined in accordance with AISI Design Specification Section B2.1 with f_1 substituted for f and k determined as follows:

$k = 4 + 2(1 - \psi)^3 + 2(1 - \psi)$ Eq. B2.3-4

COMMENTARY ON THE PRESCRIPTIVE METHOD FOR RESIDENTIAL
COLD-FORMED STEEL FRAMING

$k = 4 + 2(1 + 1)^3 + 2(1 + 1) = 24$

$h = w = \text{Web} - 2(\text{Radius} + \text{thickness}) = 3.5 - 2(0.09375 + 0.0451) = 3.222 \text{ inches.}$

$h/t = 3.222 / 0.0451 = 71.45 < 200 \text{ OK}$ $\qquad\qquad$ Section B1.2-(a)

$\lambda_{(web)} = [1.052/\sqrt{24}](71.45)\sqrt{29369/29500000} = 0.484 < 0.673$ $\qquad$ Eq. B2.1-4

$b_e = w = 3.222 \text{ "}$ $\qquad\qquad\qquad\qquad\qquad\qquad$ Eq. B2.1-1
$b_2 = 3.222 / 2 = 1.611 \text{ inches}$
$b_1 = 3.222 / (3 + 1) = 0.8055 \text{ inches.}$
$b_1 + b_2 = 1.611 + 0.8055 = 2.4165 \text{ inches}$

Compression portion of the web = y_{cg} - (Radius + thickness)
$\qquad\qquad\qquad = 1.75 - (0.09375 + 0.0451) = 1.611 \text{ inches.}$

Since $b_1 + b_2 = 2.4165" > 1.611$ in., $b_1 + b_2$ shall be taken as 1.611 in.
This verifies the assumption that the web is fully effective.

$I'_x = Ly^2 + I'_1 - Ly^2_{cg}$
$I'_x = 34.1627 + 2.796004 - 7.36956(1.75)^2 = 14.389 \text{ in.}^3$

Actual $I_x = I'_x t = (14.389)(0.0451) = 0.6489 \text{ in.}^4$

$S_e = I_x / y_{cg} = 0.6489 / 1.75 = 0.3708 \text{ in.}^3$

$M_n = S_e F_y$ $\qquad\qquad\qquad\qquad\qquad\qquad\qquad$ Eq. C.1.1-1
$\mathbf{M_n = 0.3708 * 31900 = 11829 \text{ lb.-in.} = 985.75 \text{ ft.-lb.}}$

Nominal section strength per AISI Design Specification Section C.1.3:

This section is not applicable to the wall studs in this example.

Allowable moment:

The allowable moment is the smallest nominal moment calculated per AISI Design Specification Sections C.1.1, C.1.2, and C.1.3 divided by a factor of safety of 1.67.

$\Omega = 1.67$

$M_a = M_n / 1.67$ $\qquad\qquad\qquad\qquad\qquad\qquad\qquad$ Eq. C.1-1
$\mathbf{M_a = 11,829/1.67 = 7,084 \text{ lb.-in.} = 590 \text{ ft.-lb.}}$

ii. $\qquad\qquad$ *Allowable Shear, V_a*

Shear for unpunched web:

The allowable shear V_a is calculated in accordance with Section C.2 of the AISI Design Specification .

COMMENTARY ON THE PRESCRIPTIVE METHOD FOR RESIDENTIAL
COLD-FORMED STEEL FRAMING

Calculate the depth of the flat portion of the web, h, measured along the plane of the web.

h = Web size - 2(radius + thickness) = 3.5 - 2(0.09375 + 0.0451) = 3.223 in.

$h/t = 3.223/0.0451 = 71.448$

Calculate $1.38 \sqrt{E_v}/F_y = 1.38\sqrt{29,500,000} * 5.34 * 33,000 = 95.346$
Where k_v is shear buckling coefficient = 5.34
  $F_y = 33,000$ psi   (stress increase due to cold-forming can not be used here)

$h/t = 71.448 < 95.346 \Rightarrow V_a = 0.38t^2 \sqrt{E_v F_y} \leq 0.4htF_y$ Eq. C.2-1
$V_a = 0.38(0.0451)^2 \sqrt{29,500000} * 5.34 * 33,000 \leq 0.4 * 3.223 * 0.0451 * 33,000$
$V_a = 1,762.28 \leq 1918.71$ OK.
$V_a = 1,762.28$ lbs.

Shear for punched web:

The allowable shear for webs with punchouts is calculated in accordance with the ICBO AC46 method.

Calculate the reduction factor q:
$q_s = 1 - 1.1(a/d)$ where a = web size = 3.5" and b = width of punchout = 1.5"
$q_s = 0.52857$

Allowable shear $V_a = 1762.28 * q_s$
$V_a = 931.5$ lbs.

iii.   *Axial Capacity*

The axial capacity of the member is calculated per AISI Design Specification Section C4.

$P_a = P_n / \Omega$ Eq. C4-1

Where $P_n = A_e * F_n$
A_e is the effective area at stress F_n
F_n is a function of F_e. F_e is calculated as the minimum of the elastic flexural buckling, torsional, or torsional-flexural buckling stress.
$\Omega = 1.92$ (factor of safety)

1.   Calculate F_e for sections not subject to torsional or torsional-flexural buckling per Section C4.1 of the AISI Design Specification.

$F_e = \pi^2 E / (KL/r)^2$ Eq. C4.1-1
Check $K_x L_x/r_x = 1 * 8 * 12 / 1.3973 = 69 < 200$
$F_e = (\pi^2) * 29,500,000 / (1 * 8 * 12 / 1.3973)^2 = 61,681.92$ psi

Check $K_y L_y/r_y = 1 * 4 * 12 / 0.6107 = 79 < 200$
$F_e = (\pi^2) * 29,500,000 / (1 * 4 * 12 / 0.6107)^2 = $ **47,129.67 psi**

2.   Calculate F_e for sections subject to torsional or torsional-flexural buckling per section C4.2 of the AISI Design Specification.

$F_e = 1/2\beta \, [\, (\sigma_{ex} + \sigma_t) - \sqrt{(\sigma_{ex} + \sigma_t)^2 - 4\beta\sigma_{ex}\sigma_t} \,]$
\hfill Eq. C4.2-1

$\sigma_{ex} = (\pi^2 * 29500000)/(1 * 96 / 0.6107)^2 = 61,681.92$ psi

$\sigma_t = [1/ 0.3324*(2.033)^2]/ [11,300,000 * 0.000225 + \pi^2 * 29,500,000*0.3441 / (1*48)^2]$
$\sigma_t = 33,504.95$ psi
$\beta = 1 - (x_o / r_o)^2$ \hfill Eq. C4.2-3
$\beta = 1 - (-1.3445 / 2.033)^2 = 0.5620$

$F_e = 1 / (2 * 0.5620) \, [\, (61,681.92 + 33,504.95) - \sqrt{(61,681.92 + 33,504.95)^2}$
$\quad - 4(0.562)(61,681.92)(33,504.95)$
$\mathbf{F_e = 25,572.39}$ **psi** < 47129.67 psi

$\mathbf{F_e = 25,572.39}$ **psi**

$F_y/2 = 36,500 / 2 = 18,250$ psi \hfill (Strength increase due to cold-forming is used)
$F_e = 25,572.39 > 18,250 \Rightarrow F_n = F_y (1 - F_y / 4F_e)$ \hfill Eq. C4-3

$F_n = 36,500 \, [1 - 36,500 / (4 * 25,572.39] = 23,475.7$ psi

A_e can be easily calculated to be 0.242957 in.^{2}

$P_n = 0.242957 * 23,475.7 = 5,703.28$ lbs.
$P_a = 5,703.28 / 1.92$

$\mathbf{P_a = 2,970.61}$ **lbs.**

iv. \hfill *Combined Axial and Bending*

Check combined axial and bending in accordance with AISI Design Specification Section C5

The axial force and bending moments shall satisfy the following AISI Design Specification interaction equations:

$P/P_a + (C_{mx}M_x)/(M_{ax}\alpha_x) + (C_{my}M_y) (M_{ay}\alpha_y) \le 1.0$ \hfill Eq. C5-1
$P/P_{ao} + M_x / M_{ax} + M_y / M_{ay} \le 1.0$ \hfill Eq. C5-1

In this example, $M_y = 0$ and the stresses are increased by 33% per the AISI Design Specification Section A4.4 (this is equivalent to multiplying the loads by 0.75)

$C_{mx} = 1$ \hfill conservatively taken as 1 per AISI Design Specification Section C5
$1/ \alpha_y, \, 1/\alpha_x$ are magnification factors $= 1/[1 - (\Omega_c P/P_{cr})]$ \hfill Eq. C5-4

P = Applied axial load
M_x = Applied moment = $wL^2/8$
P_a \hfill = Allowable axial load determined in accordance with AISI Design Specification Section C4.
\hfill = 2970.61 lbs.
P_{ao} \hfill = Allowable axial load determined in accordance with AISI Design Specification Section C4, with $F_n = F_y$
$P_{ao} = F_y A_e / 1.92 = 36,500 * 0.242957 / 1.92 = 4,618.21$ lbs.

COMMENTARY ON THE PRESCRIPTIVE METHOD FOR RESIDENTIAL
COLD-FORMED STEEL FRAMING

Ω_c = Factor of safety = 1.92
$P_{cr} = \pi^2 EI_b / (K_b/L_b)^2$ Eq. C5-5
I_b = Gross moment of inertia = 0.65 in.^{4}

$P_{cr} = \pi^2 (29,500,000)(0.65) / (1 / 96)^2 = 20,503.28$ lbs.
$M_{ax} = 590..31$ ft.-lb.   (previously calculated)
$\alpha_{ox} = 1 - \Omega_c P/P_{cr} = 1 - 1.92 (P/20503.28)$

There are two methods one can use to check the adequacy of the member. The first method is to check the interaction equations directly. The other method is to use the interaction equations to calculate the maximum allowable axial load, P, the member can handle for a given lateral load, and then check that load against the applied load calculated before.

Method 1:

As mentioned before, two cases will be checked: Half wind + full snow and full wind with half snow.

Case 1:

Top story: Axial load = 1036 lbs.
                                              Lateral load = 10 psf = 10 * 24" / 12" = 20 plf

                        $M_x = wL^2/8 = (20)(8)^2 / 8 = 160$ ft.-lbs.

Eq. C5-1: $1036/2970.61 + (1* 160) / (590.31)[1 - 1.92 (1036/20503.28] \leq 1.33$

                        $0.5596 \leq 1.33$ OK

Eq. C5-2: $1036 / 4,618.21 + 160 / 590.31 \leq 1.33$
                        $0.495 \leq 1.33$ OK
Bottom story: Axial load = 2316 lbs.
                        Lateral load = 10 psf = 10 * 24" / 12" = 20 plf

                        $M_x = wL^2/8 = (20)(8)^2 / 8 = 160$ ft.-lbs.

Eq. C5-1: $2316 / 2970.61 + (1* 160) / (590.31)[1 - 1.92 (2316 / 20503.28] \leq 1.33$

                        $1.105 \leq 1.33$ OK

Eq. C5-2: $2316 / 4,618.21 + 160 / 590.31 \leq 1.33$
                        $0.76 \leq 1.33$ OK

Case 2:

Top story: Axial load = 700 lbs.
                       Lateral load = 20 psf = 10 * 24" / 12" = 40 plf

                        $M_x = wL^2/8 = (20)(8)^2 / 8 = 320$ ft.-lbs.

Eq. C5-1: $700 / 2970.61 + (1* 320) / (590.31)[1 - 1.92 (700 / 20503.28] \leq 1.33$

0.865 ≤ 1.33   OK

Eq. C5-2:   700 / 4,618.21 + 320 / 590.31 ≤ 1.33
             0.69 ≤ 1.33   OK

Bottom story: Axial load = 1980 lbs.
                Lateral load = 20 psf = 20 * 24" / 12" = 40 plf

                $M_x = wL^2/8 = (40)(8)^2 / 8 = 320$ ft.-lbs.

Eq. C5-1:   1980 / 2970.61 + (1* 320) / (590.31)[1 - 1.92 (1980 / 20503.28] ≤ 1.33
             1.318 ≤ 1.33   OK

Eq. C5-2:   1980 / 4,618.21 + 320 / 590.31 ≤ 1.33
                                  0.97 ≤ 1.33   OK

Therefore, the 2 x 4 x 43 is adequate.

Method 2:

Calculate the axial capacity of the 2 x 4 x 43 for a lateral load of 20 psf:
   $M_x = wL^2/8 = (20)(8)^2 / 8 = 320$ ft.-lbs.

Use equation C5-2 to calculate P: $P/P_{ao} + M_x / M_{ax} ≤ 1.33$
                                     $P = P_{ao} (1.33 - M_x / M_{ax})$
                                     P = 4,618.21(1.33 - 320 / 590.31) = 3639 lbs.

Use equation C5-2 to calculate P: $P/P_a + (C_{mx}M_x)/(M_{ax}\alpha_x) ≤ 1.33$
                                     $P = P_a [1.33 - C_{mx}M_x / (1 - \Omega_c P/P_{cr})M_{ax}]$
                                         P=2970.61[1.33-1*320/(1 - 1.92P/20,503.28)590.31]
                                         P = 3950.91 - 1610.33 / (1 - 0.0000936P)

Use trial and error in equation C5-1 to calculate P. First iteration: P = 2000 lbs.

P = 3950.91 - 1610.33 / (1 - 0.0000936 * 2000) = 1970 lbs.

After two trial and error iterations, P = 2067.10 lbs.

Therefore, the maximum allowable axial load is 2067.10 (smaller of 2067.10 and 3639)

The applied axial loads for the 20 psf lateral loads are well below the calculated allowable axial load. Therefore, the stud is adequate.

COMMENTARY ON THE PRESCRIPTIVE METHOD FOR RESIDENTIAL
COLD-FORMED STEEL FRAMING

Similarly, one can calculate the maximum allowable axial load for 10 psf lateral load to be 2936.54 lbs. Again the applied axial loads for the 10 psf lateral loads are well below the calculated allowable axial load. Therefore, the stud is adequate.

v.   *Check Deflection Limit:*

The deflection equation used in the joist design can be utilized here:

$L/\delta = 188,800,000 * I_x / (w * L^3 * spacing)$

For 10 psf lateral load: $L/\delta = L / 997$ OK

For 20 psf lateral load: $L/\delta = L / 499$ OK

Design Header, Jack and King Studs

Design a header for an 8-foot opening and calculate the number of king and jack studs required for the example building previously described. The 8-foot opening is located on the first floor.

Vertical load acting on the header =   Roof dead load + roof live load + ceiling dead load + attic live load + second floor live and dead load + second floor wall dead load.

The attic live load and the second floor live loads are multiplied by a reduction factor based on load proportion in load and resistance factor method to account for transient loads.

Attic live load = 10 psf * 0.5/1.6 = 3.125 psf
Floor live load = 30 * 0.5/1.6 = 9.375 psf.

Uniform (vertical) load   P =   (10 psf + 21 psf + 5 psf + 10 psf + 9.375 psf) x (28'/2 + 2') + (3.125)(28/2) + 80 plf

P = 1,010 lbs./ft.

Allowable moment for a 2 x 10 x 68 mils header is 4,671.8 lbs.-ft   (From Table in Attachment A)

Conservatively, multiply this moment by two to obtain the allowable moment for a back-to-back header made out of two 2 x 10 x 68 mils C-sections.

$M_a = 4671.8 \times 2 = 9,343.6$ lbs.-ft.

Allowable shear = 5,272   (Shear value from Attachment A)

Applied moment = $wL^2 / 8 \Rightarrow L = \sqrt{8(M_a)/P}$

Maximum span due to applied moment = $L = \sqrt{8(9,343.6) / 1,010} = 8.66$ feet.

COMMENTARY ON THE PRESCRIPTIVE METHOD FOR RESIDENTIAL COLD-FORMED STEEL FRAMING

Maximum shear due to applied load = $V = PL/2 = 1{,}010 * 8.77/2 = 4{,}229$ lbs.

Maximum span due to allowable shear = $L = 2 * 5{,}272 / 1{,}010 = 10.44$ feet.

Web crippling is not a concern here because the header is connected with an angle to the king stud. The angle stiffens the web and acts as a web stiffener.

The maximum header span is, therefore, 8.6' which is greater than the opening size.

Calculate number of jack and king studs:

The number of king studs is calculated based on the actual number of full height studs required if there was no opening. The opening size is 8 feet and the spacing is 24" oc.

8 ft. / 2' = 4   (use 4 studs)

The four required studs are divided between king and jack studs with two jack studs and two king studs at each end of the header. The axial capacity of a 2 x 4 x 43 stud can be easily calculated to be 2971 lbs. with mid-height bracing (stud capacity of a fully sheathed wall is 3561 lbs.). Capacity for two jack studs is 5942 lbs. which far exceeds the applied axial load of $1{,}010 * 8/2 = 4{,}040$ lbs. Therefore, the jack studs are more than adequate to carry the axial loads from the reactions at the ends of the header (shear capacity of the header connection to the king stud is conservatively neglected).

As for lateral loads, the king studs are considered to provide the main resistance to wind loads. The jack studs, cripples, lintels, and window framing will act as a system in resisting and transferring the lateral loads to the adjacent king studs. This framing system will resist a portion of the lateral load, thus reducing the total lateral load acting on the king studs. For simplicity, however, the king studs will be assumed to carry the full lateral load.

The maximum height of a king stud for a given lateral load can be easily calculated using standard engineering equations.

Lateral load acting on each of the king studs = 20 psf * (8')/4 = 40 plf

Considering the stud as a simply supported beam, the length of this beam cab be calculated as follows:

$M_{all} = wL^2/8 = 670.8$ ft.-lbs.   (From Attachment A)

$L^2 = 8 M_{all} / w$ $L = \sqrt{8} (670.8) / 40 = 11.58$ ft.

For a stud with mid-height spacing, the maximum height can also be found by dividing the lateral load by two to account for two king studs at each end, and than calculating the capacity of each king stud for a 60"spacing [(the width of the opening + the spacing to adjacent stud)/2].

Lateral load at each end of header = 20 psf / 2 = 10 psf

For 10 psf lateral load and 60" spacing the stud height can be calculated to be 9'-5".

Maximum height of a 2 x 4 x 43 mils king stud to resist 40 plf lateral load = 9'-5" which is greater than the maximum height of the wall studs. Therefore, the king studs are capable of resisting the lateral loads with adequate reserve capacity to handle the portion of the axial loads applicable to them.

COMMENTARY ON THE PRESCRIPTIVE METHOD FOR RESIDENTIAL
COLD-FORMED STEEL FRAMING

This method has significant conservatism built into it. The jack studs will resist portion of the lateral loads. The 2 x 4 x 43 jack studs have approximately 1902 lbs. reserve axial capacity that allows them to resist approximately 35 psf lateral load. This contribution by the jack studs will greatly reduce the lateral loads acting on the king studs. Furthermore, structural sheathing around the opening as required in the *Prescriptive Method* will also help in distributing the lateral load to the different elements of the framing.

COMMENTARY ON THE PRESCRIPTIVE METHOD FOR RESIDENTIAL COLD-FORMED STEEL FRAMING

C7.0 NON-STRUCTURAL WALLS

C7.1 Non-Load Bearing Studs

For interior non-load bearing partition walls, a minimum of 18 mil studs are specified in the *Prescriptive Method*. Consideration should be given for acceptable serviceability (rigidity) at interior wall openings for doors and other purposes. ASTM C 645 is specified as the reference standard for non-load bearing studs. This is what currently is used in the industry. The *Prescriptive Method* provides a table for limiting heights for selected non-load bearing studs subjected to a lateral load of 5 psf. The table is provided for guidance only and should only be used where specific manufacturers tables or other applicable instructions and/or recommendations are not available.

COMMENTARY ON THE PRESCRIPTIVE METHOD FOR RESIDENTIAL
COLD-FORMED STEEL FRAMING

C8.0   ROOF SYSTEMS

C8.1   Roof Construction

Roof construction in the *Prescriptive Method* is limited to design wind speeds less than or equal to 110 mph (fastest mile) and to Seismic Zones 0, 1, 2, 3, and 4. The roof construction figure shown in the *Prescriptive method* is given to complement the requirements of the tables. For wind conditions in excess of 90 mph exposure C, uplift straps must be deigned by a design professional.

C8.2   Allowable Ceiling Joists

Ceiling joist tables in the *Prescriptive Method* provide the maximum allowable ceiling joist spans for two loading conditions: 10 psf and 20 psf attic live loads. Ceiling joists require stiffeners at all support locations. Joist spans are calculated based on the loading conditions, deflection criteria, and engineering methods described below.

Loading conditions:

In the design of ceiling joists, any one of several engineering criteria may control the prescriptive requirements depending on the configuration of the section, thickness of material, and member length. The analysis must include checks for:

- Yielding
- Flexural buckling
- Web crippling (not required because web stiffeners are specified)
- Shear
- Deflection

The engineering approach used in the *Prescriptive Method* considers ceiling joists dead and live load combinations as they apply to the design of joists. The objective was to maintain a simple table format without sacrificing the economies possible through case-by-case design.

The following applied loads were used in developing the joist tables:

Dead Load: Ceiling Dead Load = 5 psf

Live Loads: With Attic Storage = 20 psf
                                     Without Attic Storage = 10 psf

These loads are widely accepted by the building industry, ASCE 7-93, and are specified in the three building codes: CABO, SBCCI, and BOCA.

The section properties, moment and shear capacities were determined per the AISI Specification, as shown in a previous section. All joists are assumed punched with 1-1/2" wide x 4" long hole. The compression flange of the ceiling joists are assumed to be laterally at mid-point, third point, and unbraced.

The provisions for the combined bending and web crippling and combined bending and shear were taken directly from the AISI Design Specification.

COMMENTARY ON THE PRESCRIPTIVE METHOD FOR RESIDENTIAL
COLD-FORMED STEEL FRAMING

Deflection Limit:

Deflection limits are primarily established with regard to serviceability concerns. The intent is to prevent excessive deflection which might result in cracking of finishes. For ceiling joists, most codes generally agree that L/240 represents an acceptable serviceability limit for deflection. Therefore, ceiling finishes in the *Prescriptive Method* are limited to L/240 deflection.

C8.3   Ceiling Joist Bracing

Bracing of the bottom flanges specified in the *Prescriptive Method* is based on industry practice and engineering judgement. Gypsum board (i.e. finished ceilings) is considered to be adequate bracing for the tension flanges. Ceiling joist tables provide spans for braced, as well as unbraced, top flanges. For braced compression flanges it is necessary for steel strapping to have blocking (or bridging) installed at a maximum spacing of 12 feet and at the termination of all straps. Moreover, the ends of steel straps are designed to be fastened to a stable component of the building.

C8.4   Rafters

The rafter span table was designed from gravity loads only, hence the rafter spans are reported on the horizontal projection of the rafter, regardless of the slope. The gravity loads consist of a 10 psf dead load and the greater of a 16 psf live load or the applied snow load. Applied snow loads are calculated by multiplying the ground snow load by 0.7 (no further reductions or increases are made for special cases). Ground snow load intervals are as follows: 20 psf, 30 psf, 50 psf, and 70 psf.

Wind load effects are correlated to equivalent snow loads as shown in the *Prescriptive Method*. Wind pressures were calculated using the ASCE 7-93 Components and Cladding method. Wind loads acting perpendicular to the plane of the rafter were adjusted to represent wind loads acting orthogonal to the horizontal projection of the rafter. Wind loads were examined for both uplift and downward loads and the worst case was correlated to a corresponding snow load.

Permissible roof slopes range between 3:12 through 12:12 and more importantly, the roof system must consist of both ceiling joists (i.e. acting as rafter ties) and rafters combined. The *Prescriptive Method* does not currently address cathedral ceilings. The ridge board connection is made with a 2 inch clip angle, of equivalent size and thickness as the rafter member. The clip angle connects the rafter to the ridge board with #10 screws in each leg.

The heel joint connection, which connects the ceiling joist to the rafter, uses # 10 screws. Lapped ceiling joists shall be connected with the same screw size and number (or more) as the heel joint connection.

C8.5   Rafter Bottom Flange Bracing

The bracing requirements provided in the *Prescriptive Method* are self explanatory and are commonly used in residential steel construction.

C8.6   Splicing

Splicing of structural members is not a typical application, but some situations may arise where splicing requirements would be useful. Applications may include repair of damaged members, simplified details for dropped ceilings, and others.

Splices, generally, are required to transfer shear, bending moments, and axial loads. Some splices may occur over points of bearing, and splicing may only be required to transmit nominal axial loads. The ceiling joists and rafter spans provided in the *Prescriptive Method* are based on the assumption that the members are continuous, with no splices (except for punchouts and hole stiffeners). Therefore, ceiling joists and rafter members shall not be spliced without an approved design. Splicing details for non-structural members are shown in the *Prescriptive Method*.

C8.7   Roof Trusses

The *Prescriptive Method* allows the usage of any engineered truss system provided that its design is within the applicability limits table. Any pre-engineered truss member can not be cut, notched, or altered unless so designed.

C8.8   Uplift Resistance and Gable Endwall Bracing

In wind conditions greater than 90 mph exposure C and in Seismic Zones 3 and 4, uplift straps and endwall bracing must be designed by a design professional.

COMMENTARY ON THE PRESCRIPTIVE METHOD FOR RESIDENTIAL
COLD-FORMED STEEL FRAMING

C8.9   Design Example

Ceiling Joist Design Example

The following is an example for selecting and designing a ceiling joist member for the following building configuration and loading conditions:

Roof width = 30 ft   Wind speed= 80 mph
Attic D.L.= 5 psf   Exposure C
Ground snow load = 30 psf   Roof slope= 6:12
Spacing o.c. = 24 in   Roof D.L.= 10 psf
Load bearing wall at mid-span

First, select the proper ceiling joist from the ceiling joist tables in the *Prescriptive Method*:

With an interior bearing wall the clear span would be 14'-6" (30'/2 = 15' less 6", i.e. 4" for outside wall and 2" for inside wall), select a ceiling joist at 24 "oc, dead load 5 psf and live load of 10 psf. One feasible solution is a twin span 2 x 6 x 33 with a maximum allowable span of 14'-10".

The section properties for this member can be calculated using the AISI Design Specification (refer to the example in Section 2.11 for calculating section properties). The section properties listed below are taken from Attachment A.

F_y = 33 ksi, S_x = .5054 in^3 I_x = 1.4506 in^4 D.L. = 15 psf
F_{ya} = 33 ksi r_y = .5874 in x_o = -1.1365 in r_x = 2.1096 in
G = 11,300 ksi r_o = 2.4672 in   J = .00013 in^4 Spacing = 24" oc
E = 29,500 ksi C_W = .702 in^6 A_{gross} = .326 in^2 $I_{x\ gross}$ = 1.4506 in^4
$I_{y\ gross}$ = .1125 in^4 Def.$_{LL}$= L/240   Def.$_{TL}$= L/180   K = 1
L.L. = 10 psf   D.L. = 5 psf   L.L. = 10 psf

The allowable span shall be the shorter of:

    1. maximum span due to bending
    2. maximum span due to shear
    3. maximum span due to deflection
    4. maximum span due to bending and shear combined

<u>1.</u>   <u>Bending per AISI Design Specification Section C, "Flexural Members"</u>
        (all referenced equations and sections are those of the AISI Design Specification)

$M_a = M_n/\Omega_f$
M_n = smallest nominal moment of Sections C.1.1, C.1.2, and C.1.3
Ω_f = 1.67 (factor of safety)

C.1.1 Nominal Section Strength
$M_n = S_e F_y$
$S_e = S_x$
$M_n = (.5054in^3)x(33ksi) = 16.678$ kip-in

C3.1.2 Lateral Buckling Strength

COMMENTARY ON THE PRESCRIPTIVE METHOD FOR RESIDENTIAL
COLD-FORMED STEEL FRAMING

———

$M_n = S_c (M_c / S_f)$ Eq. C.1.2-1
$S_f = I_x / Y_{cg} = 1.4506 / 2.75 = 0.52749$ in.^{4}

$M_c = M_y (1 - M_y / 4M_e)$ Eq. C.1.2-2
$M_y = S_f F_y$ Eq. C.1.2-4
$M_y = 0.52749 * 33,000 = 17,407.2$ lb.-in.
$C_b = 1.75$ ($M_1 = 0$, Section C.1.2)

$M_e = C_b r_o A \sqrt{\sigma_{ey}\sigma_t}$ Eq. C.1.2-5
$\sigma_{ey} = (\pi^2 E)/(K_y L_y/r_y)^2$ Eq. C.1.2-8
$\sigma_t = (1/Ar_o^2)/[GJ + \pi^2 E C_w / (K_t L_t)^2]$ Eq. C.1.2-9

$\sigma_{ey} = (\pi^2 * 29500000)/(1 * 89 / 0.5874)^2 = 12,682.62$ psi

$\sigma_t = [1 / 0.326 * (2.4672)^2] / [11,300,000 * 0.00013 + \pi^2 * 29,500,000 * 0.702 / (1 * 89)^2]$
$\sigma_t = 13,744.1$ psi

$M_e = 1.75* 2.4672 * .326 \sqrt{(12,682.62)(13,744.1)} = 18,583.29$ lb.-in. $> 0.5 M_y = 8,703.6$ lb.-in.

$M_c = 17,407.2 (1 - 17,407.2 / 4 * 18583.29) = 13,330.81$ lb.-in.

Determine Sc, the elastic section modulus of the effective section, calculated at a stress of M_c / S_f in the extreme compression fiber. $M_c / S_f = 13,330.81 / 0.527491 = 25,272.12$ psi

Compression Flange:

Determine the effective width of the compression flange:

w = Flange width - 2(radius + Thickness) = 1.3683 "
w/t = 1.3683/.0346 = 39.55 <60 O.K.

$S = 1.28\sqrt{(E/f)} = 1.28\sqrt{(29500000/25,272.12)} = 43.732$ Eq. B4-1

$S/3 = 14.577 < w/t = 39.55 < S = 43.732$

$I_a = 399t^4\{[(w/t)/S] - 0.33\}^3$ Eq. B4.2-6
$I_a = 399(.0346)^4\{[(39.55)/43.732] - 0.33\}^3 = 0.000108$ in.^{4}
n = 1/2

D/w = 0.5 / 1.3683 = 0.36541, $0.25 < D/w = 0.36541 < 0.8$

$k = [4.82 - 5(D/w)](Is/Ia)^n + 0.43 \le 5.25 - 5(D/w)$ Eq. B4.2-9

d = lip-radius-thickness = .37165 in

$I_s = d^3t/12 = (0.37165)^3 (0.0346)/12 = 0.000148$ in.^{4}

$k = [4.82 - 5(0.36541)](0.000148 / .000108)^{0.5} + 0.43 \le 5.25 - 5(0.3654169)$
k = 3.4229153

COMMENTARY ON THE PRESCRIPTIVE METHOD FOR RESIDENTIAL COLD-FORMED STEEL FRAMING

Effective width of the flange shall be calculated according to AISI Design Specification Section B2.1 using k as calculated above, and $f = M_c / S_f$.

Since Is > Ia and D/w < 0.8, the stiffener is not considered as a simple lip.

w/t = 39.55 < 90 OK   Section B1.1-(a)-(3)

$\lambda = [1.052/\sqrt{k}](w/t) \sqrt{f/E}$ Eq. B2.1-4
Where $f = M_c / S_f$ = 25,272.12 psi
$\lambda = [1.052/\sqrt{3.4229153}](39.55) \sqrt{25,272.12/29500000}$ = 05043702
$\lambda \leq 0.673 \Rightarrow b = w$ Eq. B2.1-1

Effective width **b = 1.3683 inches**   (i.e. compression flange is fully effective)

Compression Stiffener:

Determine the effective width of the compression stiffener (lip):
D = 0.5 inches
d = Lip size - radius - thickness = 0.5 - 0.09375 - 0.0346 = 0.37165 inches.
d/t = 0.37165/.0346 = 10.741 < 60   OK   Section B1.1-(a)-(3)

D/w = 0.5 / 1.3683 = 0.365416,   0.25 < D/w = 0.365416 < 0.8

Calculate the effective width for the lip per AISI Design Specification Section B2.1 :

$\lambda = [1.052/\sqrt{k}](w/t) \sqrt{f/E}$ Eq. B2.1-4
Where $f = M_c / S_f$ = 25,272.12 psi and k = 0.43 per AISI Design Specification Section B3.1

$\lambda = [1.052/\sqrt{0.43}](810.741) \sqrt{25,272.12/29500000}$ = 0..5043702
$\lambda \leq 0.673 \Rightarrow d'_s = d = 0.37165"$ Eq. B2.1-1
$d_s = d'_s (I_s/I_a) \leq d'_s$ Eq. B4.2-11

Effective width of lip d_s = **0.37165 inches** (i.e. compression stiffener is fully effective)

Thus, one assumes that the section is fully effective. Y_{cg} = 5.5 / 2 = 2.75 in. (from symmetry)

Web:

Check if web is fully effective: (AISI Design Specification Section B2.3)

Assume the web is fully effective, and top fiber stress is 25,272 psi:

Element	Effective length (in)	Distance from top fiber, y (in.)	Ly in.^{2}	Ly^2 in.^{3}	I$_1'$ About own axis in.^{3}
Web	5.2433	2.75	14.4191	39.6525	12.0125
Tension flange	1.3683	5.4827	7.5020	41.1311	-
Comp. Flange	1.3683	.0176	.0237	.0004	-

59

COMMENTARY ON THE PRESCRIPTIVE METHOD FOR RESIDENTIAL
COLD-FORMED STEEL FRAMING

Upper corners	0.3487^1	.0576	.0201	.0012	-
Lower corners	0.3487^1	5.4424	1.8977	10.3283	-
Upper stiffener	0.3717	.3142	.1168	.0367	.0043
Lower stiffener	0.3717	5.1858	1.9273	9.9947	.0043
Sum	9.4206	19.25	25.9066	101.145	12.02104

[1] Where corner length = Length of arc = $1.57(R + t/2) = 0.1743485$. Two corners = 0.348697"

Distance from top fiber to x-axis is $Y_{cg} = 25.9066 / 9.4206 = 2.75$ in

$f_1 = +f (Y_{cg} - t - R)/Y_{cg} = 25,272.12(2.75 - 0.0346 - 0.09375)/2..75 = 24,092.6$ psi

$f_2 = -f (\text{Web size} - Y_{cg} - t - R)/Y_{cg} = 25,272.12(5.5 - 2.75 - 0.0346 - 0.09375)/2.75$
$f_2 = -24,092.6$ psi (tension)

The effective widths are calculated per AISI Design Specification Section B2.3 (a) for load capacity determination.

$\Psi = f_2 / f_1 = -1 \leq -0.236$ Section B2.3
$b_1 = b_e / (3 - \Psi)$ Eq. B2.3-1
$b_2 = b_e / 2$ Eq. B2.3-2
$b_1 + b_2$ shall not exceed the compression portion of the web calculated on the basis of effective section.

b_e = Effective width b determined in accordance with Section B2.1 with f_1 substituted for f and k determined as follows:
$k = 4 + 2(1 - \Psi)^3 + 2(1 - \Psi)$ Eq. B2.3-4

$k = 4 + 2(1 + 1)^3 + 2(1 + 1) = 24$
$h = w = \text{Web} - 2(\text{Radius} + \text{thickness}) = 5.5 - 2(0.09375 + 0.0346) = 5.2433$ inches.
$h/t = 5.2433 / 0.0346 = 151.54 < 200$ OK   Section B1.2-(a)

$\lambda_{(web)} = [1.052/\sqrt{24}](151.54) \sqrt{24,092.6/29500000} = 0.929968 > 0.673$ Eq. B2.1-4
$\rho_{(web)} = (1 - .22/\lambda)/\lambda = (1 - .22/.929968)/.929968 = .820922$

$b_e = \rho_{(web)} w = 4.304338$ "   Eq. B2.1-1
$b_2 = b_e / 2 = 2.1521691$ inches
$b_1 = b_e / (3 + 1) = 1.0760846$ inches.
$b_1 + b_2 = 1.0760846 + 2.1521691 = 3.2282537$ inches

Compression portion of the web = $y_{cg} - (\text{Radius} + \text{thickness})$
                    $= 2.75 - (0.09375 + 0.0346) = 2.62165$ inches.

Since $b_1 + b_2 = 3.2282537 > 2.62165$ in., $b_1 + b_2$ shall be taken as 2.62165 in.
This verifies the assumption that the web is fully effective.

$I'_x = Ly^2 + I'_1 - Ly^2_{cg}$
$I'_x = 101.1448 + 12.02104 - 19.25(2.75)^2 = 41.92257$ in.^{3}

Actual $I_x = I'_x t = (41.92257)(0.0346) = 1.450521$ in.^{4}

$S_e = I_x / y_{cg} = 1.450521 / 2.75 = 0.527462 \text{ in.}^3$

$M_n = S_e F_y$
M_n $= 0.527462 * 25,272.12 = $ **13,330.09 lb-in**

Eq. C.1.1-1

C.1.3 "Beams Having One Flange Through-Fastened to Deck or Sheathing" does not apply.

Allowable moment:

The allowable moment is the smallest nominal moment calculated per AISI Design Specification Sections C.1.1, C.1.2, and C.1.3 divided by a factor of safety of 1.67.

$\Omega = 1.67$
$M_a = M_n / 1.67$
M_a $= 13,330.09/1.67 = 7,982.09 \text{ lb.-in.} = $ **665.17 ft.-lb.**

Eq. C.1-1

$M_{max} = (9/128)\omega L^2$ (twin span beam, with web stiffeners, at the mid span where bottom flange is unbraced)

$\omega = 15 \text{ psf (10 psf Attic LL + 5 psf DL)}$
$\omega = 15 \text{ psf x 2 ft O.C.} = 30 \text{ plf}$

$L^2 = (128/9 \times 7,982.09 \text{lb-in}) / (30/12 \text{ lb/in})$
$L = 213.09 \text{ in}$
$L = 17 \text{ ft - 9 in}$

2.   <u>C.2 Strength for shear (with ICBO AC46 method of calculating shear with punchout)</u>

Calculate the depth of the flat portion of the web, h.
h = web depth - 2(r+t)
$h = 5.5-2(.09375-.0346) = 5.2433 \text{ in}$

$h/t = 151.54$
Calculate $1.38(E_v/F_y)^{.5}$
$1.38(Ek_v / F_y)^{.5} = 1.38(29,5000 \times 5.34/33)^{.5} = 95.35$

Since $h/t > 1.38(Ek_v/F_y)^{.5}$
$V_a = .53(Ek_v t^3/h)$
$V_a = .53(29,500,000 \times 5.34 \times .00346^3 /5.2433) = 659.6 \text{ lb}$ (without holes)
ICBO method for calculating shear capacity with punchouts
$q_s = 1 - 1.1$(punchout depth / web depth)   (q_s is the shear reduction factor)
$q_s = 1 - 1.1(1.5/5.5) =.7$
$V_{a(hole)} = V_a \times q_s = 461.7 \text{ lb}$

$V = 5/8 \omega L$ (simply supported twin span with web stiffeners)
$L =(5/8) \times V_{a(hole)} / \omega$
$L = (8/5) \times 461.7 / 30$
$L = 24.62 \text{ ft}$
$L= 24'-7.5"$

3.   <u>Span limit due to deflection</u>

COMMENTARY ON THE PRESCRIPTIVE METHOD FOR RESIDENTIAL
COLD-FORMED STEEL FRAMING

———

$$\delta = \omega \times L^4 / (185 \times E \times I) \qquad \text{(simply supported twin span)}$$

Total load ($\delta = L/240$, $\omega = 15$ psf)
$L = [454{,}791{,}635 \times I_x / (\omega \times \text{spacing} \times \delta)]^{1/3}$
$L = 19.69$ ft
$L = 19'\text{-}8"$

4.   Calculate the span limit due to combined bending and shear at the intermediate support.
      (AISI Design Specification Section C.3   Strength for Combined Bending and Shear)

$M/M_{ado} \le 1$ and $V/V_a \le 1$

M_{ado} = Allowable moment about the centroidal axis per AISI Design Specification Section C.1.1
$M_{axo} = 832.25$ ft-lb (as calculated before)

$V_a = 659.6$ lb   (Va without punched holes because punchouts are not allowed
within 12" of a bearing point)
$V = (5/8)\,\omega L$ (maximum shear at the intermediate support)

$M = \omega L^2/8$ (Maximum moment at intermediate support considering the
member fully braced at both flanges)

$[(\omega L^2/8) / 832.25.63] \le$ 1
$[(\,(5/8)\,\omega L) / 659.6] \le$ 1

Solve for L using worst case (shear and moment):

$L = 14.88$ ft when $M = M_{ado}$
$L = 14'\text{-}10.56"$

Therefore, at $L = 14.88$ ft, $V = (5/8)\,\omega L = (5/8) \times 30 \times 14.88 = 279$ lb

Since V/V_a (279/659.6) $= 0.42 < 0.7$ the interaction equation need not be checked.
Hence, $L = 14' = 10.56"$

Allowable rafter span is controlled by AISI Design Specification Section C.3 $= \mathbf{L = 14'\text{-}10"}$

Rafter Design Example

The following is an example for selecting and calculating the capacity for the rafters required for the example
building configuration and loading condition shown in the ceiling joist example above.

Rafter Selection:

The horizonal rafter projection is just under 15 feet. From the "Allowable Spans for High Slope Rafters"
table, select a rafter at 24 " oc and the worst case of a ground snow load of 30 psf or an equivalent load to a
wind speed of 80 mph. First, convert the 80 mph wind speed to an equivalent ground snow load. For a slope
of 6:12, an 80 mph Exposure C converts to a 20 psf ground snow load. Therefore, the 30 psf ground snow
load should be used because it is higher. Two feasible solutions are 2 x 8 x 54 spanning 17'-1" or a 2 x 10 x
43 spanning 16'-0".

COMMENTARY ON THE PRESCRIPTIVE METHOD FOR RESIDENTIAL
COLD-FORMED STEEL FRAMING

Select a ridge board and its connection.

The ridge board is a built up section composed of a joist and track of equivalent size and thickness to the rafter selection. The rafter is connected to the ridge board with a clip angle with # 10 screws in each leg. For a 32 foot wide roof and 30 psf ground snow load, three #10 screws are required in each leg of the clip angle.

Select number of # 10 screws for heel joint connection

A 6:12 roof slope with 30 psf ground snow load and 32 foot wide house requires five # 10 screws at the heel joint.

Calculate the capacity of the members selected to prove their structural adequacy.

i.   Calculate the wind load due to the 80 mph, Exposure C, wind speed:

Both uplift and inward acting wind loads must be examined and the worst case converted to an equivalent snow load effect. ASCE 7-93 components and cladding coefficients are used to calculate the wind pressures.

Downward load:

$q = .00256 \times K_z \times (GC_p + GC_{pi}) \times (V \times I)^2$
$K_z = .87$ for exposure C
$GC_p = 0$ (Zone 1, Tributary area = 75 Ft^2, pg.18 ASCE 7-93)
$GC_{pi} = .25$ ($\pm .25$ whichever is worst)
$I = 1.0$ (non-costal)
$q = .00256 \times .87 \times (0+.25)(80 \times 1)^2$
$q = 3.56$ psf   (acting perpendicular to the plane of the rafter)
$q_{y\text{-slope}} = q/\cos\theta$ (converts load to an equivalent load acting in the y direction along the rafter plane)
$q_{y\text{-slope}} = 3.56 / \cos(26.56) = 3.98$ psf
$q_{horizontal} = q_{y\text{-slope}}/\cos\theta$ (converts load to an equivalent load acting perpendicular to the horizontal projection)
$q_{horizontal} = 3.98 / \cos(26.56) = 4.45$ psf

Upward load:

$q = .00256 \times K_z \times (GC_p + GC_{pi}) \times (V \times I)^2$
$K_z = .87$ for exposure C
$GC_p = -1.2$ (Zone 1, Tributary area = 75 Ft^2, pg.18 ASCE 7-93)
$GC_{pi} = -.25$ ($\pm .25$ whichever is worse)
$I = 1.0$ (non-costal)
$q = .00256 \times .87 \times (-1.2-.25)(80 \times 1)^2$
$q = -20.67$ psf   (acting perpendicular to the plane of the rafter)
$q_{y\text{-slope}} = q/\cos\theta$ (converts load to an equivalent load acting in the y direction along the rafter plane)
$q_{y\text{-slope}} = -20.67 / \cos(26.56) = -23.11$ psf
$q_{y\text{-slope}} + DL = -23.11 + 10 = 13.11$ (compensate for DL while loads are in the same direction and plane)
$q_{horizontal} = (q_{y\text{-slope}} + DL)/\cos\theta$ (converts load to an equivalent load acting perpendicular to the horizontal projection)
$q_{horizontal} = -13.11 / \cos(26.56) = -14.66$ psf

ii   Now let us apply these loads to the rafter members to calculate the max. allowable span.

COMMENTARY ON THE PRESCRIPTIVE METHOD FOR RESIDENTIAL
COLD-FORMED STEEL FRAMING

From the Section Property Table (Attachment A)

F_y = 33 ksi S_x = 1.4125 in^3 I_x = 7.9814 in^4 D.L. = 15 psf
S.L.$_{.ground}$= 30 psf   Def.$_{TL}$= L/180   Def.$_{LL}$= L/240   Spacing = 24" oc

The allowable span shall be the shorter of:
     1. maximum span due to bending
     2. maximum span due to shear
     3. maximum span due to deflection

1.   Bending per AISI Design Specification Section C, "Flexural Members"
     (all referenced equations and sections are those of the AISI Design Specification)

$M_a = M_n/\Omega_f$
M_n = smallest nominal moment of C.1.1, C.1.2, C.1.3
Ω_f = 1.67 (factor of safety)

C.1.1 Nominal Section Strength

$M_n = S_e F_y$
$S_e = S_x$
M_n = (1.4125in^3)x(33ksi) = 46.6 kip-in

C.1.2 Lateral Buckling Strength

This section does not apply because roof sheathing provides lateral support.

C.1.3 Beams Having One Flange Through-Fastened to Deck or Sheathing

Does not apply.
M_a= (46.6 kip-in)/1.67 = 27.9 kip-in

$M_{max} = \omega L^2 / 8$ (simply supported single span)

ω = (15 + 30 x .7)psf = 36 psf
ω = 36 psf x 2 ft O.C. = 72 plf

L^2 = (8 x 27.9 x 1000 lb-in) / (72/12 lb/in)
L = 192.8 in
L = 16'-1"

2.   C.2 Strength for Shear (with ICBO AC46 method of calculating shear with punchout)

Calculate the depth of the flat portion of the web, h.
h = web depth - 2(r+t)
h = 10-2(.09374+.0451) = 9.7223 in

h/t = 215.57

Calculate 1.38$(E_v/F_y)^{.5}$
1.38$(Ek_v/F_y)^{.5}$ = 1.38(29,5000x5.34 / 33)$^{.5}$ = 95.35

COMMENTARY ON THE PRESCRIPTIVE METHOD FOR RESIDENTIAL
COLD-FORMED STEEL FRAMING

Since $h/t > 1.38(E_v/F_y)^{.5}$
$V_a = .53(E_v t^3/h)$
$V_a = .53(29,500,000 \times 5.34 \times .0451^3 / 9.7223) = 787.8$ lb (without holes)

ICBO AC46 method for calculating shear capacity with punchouts
$q_s = 1 - 1.1$(punchout depth/web depth) shear reduction factor
$q_s = 1 - 1.1(1.5/10) = .835$
$V_{a(hole)} = Va \times q_s = 657.8$ lb

$V = \omega L/2$ (simply supported single span)
$L = 2 \times V_{a(hole)} / \omega$
$L = 2 \times 657.8 / 72$
$L = 18.27$ ft $= 18'\text{-}3''$

3.   Span limit due to deflection

$\delta = 5 \times \omega \times L^4 / (384 \times E \times I)$ (simply supported single span)

Total load ($\delta = L/180$, $\omega = 36$ psf)

$L = [188,800,000 \times I_x / (\omega \times spacing \times \delta)]^{1/3}$
$L = 21.32$ ft $= 21'\text{-}4''$

Live load ($\delta = L/240$, $\omega = 21$ psf)

$L = [188,800,000 \times I_x / (\omega \times spacing \times \delta)]^{1/3}$
$L = 23.18$ ft $= 23'\text{-}2''$

Bending controls the design, hence the allowable rafter span is **$L = 16'\text{-}1''$**

iii.   Verify the adequacy of the ridge board shear connection.

Consider the horizontal projection of a simply supported rafter.

$V_{max} = \omega L/2$ (L=15 ft, ω=72 plf)
$V_{max} = 72 \times 15 / 2 = 540$ lb

The nominal shear capacity of a # 10 screw is 789 lb. Therefore the allowable screw capacity (S_{allow}) is:

$S_{allow} = 789/3$ (use a safety factor of 3)
$S_{allow} = 263$ lb

Number of screws needed $= 540/263 = 2.05$ screws

This number is rounded up to 3 screws.

iv.   Verify the adequacy of the heel joist connection

SL = 21 psf (30 x 0.7)   House Width = 32 ft (use 32' to match table intervals)

COMMENTARY ON THE PRESCRIPTIVE METHOD FOR RESIDENTIAL
COLD-FORMED STEEL FRAMING

Roof DL = 10 psf

Spacing = 24 in

Ceiling DL = 5 psf

Screw Capacity = 243 lb (w/ a safety factor of 3)

Roof Loading Diagram

Find the reactions.

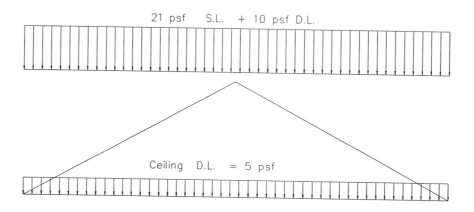

$$R = \omega L / 2$$
$$R = ((21+10+5) \times 2) \times 32 / 2 = 1,152 \text{ lb}$$

A distributed load of [(21+10) x2] psf or 62 plf is spread across the length of the building. One fourth of this load will be concentrated at the end walls, joints 1 & 2, and half the load will be concentrated at the ridge board connection, joint 3. Similarly, a distributed load of 5 psf or 10 plf is spread across the length of the ceiling joist. This load will be divided equally at each end of the wall (i.e. at joints 1 and 2).

Free Body Diagram

COMMENTARY ON THE PRESCRIPTIVE METHOD FOR RESIDENTIAL
COLD-FORMED STEEL FRAMING

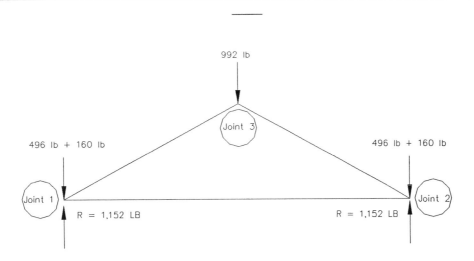

Joint 1 is examined to see which force has a greater magnitude (the compression in F_{13} or the tension in F_{12}).

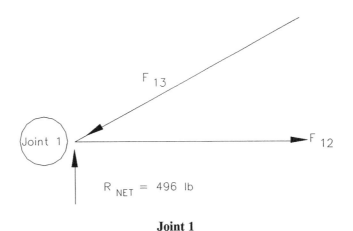

Joint 1

$\theta = \tan^{-1} (6/12) = 26.565$ degrees
$F_{13} = 496 / \sin(26.565) = 1109.01$ lb

$F_{12} = 496 \times \cot(26.565) = 992.21$ lb

Heel joint connection shall be designed from the compression in F_{13} since it represents the worst case.

Number of screws = 1109.01/243 = 4.56

Round the number of screws up to 5.

For both the heel joint and the ridge board connection analysis, the factor of safety for the number 10 screws is 3.

COMMENTARY ON THE PRESCRIPTIVE METHOD FOR RESIDENTIAL
COLD-FORMED STEEL FRAMING

___

C9.0   MECHANICAL & UTILITIES

This section provides references to the applicable building codes that must be followed when installing mechanical and utility equipment or services in residential steel framed houses.

C10.0   CONSTRUCTION GUIDELINES

The construction guidelines are provided to supplement the requirements of the *Prescriptive Method* and are considered good construction practices. These guidelines should not be considered comprehensive. Manufacturer's catalogs, recommendations, and other technical literature should also be consulted.

Driveability of screws in different thicknesses of steels have been an issue with end users. This issue is addressed in the *Prescriptive Method* by providing a table listing the maximum steel thicknesses that a particular size screws can be driven into. The maximum steel thickness represents the total thickness of the connected parts. This information is provided in the *Prescriptive Method* in the General Guidelines Section because the driveability of certain screws vary from one manufacturer to another, and the user should always consult with screw manufacturers for proper fastener application.

Thermal concerns have also been an issue with end users. Guidance for insulation of steel framed walls in various climates are found in AISI publication RG-9405 (ref. 11.15). For a complete energy analysis a design professional that is familiar with steel framing may be consulted.

COMMENTARY ON THE PRESCRIPTIVE METHOD FOR RESIDENTIAL COLD-FORMED STEEL FRAMING

COMMENTARY ON THE PRESCRIPTIVE METHOD FOR RESIDENTIAL COLD-FORMED STEEL FRAMING

C11.0   REFERENCES

11.1   *Specification For The Design Of Cold Formed Steel Structural Members*, August 19, 1986 Edition with December 11, 1989 Addendum.

11.2   *Minimum Design Load for Buildings and Other Structures*, ASCE 7-93. American Society of Civil Engineers, New York, NY. 1993.

11.3   *CABO One and Two Family Dwelling Code*, 1995 Edition.

11.4   Building Officials & Code Administrators International, Inc. *The BOCA National Building Code*, 1993 Edition.

11.5   International Conference of Building Officials *Uniform Building Code*, 1994 Edition.

11.6   Southern Building Code Congress International, Inc. *Standard Building Code*, 1993 Edition.

11.7   ICBO Evaluation Service, Inc. No. AC46 *Acceptance Criteria for Steel Studs and Joists*, Approved July, 1992; Effective April, 1993.

11.8   WJE report number 73345 *Report of Laboratory Tests and Analytical Studies of Structural Characteristics of Cold-Formed Steel-Joists Floor Systems*, December 1977.

11.9   ASTM Standards:

   A 307 - 93a *Standard Specification for Carbon Steel Bolts and Studs, 60000 PSI Tensile Strength*.

   A 653 - 1994 *Specification for Steel Sheet, Zinc-Coated (galvanized) or Zinc-Iron Ally Coated (Galvanized) by the Hot-Dip Process*.

   A 875-1994 *Specification for Steel Sheet, Zinc-5% Aluminum Alloy Metallic-Coated by the Hot-Dip Process*.

   A 792-1994 *Specification for Steel Sheet, Aluminum-Zinc Alloy-Coated by the Hot-Dip Process, General Requirements*.

   C 645 - 1988 *Specification for Non-Load (Axial) Bearing Steel Studs, Runners (Tracks), and Rigid Furring Channels for Screw Application of Gypsum Board*.

11.10   Society of Automotive Engineers SAE J-78-1979 *Steel Self Drilling Tapping Screws*.

11.11   *Cold-Formed Steel Design*, Second Edition; By: Wei-Wen Yu; John Wiley & Sons, Inc.

11.12   CCFSS Technical Bulletin Vol. 2, No. 1, February 1993, *AISI Specification Provisions for Screw Connections*.

COMMENTARY ON THE PRESCRIPTIVE METHOD FOR RESIDENTIAL
COLD-FORMED STEEL FRAMING

11.13   Australian Institute of Steel Construction (AISC) *"Structural Performance Requirements For Domestic Steel Framing"* Australia.

COMMENTARY ON THE PRESCRIPTIVE METHOD FOR RESIDENTIAL
COLD-FORMED STEEL FRAMING

ATTACHMENT B

Metric Conversion

COMMENTARY ON THE PRESCRIPTIVE METHOD FOR RESIDENTIAL COLD-FORMED STEEL FRAMING

Metric Conversion

1 inch   = 25.4 millimeters (mm.)

1 mil = 1/1000 inch = 0.0254 millimeter (mm.)

1 foot   = 304.8 mm. = 3.05 meters (m)

1 pound/foot^2 (psf) = 0.0479 kN/m^2 = 48 Pascals (Pa)

1 mile = 5280 feet = 1.609 kilometers (km)

1 mile/hr. (mph) = 1.467 feet/sec. (ft/sec) = 1.609 kilometers/hr. (km/hr)

1 mile/hr. (mph) = 0.02682 kilometers/min. = 49 meters/sec. (m/sec.)

1 pound force (lbf) = 4.5 Newtons (N)

1 pound/inch^2 (psi)   = 703 kilograms/meter^2 (kgs./m^2) = 6.9 kilo-Pascals (kPa)

D

EZFRAME User's Manual

EZFRAME User's Manual
VERSION 2.0
California Energy Commission's Automated Procedure for Calculating U-values
of Framed Envelope Assemblies for Residential and Nonresidential Buildings.

SEPTEMBER 1994

CALIFORNIA
ENERGY
COMMISSION

ACKNOWLEDGEMENTS

EZFRAME is a computer program developed by the California Energy Commission to automate the calculation of the overall U-value and the overall R-value of wood and metal frame envelope assemblies for showing compliance with residential and nonresidential standards. The EZFRAME program uses the thermal and physical characteristics of the layers in an envelope assembly to calculate the overall thermal characteristics (R-value and U-value) for the assembly. This program was designed and developed by Soheil Loghmanpour, with support from Jim Trowbridge, technical staff in the Efficiency Technology Office.

TABLE OF CONTENTS

SECTION 1.   INTRODUCTION

New Features of EZFRAME Version 2.0

If you have used EZFRAME version 1.0, you will find many improvements and new features in EZFRAME version 2.0. These features include changes to file operations and data entry which make the program more efficient and user friendly. The files that you have created using version 1.0 are fully compatible with version 2.0.

EZFRAME version 2.0 has the following new features:

▸   The ability to list files located in the current directory or selected directory and retrieve the desired file into the EZFRAME program.

▸   Additional dialog boxes to ensure that you do not accidently overwrite a saved file or discard your current unsaved information.

▸   The ability to select assembly layers from the library. The library has most of the common materials used in construction today. The user can add, delete or modify the information in the library if needed.

▸   New on-line "Help" system which allows the user to easily access the instructions for entering the required information for modeling an assembly.

▸   The name of the current file appears on every screen to prevent accidental overwrite.

▸   Hot keys for entering Frame Type, Envelope Type and the layer information.

▸   The Layer Description field has been increased from 14 characters to 18 characters to allow longer description for clarity.

▸   The calculation module has been modified to improve the accuracy of the results. In version 1.0, when the air space in the assembly is adjacent to the framing member, the program uses the isothermal planes method to calculate the overall U-value. For small frame spacing (under four feet), this method may be overly conservative by up to 25 percent. In version 2.0, a combination of the zone method and the isothermal planes method is used.

Application of the EZFRAME Program

Steel framing systems for homes and commercial buildings have gained popularity in recent years. The use of steel in building envelope components can have a significant impact on the thermal performance of the building. Thermal conductivity of a steel stud is 310 times greater than that of typical wood stud. If the wood framing members of an assembly is replaced with steel stick for stick, the overall R-value of the assembly is reduced by more than 50 percent.

When showing compliance with the building energy efficiency standards, it becomes important that the energy analyst accurately account for the thermal effects of the steel framing. The energy analyst will have to show if an envelope assembly meets the U-value requirement of the standards. For medium-mass and heavy-mass assemblies, residential FORM-3RM 'Residential Masonry Wall Assembly' or nonresidential ENV-3 Form 'Proposed Masonry Wall Assembly' must be used. For light-mass framed assemblies, the energy analyst may use one of the following options:

    1.   Use the tabulated default values included in the residential and nonresidential manuals,

    2.   Use nonresidential ENV-3 Form for metal framed assemblies for both residential and nonresidential buildings,

    3.   Hand calculate the U-value using the ASHRAE zone method and accounting for the appropriate framing percentage, knock-outs in the web and appropriate assumptions for air spaces, metal skin, interior and exterior insulations in the assembly,

    4.   Use the EZFRAME program. This program requires minimal information about the assembly being analyzed. After the assembly is modeled, the EZFRAME program automatically applies the appropriate method and assumptions for calculating the overall R-value and U-value of the assembly. The application of the EZFRAME program is limited to light-mass assemblies (heat capacity less than 7.0 Btu/ft²-°F) and it can be used for both wood and steel framed construction assemblies. Typically, if an envelope assembly includes a brick, clay or masonry layer with a thickness greater than or equal to 2.5 inches, that assembly will have a heat capacity of 7.0 Btu/ft²-°F or greater and therefore considered to be a medium-mass or a heavy-mass assembly.

The ASHRAE procedures for calculating the overall U-value for metal frame envelope assemblies are especially complex and lengthy which can lead to significant errors and inaccuracies in the calculations as well as making it difficult for the building official to review these calculations. There are many different metal frame systems available to the building community and new systems are continuously entering the construction market. The calculation procedure for metal frame assemblies require not only the thermal properties of the assembly layers, but also the physical properties and the order in which the layers are assembled. In order to properly and accurately calculate the U-value of these assemblies, the energy analyst needs to know the appropriate procedure and assumptions in every case and situation. Variations in metal frame systems and the complexity of the calculations will often prohibit the building official from checking the calculations in any detail because of the limited time available to review the submittal.

The Commission has developed the EZFRAME program to automate ASHRAE procedures in order to help energy analysts and others in the building community calculate the U-values of wood and especially metal frame assemblies with higher accuracy and efficiency. This program will also be a useful tool for the building official who needs to quickly verify the calculated values in a submittal. The EZFRAME program can be used for demonstrating envelope compliance for both residential and nonresidential buildings.

Organization of This Manual

This manual is organized in eight sections:

SECTION 1.   Introduction -- a general discussion about EZFRAME program and the system requirements.

SECTION 2.   Technical Approach -- the techniques and assumptions that the program uses to calculate the assembly U-value.

SECTION 3.   Installation -- system requirements and installation procedure.

SECTION 4.   Program Inputs -- instructions on how to model and how to enter parameters for an envelope assembly.

SECTION 5.   Library Operations -- instructions on how to retrieve, exit, add, delete and modify the library information.

SECTION 6.   File Operations -- describes how you can save and retrieve a file.

SECTION 7.   Running EZFRAME -- describes how to use the on-line help system, calculate results, print FORM-3R/ENV-3 form and how to exit the program.

SECTION 8.   Examples -- describes how to set up and run typical and special envelope assemblies.

Most of the information needed to model an envelope assembly such as the thermal properties of the assembly layers are included in ASHRAE Handbook of Fundamentals. Additional information may be found in the Nonresidential Manual or in the Residential Manual. These manuals are available at the Energy Commission's Publication Office.

	Publication #	Price
Nonresidential Manual	P400-92-005	$28.00
Residential Manual	P400-92-002	$34.00

You may order these publications by sending your check or money order to

California Energy Commission
Publications MS-13
P.O. Box 944295
Sacramento, California 94244-2950

Questions about the publications or this program may be directed to the Energy Hotline at (800) 772-3300 or (916) 654-5106.

The EZFRAME program can be used with any IBM-compatible personal computer.

SECTION 2.   TECHNICAL APPROACH

Overview

Heat transfer across an envelope assembly is the result of convection on both interior and exterior surfaces and conduction through the layers. In a framed assembly where the framing member penetrates the insulation in the cavity, a greater percentage of heat is conducted through the framing member because it is a much better conductor than the insulation it penetrates. This is referred to as a "thermal bridge" in the ASHRAE Handbook of Fundamentals. Therefore, more conductive framing members transfer more heat and degrade the thermal performance of the assembly. For example, when metal framing is used, nearly fifty percent more heat is conducted through the envelope assembly compared to the same assembly with wood framing.

The Energy Efficiency Standards require that ASHRAE methods be used for calculating the U-value of framed envelope assemblies with a heat capacity of less than 7.0 Btu/ft^2-°F. ASHRAE uses the parallel flow method for calculating the overall U-value of wood framed assemblies. The parallel flow method assumes unidirectional transmission of heat through the envelope assembly. For metal framed assemblies, ASHRAE uses the zone method because this method assumes bi-directional heat transfer. Another ASHRAE technique used for metal framed assemblies is isothermal planes method. The calculation details for determining the effective U-value of metal framed assemblies are included in Chapter 22 of 1989 ASHRAE Handbook of Fundamentals. The zone method and isothermal planes method are briefly described in the following paragraphs.

1.   **Isothermal Planes Method:** The Isothermal Planes method treats the core layer that consists of insulation between framing members and the framing member itself as having the thermal properties of a homogeneous material with an equivalent R-value which is the weighted average of metal member and cavity insulation R-values based on the framing flange area and the cavity area. After calculating the equivalent R-value of the core layer, it is directly added to the R-values of the remaining assembly layers.

2.   **Zone Method:** The Zone method divides the assembly into two separate assemblies (zones) that act in parallel for transferring heat. The two zones are the cavity zone which is unaffected by thermal bridging and framing zone which is the assembly that includes the metal framing and is affected by thermal bridging. The area transmittance (UA) of these two zones are calculated separately. The effective R-value of the assembly is calculated by weight averaging of the area transmittances of the two zones based on the zone areas.

Based on the information and the parameters that are entered into the EZFRAME program for describing the assembly, the program uses the appropriate combination of these methods and assumptions for calculating the U-value of that assembly.

Assumptions

The EZFRAME program uses additional assumptions when calculating the U-value of **metal** framed assemblies. These assumptions are used when : (1) the assembly layers are not directly attached to the framing member creating an air space between the framing and the assembly layers, (2) a layer of air is present anywhere in the assembly, (3) an insulative sheathing is used as an assembly layer, or (4) when metal skin is used in the assembly.

The zone method assumes that heat is conducted in both longitudinal and lateral (parallel to the assembly surface) directions. For that reason, the framing zone area is determined by drawing lines which are at 45 degree angles relative to the flange surface indicating that conduction occurs at the same rate in both lateral and longitudinal directions . This assumption is relatively accurate when the assembly layers have similar thermal conductivity values which will prevent accumulation of heat on one layer when the adjacent layer is too conductive (such as metal) or too resistive (such as insulative sheathing). When insulative sheathing, metal or air space is used as a layer in an assembly, the 45 degree lines that define the zones will not properly describe the flow path of heat in that assembly.

The zone method is sensitive to the thickness of the assembly layers. This sensitivity is more obvious for a layer that has a high R-value such as an insulative sheathing. Calculations have shown that if an insulation layer that is not penetrated by a framing member such as an interior or an exterior insulative sheathing is included in the zone calculations, the ability of the insulation layer to insulate may be misrepresented. For instance, when a thin layer of insulative sheathing (about 1/2") is used, the zone method over-estimates the R-value of the sheathing and when a thick layer of insulative sheathing (about 2") is used, the zone method under-estimates the R-value of the sheathing. For that reason, the EZFRAME program initially ignores the insulative sheathing. After the effective R-value of the assembly without the insulative sheathing is calculated, the program adds the R-value of the sheathing to the calculated R-value to obtain the effective R-value of the complete assembly.

When air space separates the assembly layers (ie, some or all of the assembly layers are attached to the framing via furring), a combination of the zone method and the isothermal planes method is used. The program separates the assembly into two sub-assemblies. Sub-assembly one consists of the framing member, cavity insulation and all of the assembly layers except the air space(s) and layers that are connected to the framing via furring (air space) and sub-assembly two that consists of the air space(s) and layers not considered in sub-assembly one. The program calculates the R-value of sub-assembly one using the zone method and adds it to the resistances of the layers in sub-assembly two using the isothermal planes method. A similar procedure is also used for cases where a layer of steel is used in the assembly.

For wall assemblies, an additional 5 percent of framing is assumed in order to account for the effect of the additional framing members in the window area, cross members and wall header and footer.

Therefore, the program reduces the spacing between studs to account for the additional framing members. Spacing between joists and rafters are not adjusted because it is assumed there are no skylights or cross members in the floors, ceilings and roofs.

SECTION 3.   INSTALLATION

You can run the EZFRAME program from your hard disk or a floppy disk. You should copy this program and the library into a separate directory. This will prevent your root directory from cluttering and will allow easy identification of your saved program files.

To load the program onto the hard disk, change to the directory where you want to load the EZFRAME program, insert the diskette into drive A: or B: and at the C:\> prompt type:

COPY A:*.*   *.*

This command will copy both the EZFRAME.EXE program and the library of materials (LIBRARY.RSC file) into the directory that you specified.

If you would like to run this program from a floppy drive, the installation to your hard disk is not required. To run the program, at DOS prompt (A:\> or B:\>) type EZFRAME and follow the instructions for entering the envelope parameters.

Whether you are running the program from the hard drive or floppy drive, you must have the library in the current directory.

SECTION 4. PROGRAM INPUTS

The information describing an envelope assembly is entered into the EZFRAME program by completing the four worksheets (Windows) in the program. You can pull up one Window at a time by pressing and holding the **Alt** key and pressing the Window number. The following paragraphs will describe the entries for these Windows.

Window #1. Frame Type & Layers Descriptions -- This Window is automatically called when you invoke the program and press a key. You can bring up this Window at any point within the program by pressing **Alt-1**. You must describe all of the layers in your assembly because you will not be able to enter the information about layers that have not been described in this Window. Do not enter descriptions such as 'NONE', 'N/A' etc., for layers that do not exist. In Window #1, you can enter the project name, envelope assembly name, frame type and the descriptions of the assembly layers on both sides of the framing. The order of the layers is very important. Figure 1 illustrates the layers numbering convention. Please be sure that Layer #1 is the layer adjacent to the frame .

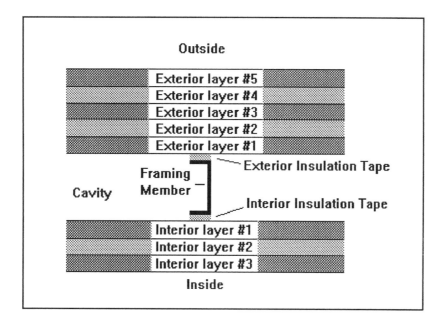

Figure 1. Envelope assembly layers numbering convention

The parameters included in this Window are:

1. *Project Name*: The project name as you would like it to appear on the FORM-3R/ENV-3 Form. The project name can include up to 40 alphanumeric characters.

2. *Envelope Name*: The name of the envelope assembly as you would like it to appear on the FORM-3R/ENV-3 Form. The envelope name can include up to 12 alphanumeric characters.

3. *Frame Type*: The frame type used in the assembly. Acceptable entries are **W** for Wood and **M** for Metal. The program assumes 0.99 h-ft^2-°F/Btu per inch for the thermal resistance of wood framing and 0.0032 h-ft^2-°F/Btu per inch for the thermal resistance of metal framing.

4. *Interior Layer #1*: The description of the first layer in the interior side of theframe (facing the inside of the building). The first layer is adjacent to the frame. You can enter up to 18 alphanumeric characters to describe this layer. Leave this entry blank if it does not exit.

5. *Interior Layer #2*: The description of the second layer in the interior side of the frame. The second layer is adjacent to layer #1. You can enter up to 18 alphanumeric characters to describe this layer. Leave this entry blank if it does not exist.

6. *Interior Layer #3*: The description of the third layer in the interior side of the frame. The third layer is adjacent to layer #2. You can enter up to 18 alphanumeric characters to describe this layer. Leave this entry blank if it does not exist.

7. *Exterior Layer #1*: The description of the first layer in the exterior side of the frame (facing the outside of the building). The first layer is adjacent to the frame. You can enter up to 18 alphanumeric characters to describe this layer. Leave this entry blank if it does not exist.

8. *Exterior Layer #2*: The description of the second layer in the exterior side of the frame. The second layer is adjacent to layer #1. You can enter up to 18 alphanumeric characters to describe this layer. Leave this entry blank if it does not exist.

9. *Exterior Layer #3*: The description of the third layer in the exterior side of theframe. The third layer is adjacent to layer #2. You can enter up to 18 alphanumeric characters to describe this layer. Leave this entry blank if it does not exist.

10. *Exterior Layer #4*: The description of the fourth layer in the exterior side of the frame (facing the outside of the building). The fourth layer is adjacent to layer #3. You can enter up to 18 alphanumeric characters to describe this layer. Leave this entry blank if it does not exist.

11.   *Exterior Layer #5*: The description of the fifth layer in the exterior side of the frame. The fifth layer is adjacent to layer #4. You can enter up to 18 alphanumeric characters to describe this layer. Leave this entry blank if it does not exist.

The program has a built-in library with complete modeling information for common construction materials. You can call up the library by pressing **Alt-L**. The library is accessible only when you are in Window #1 and the cursor is on Interior Layers #1 through #3 or Exterior Layers #1 through #5. When you are in the library Window, use the Up and Down arrow keys or the Enter key to move the scroll bar up and down. To select a material from the library, place the cursor on the layer that you want to describe and press **Alt-L**. The library Window will appear. While you are in the library, move the scroll bar to the desired material, press the Space Bar and the program will automatically retrieve the information about the selected material into the program and assign it to the layer that you wanted to describe.

If you delete a layer description in Window #1, the related information in Windows #3 or #4 such as its R-value and thickness will be automatically deleted. For more information about the library features and operations, please refer to Section 5.

Window #2.   **Frame & Cavity Information** -- In this Window, you can enter the information about the envelope type (wall, floor, ...), frame size, spacing between framing members and the cavity insulation R-value. You can bring up this Window at any point within the program by pressing **Alt-2**. The parameters included in this Window are:

1.   *Envelope Type*: Enter the envelope type of your assembly. Acceptable entries are **W** for Wall, **F** for Floor, **R** for Roof or **C** for Ceiling. For insulated ceilings with vented attic space, the attic air must be treated as outside air and the ceiling becomes the exterior envelope. For insulated ceilings you must use the Ceiling option for a horizontal ceiling and the Roof option for a vaulted ceiling regardless of the slope of the roof. The attic air and the layers beyond it are not included in the calculations. For non-vented attics with horizontal ceiling, use the Ceiling option and assume that the roof is parallel to the horizontal ceiling. Include all layers from inside to the roof line. For the thickness of the attic air, you may use the average of the height of the attic space. For vaulted ceilings, use the Roof option and include all layers from inside to the roof line.

2.   *Frame Depth*: Enter the actual depth of the framing member in inches. See Figure 2.

3.   *Flange Width (for metal frames)*: Enter the actual width of the flange face for metal frame in inches. See Figure 2. The framing percentage is automatically calculated by the program using the flange width and the frame spacing.

4.   *Frame Width (for wood frames)*: Enter the actual width of the wood frame in inches. See Figure 2. The framing percentage is automatically calculated by the program using the frame width and the frame spacing.

5.   *Frame Spacing*: Enter the center to center distance between framing members in inches.

6.   *Web Thickness (for metal frames)*: Enter the web thickness of the frame in inches. See Figure 2. The frame thickness is usually specified in Gauge, but you should be able to obtain the thickness in inches from the steel manufacturer or you may use the following default Table:

GAUGE	THICKNESS (Inches)
12	0.1046
14	0.0747
16	0.0598
18	0.0478
20	0.0359
22	0.0299
24	0.0239
26	0.0179
28	0.0149

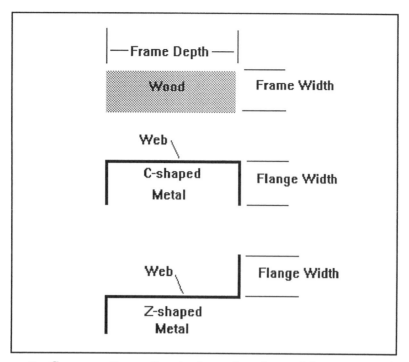

Figure 2. *Cross-sections of typical wood and metal framing members*

7.   *Percent Knock-out in Web (for metal frames)*: Enter the percentage of the length of the web which does not conduct heat because of the knock-outs. The knock -out percentage is typically 15 percent. For regularly spaced knock-outs, it is calculated by dividing the length of the knock-out by the spacing between them. Figure 3 illustrates how the knock-out percentage is calculated for a typical metal frame.

8.   *Cavity Insulation R-value*: Enter the R-value of the insulation in the cavity (between framing members). If your assembly does not have any insulation between the framing members, you must enter the resistance of the cavity air space. If the cavity insulation in compressed, enter the compressed R-value.

9.   *R-value of Insulation Tape on Interior Flange (for metal frames)*: Enter the R-value of the insulation tape placed on the flange facing the inside of the building.

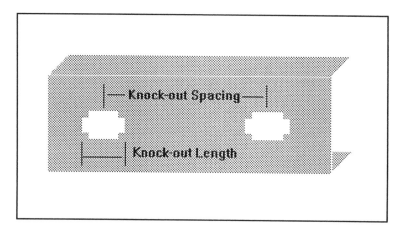

Figure 3. *Knock-outs in a metal frame web*

10.   *Thickness of Insulation Tape on Interior Flange (for metal frames)*: Enter the thickness of the insulation tape placed on the flange facing the inside of the building. The thickness must be expressed in inches.

11.   *R-value of Insulation Tape on Exterior Flange (for metal frames)*: Enter the R-value of the insulation tape placed on the flange facing the outside of the building.

12.   *Thickness of Insulation Tape on Exterior Flange (for metal frames)*: Enter thethickness of the insulation tape placed on the flange facing the outside of the building. The thickness must be expressed in inches.

Window #3.   **Interior Layers Information** -- In this Window, you can enter the information about the assembly layers that are in the interior side of the assembly. For the described layers you must enter the layer R-value, the layer thickness and whether this layer is an air space, a sheet of metal or interior insulation. Interior insulation refers to any interior insulation that is not penetrated by framing members. You can enter up to three layers excluding the inside air film. You can bring up this Window at any point within the program by pressing **Alt-3**. Depending on the envelope type the program automatically selects and uses the appropriate inside air film resistance for the assembly. For each interior layer, the parameters included in this Window are:

1.   *Interior Insulation (for metal frames)*: If the layer is an interior insulation you must respond **Y** for Yes, otherwise enter **N** for No.

2.   *Air Space or metal (for metal frames)*: If the layer is an air space or sheet of metal

enter **Y** for Yes, otherwise enter **N** for No. In assemblies where metal furring is used to attach the assembly layers to widely spaced framing members, layer #1 becomes an air space with a thickness equal to the thickness of the furring.

3.   *R-value*: Enter the R-value of the layer in h-ft^2-°F/Btu.

4.   *Thickness*: Enter the thickness of the layer in inches.

If the library is used to enter the information for an assembly layer, the required information for this Window is automatically entered from the library and no additional action will be required by the user.

Window #4.   **Exterior Layers Information** -- In this Window, you can enter the information about the assembly layers which are in the exterior side of the assembly. You can enter the layer R-value, the layer thickness and whether this layer is an air space, sheet of metal or exterior insulation. You can enter up to five layers excluding the outside air film. You can bring up this Window at any point within the program by pressing **Alt-4**. Depending on the envelope type the program automatically selects and uses the appropriate outside air film resistance for the assembly. For each exterior layer, the parameters included in this Window are:

1.   *Exterior Insulation (for metal frames)*: If the layer is an exterior insulation you must respond **Y** for Yes, otherwise enter **N** for No.

2.   *Air Space or metal (for metal frames)*: If the layer is an air space or sheet of metal enter **Y** for Yes, otherwise enter **N** for No. In assemblies where metal furring is used to attach the assembly layers to widely spaced framing members, layer #1 becomes an air space with a thickness equal to the thickness of the furring.

3.   *R-value*: Enter the R-value of the layer in h-ft^2-°F/Btu.

4.   *Thickness*: Enter the thickness of the layer in inches.

If the library is used to enter the information for an assembly layer, the required information for this Window is automatically entered from the library and no additional action will be required by the user.

SECTION 5.   LIBRARY OPERATIONS

Alt-L.   Library -- You can open the Library only when you are in Window #1 describing the layers. After the library is opened, the program allows the user to add, delete or modify materials in the library. Again, the library can be accessed only when you are in Window #1 and the cursor is at the location where you describe layers. After you select and retrieve an item from the library, the required information about that layer is automatically retrieved by the program and placed in the appropriate Window and layer location. The information about an assembly layers that have been selected from the library can be modified if needed and saved for that particular assembly. The library information will not change unless you edit the library when you are in the library Window.

You can run the EZFRAME program from any directory as long as you specify the path to the directory where the program is located. However, if you want to use the materials library, be sure that the library is in the **current** directory.

Alt-I.   Insert an Item in the Library -- While you are in the library Window, move the scroll bar to the position just above where you want to insert a new item and press **Alt-I**. A small window will appear in the bottom of the screen where you can enter the information about the new item. Be sure that you complete the required information or you will not be able to save the new item. Press Enter to insert the new item. The Esc key will discard the information about the new item and will place you back in the library Window. The user can store up to 600 items in the library.

Alt-R.   Remove an Item from the Library -- While you are in the library Window, move the scroll bar to the item that you would like to remove and press **Alt-R**.

Alt-M.   Modify an Item in the Library -- While you are in the library Window, move the scroll bar to the item that you would like to modify and press **Alt-M**. A small window will appear in the bottom of the screen where you can edit the item that you have selected. Press Enter to insert back the modified item. The Esc key will discard the edits and will place you back in the library Window with no changes to that item.

Space Bar.   Select and Retrieve an Item from the Library -- Use the Up and Down arrow keys or the Enter key to move the scroll bar to the desired item. Press Space Bar to retrieve the information about that item from the library.

Esc.   Exit the Library -- After you are through with the library, press the Esc key to close the library Window. At this time, all of the changes that you have made to the library is saved in LIBRARY.RSC file for future use.

SECTION 6.   FILE OPERATIONS

F1.   Retrieve -- You can retrieve a previously saved file by pressing **F1** at any point within the program. The program automatically shows the path for the current directory. Any file in that directory can be retrieved by specifying the file name after the path name. If the file that you want to retrieve is located in a different directory, you must change the path name to where the desired file is located. Before the new file is retrieved, the program will ask you if you want to save the current information. Respond **Y** for Yes or **N** for No and press Enter and follow instructions to save or discard the current information. After loading the new file, the program will automatically place you in Window #1.   You can cancel the **Retrieve** command by pressing the Esc key.

Alt-F1.   List Files & Retrieve --   You can retrieve a previously saved file from the directory by selecting it from the list of directory files. Press **Alt-F1** at any point within the program to list files in the current directory. You can list files from any directory by changing the path name. You can also request only certain pattern files to be shown on the screen. Move the scroll bar up or down to the desired file and press Space Bar to retrieve it into the program.   Before the new file is retrieved, the program will ask if you want to save the current work. Respond **Y** for Yes or **N** for No and press Enter and follow instructions to save or discard the current information. After loading a new file, the program will automatically place you in Window #1.   You can cancel the **List Files & Retrieve** command by pressing the Esc key.

F2.   Save -- You can save the current information by pressing **F2** at any point within the program. The program automatically shows the path for the current directory and any saved file will be placed in this directory. If you want to save the file in another directory, you must change the path name to the desired directory where you wish to save the file. If the file name you have entered refers to an existing file, the program will ask if you want to replace it. A **Y** (Yes) response will replace the existing file with the new file and a **N** (No) response will prompt you for a different file name. You can cancel the **Save** command by pressing the Esc key.

Alt-Z.   Clear -- If you are manually entering information for a new file, it is recommended that you first clear the Windows. Press **Alt-Z** and the program will reset all of the current parameters and place you in Window #1. Before the parameters are reset, the program will ask you if you want to save the current work. Enter the file name and press the Return key to save the current information or press the Esc key to discard it.   When you retrieve a file, the program automatically resets the parameters before the file is loaded.

SECTION 7.   RUNNING EZFRAME

Alt-C. Calculate -- After you have completed all four Windows, you can calculate the overall R-value and the overall U-value of the assembly by pressing **Alt-C**. You can review the results on the screen and if you need to change any of the parameters, simply press **Alt-(window number)** for the Window you want to call up. If you have not entered all of the required information for performing the calculations, an error message will appear on the screen informing you which parameters are needed but missing in order to perform the calculations. The program will automatically put you in the appropriate Window for entering that information. If you update any information in any of the four Windows, the calculation results will include the updated information.

The calculated results include both inside and outside air film resistances. If you are using the results of this program as an input to a compliance program that requires entering the R-value or the U-value of the envelope assembly without the air film resistances (R-values), you must subtract the air films from the calculated overall R-value.   The R-values of the inside and outside air films are printed in the Construction Components section of the FORM-3R/ENV-3 Form printed by the program. The new U-value will then be the inverse of the newly calculated overall R-value.

Alt-P. Print -- When you are satisfied with the results (calculated total U-value), you can print the FORM-3R/ENV-3 From by pressing **Alt-P**. The program uses the printer's default font to print the form.   It is recommended that the printer be set to a 10-pitch font (10 characters per inch) which will result in equal left and right margins. This form can be used for showing compliance with light-mass wood or metal framed assemblies for residential and nonresidential buildings.

When printing the form, the program will look for LPT1 printer port. If the default printer port in your computer is other than LPT1 (such as LPT2, LPT3, ..., COM1, COM2, ...), you must redirect the program output to your default printer port before starting the EZFRAME program by using the MODE command.

MODE LPT1 := [default printer port] :

For example, if your default printer port is COM1, the appropriate command will be

MODE LPT1 := COM1:

It will be more convenient to include this command in your AUTOEXEC.BAT file after the MODE command that configures the printer.   The redirection command is automatically executed when your computer is turned on or rebooted. Redirecting the program output to your default printer will not affect the operation of your computer or interrupt other application programs in your computer.   For more information about the MODE command, please refer to your DOS User's Reference.

F7.   Help -- The on-line "Help" provides the user with instructions for each entry of every Window. Place the cursor on the desired location on the screen and press F7. The instructions for that entry will appear on the screen. You can terminate the on-line "Help" by pressing any key while you are in the "Help" Window.

Alt-E.   Exit -- Before exiting the program, you will be asked if you want to save your information. If you have not saved your current information, this will be your last chance. Once you leave the program, any information that was not saved will be lost. Press the Esc key to exit the program without saving the current information. You can exit by pressing **Alt-E** at any point in the program.

SECTION 8.   EXAMPLES

Definitions

Thickness -- Thickness of an assembly layer is the length in inches of the conduction path of the layer. Longer conduction path (thicker layer) will increase the resistance (R-value) of the layer. Thickness is shown by the letter (L).

Area -- Area of a layer is the area in square feet of the surface that is perpendicular to the direction of the heat flow in the layer. Area is shown by the letter (A).

Conductivity -- Thermal conductivity (conductivity) of a layer is a measure of heat transfer rate through the layer per inch of thickness. For a given material, the conductivity is a constant value. The conductivity of a layer is the rate of energy in Btu per hour that crosses the layer that is one inch thick and one square foot in area with one degree Fahrenheit temperature difference across the two sides (Btu-in/h-ft^2-°F). Conductivity is shown by the letter (K).

Conductance -- Conductance of a layer is the thermal conductivity for a given thickness of the layer. The conductance of a layer is the rate of energy in Btu per hour that crosses the layer that has one square foot of area with one degree Fahrenheit temperature difference across the two sides (Btu/h-ft^2-°F). Conductance of a layer is calculated by dividing the conductivity by the thickness of the layer (K/L). Conductance is shown by the letter (C). Conductance is also referred to as Transmittance or U-value.

Resistance -- Resistance of a layer is the inverse of conductance of that layer (h-ft^2-°F/Btu). Resistance of a layer is calculated by dividing the thickness of the layer by the conductivity (L/K). Resistance is shown by the letter (R). Resistance ia also referred to as R-value.

Examples

This section will illustrate how an envelope assembly is modeled in the EZFRAME program. Since modeling the entire assembly is not practical, a representative area of the assembly is modeled. The representative area of an assembly is a section of the assembly that measures one foot along the framing member and has a width equal to the spacing between framing members.

Example #1 -- Wall assembly #1 has the following construction:

Interior layer #1 -- 5/8" Gyp Board $\qquad$ L/K = R = .56 h-ft^2-°F/Btu
Exterior layer #1 -- 1" Foam Type Sheathing $\qquad$ L/K = R = 4.16 $\qquad$ h-ft^2-°F/Btu
Exterior layer #2 -- 5/8" Plywood Exterior Facing $\quad$ L/K = R = .78 $\qquad$ h-ft^2-°F/Btu
Cavity insulation -- R-13 Batt Insulation $\qquad$ L/K = R = 13.00 $\qquad$ h-ft^2-°F/Btu
Framing information -- (2 × 4) Steel Framing

Spacing	= 16	inches O.C.
Depth (actual)	= 3.625	inches
Flange Width (actual)	= 1.625	inches
Thickness	= .060 inches	

The representative area of this wall assembly is a section that is one foot high and 16 inches wide.

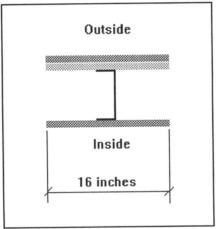

```
+---------- Window #1: Frame Type & Layers Descriptions ----------+
|                                                                 |
| Project Name : ABC Commercial Building                          |
| Envelope Name : wall-1                                          |
|                                                                 |
| Frame Type (Metal or Wood) ........ : Metal                     |
|                                                                 |
| Interior layer #1 (adjacent to frame)..... : 5/8" gyp board     |
| Interior layer #2 (adjacent to layer #1).. :                    |
| Interior layer #3 (adjacent to layer #2).. :                    |
|                                                                 |
| Exterior layer #1 (adjacent to frame)..... : 1" foam board      |
| Exterior layer #2 (adjacent to layer #1).. : 5/8" plywood       |
| Exterior layer #3 (adjacent to layer #2).. :                    |
| Exterior layer #4 (adjacent to layer #3).. :                    |
| Exterior layer #5 (adjacent to layer #4).. :                    |
|                                                                 |
| EXAMPLE1.INP          Up/Down/Enter              F7: Help        |
+-----------------------------------------------------------------+
```

```
+--------- Window #2: Frame & Cavity Information ----------+
|                                                         |
| Envelope type (Wall, Floor, Roof or Ceiling) : Wall     |
| Frame depth (inches) ...................... : 3.625     |
| Flange width (inches) ..................... : 1.625     |
| Frame spacing (inches) .................... : 16        |
| Web thickness (inches) .................... : .06       |
| Percent knock-out in web .................. : 14        |
| Cavity Insulation R-value ................. : 13        |
|                                                         |
|                                                         |
| R-value   of insulation tape on interior flange :       |
| Thickness of insulation tape on interior flange :       |
| R-value   of insulation tape on exterior flange :       |
| Thickness of insulation tape on exterior flange :       |
|                                                         |
|                                                         |
| EXAMPLE1.INP          Up/Down/Enter            F7: Help |
+---------------------------------------------------------+
```

```
+---------- Window #3: Interior Layers Information ------------+
|                                                             |
| 1. 5/8" gyp board    ; interior insulation (Y/N) . : NO     |
|                      ; air space or metal (Y/N) .. : NO     |
|                      ; R-value (h.ft2.F/Btu) ..... : .56    |
|                      ; thickness (inches) ........ : .625   |
| 2. No Description    ; interior insulation (Y/N) . :        |
|                      ; air space or metal (Y/N) .. :        |
|                      ; R-value (h.ft2.F/Btu) ..... :        |
|                      ; thickness (inches) ........ :        |
| 3. No Description    ; interior insulation (Y/N) . :        |
|                      ; air space or metal (Y/N) .. :        |
|                      ; R-value (h.ft2.F/Btu) ..... :        |
|                      ; thickness (inches) ........ :        |
|                                                             |
| EXAMPLE1.INP          Up/Down/Enter            F7: Help     |
+-------------------------------------------------------------+
```

```
+---------- Window #4: Exterior Layers Information ------------+
| 1. 1" foam board     ; exterior insulation (Y/N) . : YES    |
|                      ; air space or metal (Y/N) .. : NO     |
|                      ; R-value (h.ft2.F/Btu) ..... : 4.16   |
|                      ; thickness (inches) ........ : 1      |
| 2. 5/8" plywood      ; exterior insulation (Y/N) . : NO     |
|                      ; air space or metal (Y/N) .. : no     |
|                      ; R-value (h.ft2.F/Btu) ..... : .78    |
|                      ; thickness (inches) ........ : .625   |
| 3. No Description    ; exterior insulation (Y/N) . :        |
|                      ; air space or metal (Y/N) .. :        |
|                      ; R-value (h.ft2.F/Btu) ..... :        |
|                      ; thickness (inches) ........ :        |
| 4. No Description    ; exterior insulation (Y/N) . :        |
|                      ; air space or metal (Y/N) .. :        |
|                      ; R-value (h.ft2.F/Btu) ..... :        |
|                      ; thickness (inches) ........ :        |
| 5. No Description    ; exterior insulation (Y/N) . :        |
|                      ; air space or metal (Y/N) .. :        |
|                      ; R-value (h.ft2.F/Btu) ..... :        |
|                      ; thickness (inches) ........ :        |
| EXAMPLE1.INP          Up/Down/Enter            F7: Help     |
+-------------------------------------------------------------+
```

```
+--------------------- Results ---------------------+
|                                                    |
|   Results for  EXAMPLE1.INP :                      |
|                                                    |
|   Total R-value (h.ft2.F/Btu) =      11.11         |
|                                                    |
|   Total U-value (Btu/h.ft2.F) =      0.090         |
|                                                    |
|                                                    |
|  Note: These results include inside & outside air films.  |
+----------------------------------------------------+
```

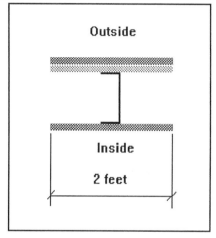

Example #2 -- Wall assembly #2 has the following construction:

Interior layer #1 -- 5/8" Gyp Board $L/K = R = .56$ h-ft^2-°F/Btu
Exterior layer #1 -- 7/16" OSB Sheathing $L/K = R = .62$ h-ft^2-°F/Btu
Exterior layer #2 -- 7/8" Masonry Stucco $L/K = R = .175$ h-ft^2-°F/Btu
Cavity Insulation $L/K = R = 30.0$ h-ft^2-°F/Btu
Steel Framing Member (Secondary)

Spacing	= 2	feet O.C.
Depth (actual)	= 8	inches
Flange Width (actual)	= 1.625	inches
Thickness	= .030	inches

Steel Framing Member (Primary)

Spacing	= 8	feet O.C.
Depth(actual)	= 8	inches
Flange Width(actual)	= 3.50	inches
Thickness	= .104	inches

The primary members are eight feet apart. There are three secondary members between primary members spaced at 24 inches (two feet) on center. Therefore, the primary and secondary members are equally spaced at 24 inches. In this type of construction the primary member and the secondary member must be modeled separately.   The total U-value will be the area weight average of the calculated U-values for the wall areas associated with the primary and the secondary members.

For the secondary member, the representative area is a section that is one foot high and two feet wide. The assembly with the secondary member will be modeled as follows:

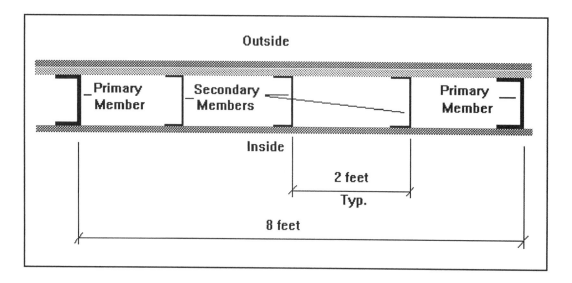

```
---------- Window #1: Frame Type & Layers Descriptions------------
¦                                                                 ¦
¦ Project Name : ABC Commercial Building                          ¦
¦ Envelope Name : wall-2 (sec)                                    ¦
¦                                                                 ¦
¦ Frame Type (Metal or Wood) ....... : Metal                      ¦
¦                                                                 ¦
¦ Interior layer #1 (adjacent to frame)..... : 5/8" gyp board     ¦
¦ Interior layer #2 (adjacent to layer #1).. :                    ¦
¦ Interior layer #3 (adjacent to layer #2).. :                    ¦
¦                                                                 ¦
¦ Exterior layer #1 (adjacent to frame)..... : 7/16" OSB Shtg     ¦
¦ Exterior layer #2 (adjacent to layer #1).. : 7/8" stucco        ¦
¦ Exterior layer #3 (adjacent to layer #2).. :                    ¦
¦ Exterior layer #4 (adjacent to layer #3).. :                    ¦
¦ Exterior layer #5 (adjacent to layer #4).. :                    ¦
¦                                                                 ¦
¦ EXAMPLE2.SEC          Up/Down/Enter              F7: Help ¦
+-----------------------------------------------------------------+
```

```
+--------- Window #2: Frame & Cavity Information ----------+
|                                                          |
| Envelope type (Wall, Floor, Roof or Ceiling) : Wall      |
| Frame depth (inches) ....................... : 8         |
| Flange width (inches) ...................... : 1.625     |
| Frame spacing (inches) ..................... : 24        |
| Web thickness (inches) ..................... : .03       |
| Percent knock-out in web ................... : 14        |
| Cavity Insulation R-value .................. : 30        |
|                                                          |
| R-value   of insulation tape on interior flange :       |
| Thickness of insulation tape on interior flange :       |
| R-value   of insulation tape on exterior flange :       |
| Thickness of insulation tape on exterior flange :       |
|                                                          |
|                                                          |
| EXAMPLE2.SEC           Up/Down/Enter            F7: Help |
+----------------------------------------------------------+
+---------- Window #3: Interior Layers Information ------------+
|                                                             |
| 1. 5/8" gyp board     ; interior insulation (Y/N) . : NO    |
|                       ; air space or metal (Y/N) .. : NO    |
|                       ; R-value (h.ft2.F/Btu) ..... : .56   |
|                       ; thickness (inches) ........ : .625  |
| 2. No Description     ; interior insulation (Y/N) . :       |
|                       ; air space or metal (Y/N) .. :       |
|                       ; R-value (h.ft2.F/Btu) ..... :       |
|                       ; thickness (inches) ........ :       |
| 3. No Description     ; interior insulation (Y/N) . :       |
|                       ; air space or metal (Y/N) .. :       |
|                       ; R-value (h.ft2.F/Btu) ..... :       |
|                       ; thickness (inches) ........ :       |
|                                                             |
| EXAMPLE2.SEC           Up/Down/Enter            F7: Help    |
+-------------------------------------------------------------+
+---------- Window #4: Exterior Layers Information ------------+
| 1. 7/16" OSB Shtg     ; exterior insulation (Y/N) . : NO    |
|                       ; air space or metal (Y/N) .. : NO    |
|                       ; R-value (h.ft2.F/Btu) ..... : .62   |
|                       ; thickness (inches) ........ : .4375 |
| 2. 7/8" stucco        ; exterior insulation (Y/N) . : NO    |
|                       ; air space or metal (Y/N) .. : NO    |
|                       ; R-value (h.ft2.F/Btu) ..... : .175  |
|                       ; thickness (inches) ........ : .875  |
| 3. No Description     ; exterior insulation (Y/N) . :       |
|                       ; air space or metal (Y/N) .. :       |
|                       ; R-value (h.ft2.F/Btu) ..... :       |
|                       ; thickness (inches) ........ :       |
| 4. No Description     ; exterior insulation (Y/N) . :       |
|                       ; air space or metal (Y/N) .. :       |
|                       ; R-value (h.ft2.F/Btu) ..... :       |
|                       ; thickness (inches) ........ :       |
| 5. No Description     ; exterior insulation (Y/N) . :       |
|                       ; air space or metal (Y/N) .. :       |
|                       ; R-value (h.ft2.F/Btu) ..... :       |
|                       ; thickness (inches) ........ :       |
| EXAMPLE2.SEC           Up/Down/Enter            F7: Help    |
+-------------------------------------------------------------+
```

```
+---------------------- Results ----------------------+
¦                                                     ¦
¦   Results for  EXAMPLE2.SEC :                       ¦
¦                                                     ¦
¦   Total R-value (h.ft2.F/Btu) =       13.56         ¦
¦                                                     ¦
¦   Total U-value (Btu/h.ft2.F) =       0.074         ¦
¦                                                     ¦
¦                                                     ¦
¦                                                     ¦
¦ Note: These results include inside & outside air films. ¦
+-----------------------------------------------------+
```

The representative area for the primary member will be an area which is one foot high and two feet
wide. This section of the assembly with the primary member will be modeled as follows:

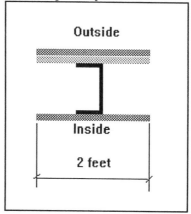

```
+---------- Window #1: Frame Type & Layers Descriptions ----------+
¦                                                                 ¦
¦ Project Name : ABC Commercial Building                          ¦
¦ Envelope Name : wall-2 (prm)                                    ¦
¦                                                                 ¦
¦ Frame Type (Metal or Wood) ........ : Metal                     ¦
¦                                                                 ¦
¦ Interior layer #1 (adjacent to frame)..... : 5/8" gyp board     ¦
¦ Interior layer #2 (adjacent to layer #1).. :                    ¦
¦ Interior layer #3 (adjacent to layer #2).. :                    ¦
¦                                                                 ¦
¦ Exterior layer #1 (adjacent to frame)..... : 7/16" OSB Shtg     ¦
¦ Exterior layer #2 (adjacent to layer #1).. : 7/8" stucco        ¦
¦ Exterior layer #3 (adjacent to layer #2).. :                    ¦
¦ Exterior layer #4 (adjacent to layer #3).. :                    ¦
¦ Exterior layer #5 (adjacent to layer #4).. :                    ¦
¦                                                                 ¦
¦ EXAMPLE2.PRM            Up/Down/Enter              F7: Help      ¦
+-----------------------------------------------------------------+
```

```
+--------- Window #2: Frame & Cavity Information ----------+
¦                                                         ¦
¦ Envelope type (Wall, Floor, Roof or Ceiling) : Wall     ¦
¦ Frame depth (inches) ....................... : 8        ¦
¦ Flange width (inches) ...................... : 3.5      ¦
¦ Frame spacing (inches) ..................... : 24       ¦
¦ Web thickness (inches) ..................... : .104     ¦
¦ Percent knock-out in web ................... : 14       ¦
¦ Cavity Insulation R-value .................. : 30       ¦
¦                                                         ¦
¦ R-value   of insulation tape on interior flange :       ¦
¦ Thickness of insulation tape on interior flange :       ¦
¦ R-value   of insulation tape on exterior flange :       ¦
¦ Thickness of insulation tape on exterior flange :       ¦
¦                                                         ¦
¦                                                         ¦
¦ EXAMPLE2.PRM        Up/Down/Enter           F7: Help ¦
+---------------------------------------------------------+
```

```
+----------- Window #3: Interior Layers Information -----------+
¦                                                             ¦
¦ 1. 5/8" gyp board    ; interior insulation (Y/N) . : NO     ¦
¦                      ; air space or metal (Y/N) .. : NO     ¦
¦                      ; R-value (h.ft2.F/Btu) ..... : .56    ¦
¦                      ; thickness (inches) ........ : .625   ¦
¦ 2. No Description    ; interior insulation (Y/N) . :        ¦
¦                      ; air space or metal (Y/N) .. :        ¦
¦                      ; R-value (h.ft2.F/Btu) ..... :        ¦
¦                      ; thickness (inches) ........ :        ¦
¦ 3. No Description    ; interior insulation (Y/N) . :        ¦
¦                      ; air space or metal (Y/N) .. :        ¦
¦                      ; R-value (h.ft2.F/Btu) ..... :        ¦
¦                      ; thickness (inches) ........ :        ¦
¦                                                             ¦
¦ EXAMPLE2.PRM        Up/Down/Enter           F7: Help ¦
+-------------------------------------------------------------+
```

```
+----------- Window #4: Exterior Layers Information -----------+
| 1. 7/16" OSB Shtg     ; exterior insulation (Y/N) . : NO     |
|                       ; air space or metal (Y/N) .. : NO     |
|                       ; R-value (h.ft2.F/Btu) ..... : .62    |
|                       ; thickness (inches) ........ : .4375  |
| 2. 7/8" stucco        ; exterior insulation (Y/N) . : NO     |
|                       ; air space or metal (Y/N) .. : NO     |
|                       ; R-value (h.ft2.F/Btu) ..... : .175   |
|                       ; thickness (inches) ........ : .875   |
| 3. No Description     ; exterior insulation (Y/N) . :        |
|                       ; air space or metal (Y/N) .. :        |
|                       ; R-value (h.ft2.F/Btu) ..... :        |
|                       ; thickness (inches) ........ :        |
| 4. No Description     ; exterior insulation (Y/N) . :        |
|                       ; air space or metal (Y/N) .. :        |
|                       ; R-value (h.ft2.F/Btu) ..... :        |
|                       ; thickness (inches) ........ :        |
| 5. No Description     ; exterior insulation (Y/N) . :        |
|                       ; air space or metal (Y/N) .. :        |
|                       ; R-value (h.ft2.F/Btu) ..... :        |
|                       ; thickness (inches) ........ :        |
| EXAMPLE2.PRM              Up/Down/Enter              F7: Help |
+-------------------------------------------------------------+

      +--------------------- Results ---------------------+
      |                                                   |
      |   Results for  EXAMPLE2.PRM :                     |
      |                                                   |
      |   Total R-value (h.ft2.F/Btu) =      9.15         |
      |                                                   |
      |   Total U-value (Btu/h.ft2.F) =      0.109        |
      |                                                   |
      |                                                   |
      |                                                   |
      | Note: These results include inside & outside air films. |
      +---------------------------------------------------+
```

Analysis summary:

Number of primary members in the representative wall section   = 1
Area of the wall associated with each primary member   = 2 ft^2
U-value of the wall associated with primary member   = .109
        U x Area x number of primary members = .109 x 2 x 1 = .218
Number of secondary members in the representative wall section   = 3
Area of the wall associated with each secondary member   = 2 ft^2
U-value of the wall associated with secondary member   = .074
U x Area x number of secondary members = .074 x 2 x 3 = .444
Total area of the representative wall section   = 8 ft^2

Total U-value = (.444 + .218) / 8 = .083   Btu/h-ft^2.F

Example #3 -- In wall assembly #3, the framing members are offset. The framing is isolated from the assembly layers by laying the insulation batt over the flange faces. The construction of this assembly is as follows:

Interior layer #1 -- 5/8" Gyp Board $L/K = R = .56$ h-ft²-°F/Btu
Exterior layer #1 -- 5/8" Plywood Exterior Facing $L/K = R = .78$ h-ft²-°F/Btu
Cavity insulation -- $L/K = R = 11.0$ h-ft²-°F/Btu
Framing information -- (2 × 4) Offset Steel Framing

Spacing	= 24	inches O.C.
Depth (actual)	= 3.625	inches
Flange Width (actual)	= 1.625	inches
Thickness	= .060 inches	

Because of the symmetrical arrangement of the framing members and the insulation, the representative section of this assembly will consist of a one foot high wall which is two feet wide (spacing between the framing members). When modeling this assembly, the insulation on the flange face will be assumed to be an interior or exterior insulation tape. The cavity insulation will be adjusted to account for the air pockets created around the framing member.

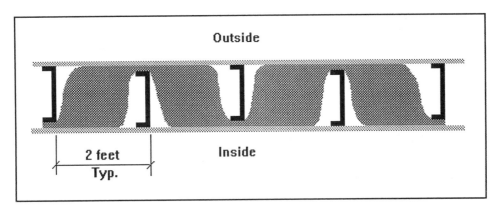

The compressed R-value of the insulation was obtained from ASHRAE Handbook of Fundamentals, Figure 2, Page 20.5. The fiber diameter for fiber glass insulation is approximately 0.00018 inches. R-11 insulation is 3.5 inches thick and has a density of approximately 0.50 lbs per ft³. From Figure 2 of ASHRAE Handbook, the maximum density for compressed insulation that has a fiber diameter of 0.00018 inches is 3 lbs per ft³. Assume that R-11 is fully compressed to (0.50 lbs per ft³ / 3 lbs per ft³) x 3.5 inches = 0.58 inches. Figure 2 shows that in the fully compressed condition, the conductivity per inch becomes .22 Btu-in/h-ft²-°F or transmittance of (.22 / 0.58") = .379 Btu/ h-ft²-°F. The compressed R-value will then be 1/.379 = 2.6 h-ft²-°F/Btu.

R$_{insulation}$ for fully compressed batt   = 2.6 h-ft^2-°F/Btu

R$_{insulation}$ for fully expanded batt   = 11.0   h-ft^2-°F/Btu

R$_{insulation}$ between fully expanded and fully compressed batt   = (2.6 + 11.0)/2
                                                                        = 6.8   h-ft^2-°F/Btu

It is further assumed that on each side of the metal member, the insulation will not fully expand for a distance from the metal member equal to the cavity depth (3.625 inches). Therefore, for the cavity insulation of the representative section, the length of fully expanded insulation will be

24" - (3.625" x 2) - 1.625 = 15.125"

The effective cavity R-value will be

$$R_{cavity} = \frac{(6.8 \times 2 \times 3.625) + (11 \times 15.125)}{15.125 + (2 \times 3.625)} = 9.6 \quad \text{h-ft}^2\text{-°F/Btu}$$

This assembly will be modeled as follows:

```
+---------- Window #1: Frame Type & Layers Descriptions ----------+
¦                                                                 ¦
¦ Project Name : ABC Commercial Building                          ¦
¦ Envelope Name : wall-3                                          ¦
¦                                                                 ¦
¦ Frame Type (Metal or Wood) ........ : Metal                     ¦
¦                                                                 ¦
¦ Interior layer #1 (adjacent to frame)..... : 5/8" gyp board     ¦
¦ Interior layer #2 (adjacent to layer #1).. :                    ¦
¦ Interior layer #3 (adjacent to layer #2).. :                    ¦
¦                                                                 ¦
¦ Exterior layer #1 (adjacent to frame)..... : 5/8" plywood       ¦
¦ Exterior layer #2 (adjacent to layer #1).. :                    ¦
¦ Exterior layer #3 (adjacent to layer #2).. :                    ¦
¦ Exterior layer #4 (adjacent to layer #3).. :                    ¦
¦ Exterior layer #5 (adjacent to layer #4).. :                    ¦
¦                                                                 ¦
¦ EXAMPLE3.INP          Up/Down/Enter              F7: Help ¦
+-----------------------------------------------------------------+
```

```
+--------- Window #2: Frame & Cavity Information ----------+
|                                                          |
| Envelope type (Wall, Floor, Roof or Ceiling) : Wall      |
| Frame depth (inches) ........................ : 3.625    |
| Flange width (inches) ....................... : 1.625    |
| Frame spacing (inches) ...................... : 24       |
| Web thickness (inches) ...................... : .06      |
| Percent knock-out in web .................... : 14       |
| Cavity Insulation R-value ................... : 9.6      |
|                                                          |
| R-value   of insulation tape on interior flange : 2.6    |
| Thickness of insulation tape on interior flange : .58    |
| R-value   of insulation tape on exterior flange :        |
| Thickness of insulation tape on exterior flange :        |
|                                                          |
|                                                          |
| EXAMPLE3.INP          Up/Down/Enter            F7: Help   |
+----------------------------------------------------------+


+----------- Window #3: Interior Layers Information ------------+
|                                                              |
| 1. 5/8" gyp board     ; interior insulation (Y/N) . : NO     |
|                       ; air space or metal (Y/N) .. : NO     |
|                       ; R-value (h.ft2.F/Btu) ..... : .56    |
|                       ; thickness (inches) ........ : .625   |
| 2. No Description      ; interior insulation (Y/N) . :        |
|                       ; air space or metal (Y/N) .. :        |
|                       ; R-value (h.ft2.F/Btu) ..... :        |
|                       ; thickness (inches) ........ :        |
| 3. No Description      ; interior insulation (Y/N) . :        |
|                       ; air space or metal (Y/N) .. :        |
|                       ; R-value (h.ft2.F/Btu) ..... :        |
|                       ; thickness (inches) ........ :        |
|                                                              |
| EXAMPLE3.INP          Up/Down/Enter            F7: Help       |
+--------------------------------------------------------------+


+----------- Window #4: Exterior Layers Information ------------+
| 1. 5/8" plywood       ; exterior insulation (Y/N) . : NO     |
|                       ; air space or metal (Y/N) .. : NO     |
|                       ; R-value (h.ft2.F/Btu) ..... : .78    |
|                       ; thickness (inches) ........ : .625   |
| 2. No Description      ; exterior insulation (Y/N) . :        |
|                       ; air space or metal (Y/N) .. :        |
|                       ; R-value (h.ft2.F/Btu) ..... :        |
|                       ; thickness (inches) ........ :        |
| 3. No Description      ; exterior insulation (Y/N) . :        |
|                       ; air space or metal (Y/N) .. :        |
|                       ; R-value (h.ft2.F/Btu) ..... :        |
|                       ; thickness (inches) ........ :        |
| 4. No Description      ; exterior insulation (Y/N) . :        |
|                       ; air space or metal (Y/N) .. :        |
|                       ; R-value (h.ft2.F/Btu) ..... :        |
|                       ; thickness (inches) ........ :        |
| 5. No Description      ; exterior insulation (Y/N) . :        |
|                       ; air space or metal (Y/N) .. :        |
|                       ; R-value (h.ft2.F/Btu) ..... :        |
|                       ; thickness (inches) ........ :        |
| EXAMPLE3.INP          Up/Down/Enter            F7: Help       |
+--------------------------------------------------------------+
```

```
+----------------------- Results ---------------------+
|                                                     |
|   Results for  EXAMPLE3.INP :                       |
|                                                     |
|   Total R-value (h.ft2.F/Btu) =        8.91         |
|                                                     |
|   Total U-value (Btu/h.ft2.F) =        0.112        |
|                                                     |
|                                                     |
|                                                     |
| Note: These results include inside & outside air films. |
+-----------------------------------------------------+
```

The R-values for fully compressed insulation batts should be obtained from the insulation manufacturer. If this information is not available, the following Table may be used:

Rated R-value	Density (lbs/ft^3)	Thickness (Inches)	Fully Compressed Thickness (Inches)	Fully Compressed R-value
11	0.50	3.50	0.58	2.6
13	0.75	3.50	0.88	4.0
15	1.30	3.50	1.52	6.9
19	0.50	6.25	1.04	4.7
21	0.80	5.50	1.47	6.7
25	0.50	8.00	1.33	6.0
30	0.50	9.50	1.58	7.2

Example #4 -- Roof assembly #1 is a vaulted ceiling and has the following construction:

Exterior layer #1 -- 2 1/2" Air Space   R = .80   h-ft²-°F/Btu
    From Table 2 on page 22.3 of the 1989 ASHRAE Handbook of Fundamentals, interpolated
    by assuming winter conditions, roof slope of 22.5 degrees, effective emittance of .82, mean
    temperature of 90 degrees, and temperature difference of 10 degrees.
Exterior layer #2 -- 7/16" OSB Sheathing   L/K = R = .62 h-ft²-°F/Btu
Exterior layer #3 -- 1/2" Roofing Tile   L/K = R = .21   h-ft²-°F/Btu
Cavity Insulation   L/K = R = 30.0   h-ft²-°F/Btu
Interior layer #1 -- 7/8" Air Space   R = .78   h-ft²-°F/Btu
    From Table 2 on page 22.2 of the 1989 ASHRAE Handbook of Fundamentals.
Interior layer #2 -- 5/8" Gyp Board   L/K = R = .56 h-ft²-°F/Btu
Steel Framing Member

Spacing	= 8	feet O.C.
Depth (actual)	= 10	inches
Flange Width (actual) = 3.50	inches	
Thickness	= .104	inches

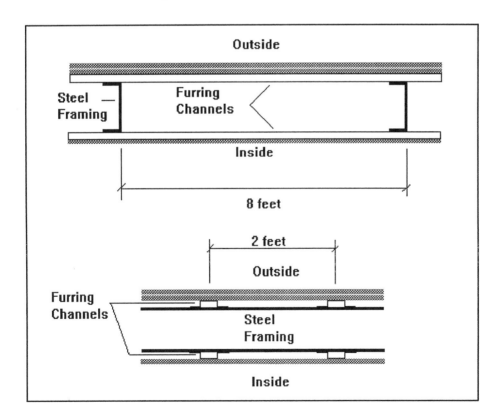

The representative section of the roof assembly is eight feet wide and one foot deep. In this assembly, it is assumed that the cavity insulation does not occupy the air spaces because the insulation thickness is less than the depth of the framing (10"). This roof assembly will be modeled as follows:

```
+---------- Window #1: Frame Type & Layers Descriptions ----------+
¦                                                                 ¦
¦ Project Name : ABC Commercial Building                          ¦
¦ Envelope Name : roof-1                                          ¦
¦                                                                 ¦
¦ Frame Type (Metal or Wood) ........ : Metal                     ¦
¦                                                                 ¦
¦ Interior layer #1 (adjacent to frame)..... : 7/8" air space     ¦
¦ Interior layer #2 (adjacent to layer #1).. : 5/8" gyp board     ¦
¦ Interior layer #3 (adjacent to layer #2).. :                    ¦
¦                                                                 ¦
¦ Exterior layer #1 (adjacent to frame)..... : 2.5" air space     ¦
¦ Exterior layer #2 (adjacent to layer #1).. : 7/16" OSB shtg     ¦
¦ Exterior layer #3 (adjacent to layer #2).. : 1/2" rfg tile      ¦
¦ Exterior layer #4 (adjacent to layer #3).. :                    ¦
¦ Exterior layer #5 (adjacent to layer #4).. :                    ¦
¦                                                                 ¦
¦ EXAMPLE4.INP          Up/Down/Enter              F7: Help ¦
+-----------------------------------------------------------------+


    +--------- Window #2: Frame & Cavity Information ----------+
    ¦                                                         ¦
    ¦ Envelope type (Wall, Floor, Roof or Ceiling) : Roof     ¦
    ¦ Frame depth (inches) ....................... : 10       ¦
    ¦ Flange width (inches) ...................... : 3.5      ¦
    ¦ Frame spacing (inches) ..................... : 96       ¦
    ¦ Web thickness (inches) ..................... : .104     ¦
    ¦ Percent knock-out in web ................... :          ¦
    ¦ Cavity Insulation R-value .................. : 30       ¦
    ¦                                                         ¦
    ¦ R-value   of insulation tape on interior flange :       ¦
    ¦ Thickness of insulation tape on interior flange :       ¦
    ¦ R-value   of insulation tape on exterior flange :       ¦
    ¦ Thickness of insulation tape on exterior flange :       ¦
    ¦                                                         ¦
    ¦                                                         ¦
    ¦ EXAMPLE4.INP          Up/Down/Enter          F7: Help ¦
    +---------------------------------------------------------+


    +---------- Window #3: Interior Layers Information -----------+
    ¦                                                            ¦
    ¦ 1. 7/8" air space    ; interior insulation (Y/N) . : NO    ¦
    ¦                      ; air space or metal (Y/N) .. : YES   ¦
    ¦                      ; R-value (h.ft2.F/Btu) ..... : .78   ¦
    ¦                      ; thickness (inches) ........ : .875  ¦
    ¦ 2. 5/8" gyp board    ; interior insulation (Y/N) . : NO    ¦
    ¦                      ; air space or metal (Y/N) .. : no    ¦
    ¦                      ; R-value (h.ft2.F/Btu) ..... : .56   ¦
    ¦                      ; thickness (inches) ........ : .625  ¦
    ¦ 3. No Description     ; interior insulation (Y/N) . :       ¦
    ¦                      ; air space or metal (Y/N) .. :       ¦
    ¦                      ; R-value (h.ft2.F/Btu) ..... :       ¦
    ¦                      ; thickness (inches) ........ :       ¦
    ¦                                                            ¦
    ¦ EXAMPLE4.INP          Up/Down/Enter          F7: Help ¦
    +------------------------------------------------------------+
```

```
+----------- Window #4: Exterior Layers Information ------------+
| 1. 2.5" air space    ; exterior insulation (Y/N) . : NO      |
|                      ; air space or metal (Y/N) .. : yes     |
|                      ; R-value (h.ft2.F/Btu) ..... : .8      |
|                      ; thickness (inches) ........ : 2.5     |
| 2. 7/16" OSB shtg    ; exterior insulation (Y/N) . : NO      |
|                      ; air space or metal (Y/N) .. : NO      |
|                      ; R-value (h.ft2.F/Btu) ..... : .62     |
|                      ; thickness (inches) ........ : .4375   |
| 3. 1/2" rfg tile     ; exterior insulation (Y/N) . : NO      |
|                      ; air space or metal (Y/N) .. : no      |
|                      ; R-value (h.ft2.F/Btu) ..... : .21     |
|                      ; thickness (inches) ........ : .5      |
| 4. No Description    ; exterior insulation (Y/N) . :         |
|                      ; air space or metal (Y/N) .. :         |
|                      ; R-value (h.ft2.F/Btu) ..... :         |
|                      ; thickness (inches) ....... :          |
| 5. No Description    ; exterior insulation (Y/N) . :         |
|                      ; air space or metal (Y/N) .. :         |
|                      ; R-value (h.ft2.F/Btu) ..... :         |
|                      ; thickness (inches) ....... :          |
| EXAMPLE4.INP              Up/Down/Enter              F7: Help |
+--------------------------------------------------------------+
```

```
+---------------------- Results ----------------------+
|                                                     |
|   Results for  EXAMPLE4.INP :                       |
|                                                     |
|   Total R-value (h.ft2.F/Btu) =     21.77           |
|                                                     |
|   Total U-value (Btu/h.ft2.F) =     0.046           |
|                                                     |
|                                                     |
|                                                     |
| Note: These results include inside & outside air films. |
+-----------------------------------------------------+
```

Example #5 -- In Roof assembly #2, the insulation fills the cavity and the air space created by furring channels because the insulation thickness is greater than the depth of the framing (6"). Air pockets, however, are created between the framing members and the roof's exterior layer. Since the metal members conduct most of the energy, it is assumed that the air space is continuous.

Exterior layer #1 -- 7/8" Air Space $L/K = R = .78$ h-ft^2-°F/Btu
          (7/8" Furring channel, 24 inches on center)
Exterior layer #2 -- 3/8" Metal Roof $L/K = R = .001$ h-ft^2-°F/Btu
Cavity Insulation $L/K = R = 22.0$ h-ft^2-°F/Btu
Interior layer #1 -- 5/8" Gyp Board $L/K = R = .56$ h-ft^2-°F/Btu
(2×6) Steel Framing Member

        Spacing   = 16   inches O.C.
        Depth (actual)   = 6   inches
        Flange Width (actual) = 1.625   inches
        Thickness   = .060 inches

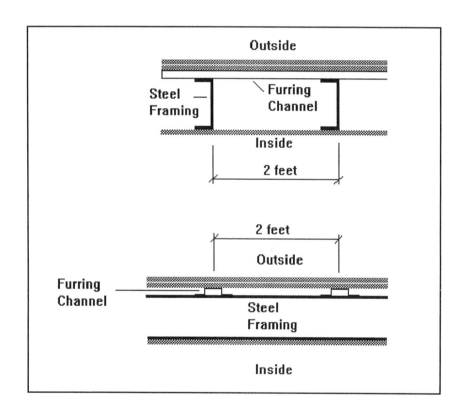

This assembly will be modeled as follows:

```
+---------- Window #1: Frame Type & Layers Descriptions ----------+
|                                                                 |
| Project Name : ABC Commercial Building                          |
| Envelope Name : roof-2                                          |
|                                                                 |
| Frame Type (Metal or Wood) ........ : Metal                     |
|                                                                 |
| Interior layer #1 (adjacent to frame)..... : 5/8" gyp board     |
| Interior layer #2 (adjacent to layer #1).. :                    |
| Interior layer #3 (adjacent to layer #2).. :                    |
|                                                                 |
| Exterior layer #1 (adjacent to frame)..... : 7/8" air space     |
| Exterior layer #2 (adjacent to layer #1).. : 3/8" metal rfg     |
| Exterior layer #3 (adjacent to layer #2).. :                    |
| Exterior layer #4 (adjacent to layer #3).. :                    |
| Exterior layer #5 (adjacent to layer #4).. :                    |
|                                                                 |
| EXAMPLE5.INP          Up/Down/Enter              F7: Help       |
+-----------------------------------------------------------------+


       +--------- Window #2: Frame & Cavity Information ----------+
       |                                                         |
       | Envelope type (Wall, Floor, Roof or Ceiling) : Roof     |
       | Frame depth (inches) ...................... : 6         |
       | Flange width (inches) ..................... : 1.625     |
       | Frame spacing (inches) .................... : 24        |
       | Web thickness (inches) .................... : .06       |
       | Percent knock-out in web .................. :           |
       | Cavity Insulation R-value ................. : 22        |
       |                                                         |
       | R-value   of insulation tape on interior flange :       |
       | Thickness of insulation tape on interior flange :       |
       | R-value   of insulation tape on exterior flange :       |
       | Thickness of insulation tape on exterior flange :       |
       |                                                         |
       |                                                         |
       | EXAMPLE5.INP          Up/Down/Enter         F7: Help    |
       +---------------------------------------------------------+


     +---------- Window #3: Interior Layers Information ----------+
     |                                                           |
     | 1. 5/8" gyp board   ; interior insulation (Y/N)  : NO     |
     |                     ; air space or metal (Y/N) .. : no    |
     |                     ; R-value (h.ft2.F/Btu) ..... : .56   |
     |                     ; thickness (inches) ........ : .625  |
     | 2. No Description   ; interior insulation (Y/N) . :       |
     |                     ; air space or metal (Y/N) .. :       |
     |                     ; R-value (h.ft2.F/Btu) ..... :       |
     |                     ; thickness (inches) ........ :       |
     | 3. No Description   ; interior insulation (Y/N) . :       |
     |                     ; air space or metal (Y/N) .. :       |
     |                     ; R-value (h.ft2.F/Btu) ..... :       |
     |                     ; thickness (inches) ........ :       |
     |                                                           |
     | EXAMPLE5.INP          Up/Down/Enter            F7: Help   |
     +-----------------------------------------------------------+
```

```
+---------- Window #4: Exterior Layers Information -----------+
¦ 1. 7/8" air space    ; exterior insulation (Y/N) . : NO       ¦
¦                      ; air space or metal (Y/N) .. : YES      ¦
¦                      ; R-value (h.ft2.F/Btu) ..... : .78      ¦
¦                      ; thickness (inches) ........ : .875     ¦
¦ 2. 3/8" metal rfg    ; exterior insulation (Y/N) . : NO       ¦
¦                      ; air space or metal (Y/N) .. : yes      ¦
¦                      ; R-value (h.ft2.F/Btu) ..... : .001     ¦
¦                      ; thickness (inches) ........ : .375     ¦
¦ 3. No Description    ; exterior insulation (Y/N) . :          ¦
¦                      ; air space or metal (Y/N) .. :          ¦
¦                      ; R-value (h.ft2.F/Btu) ..... :          ¦
¦                      ; thickness (inches) ........ :          ¦
¦ 4. No Description    ; exterior insulation (Y/N) . :          ¦
¦                      ; air space or metal (Y/N) .. :          ¦
¦                      ; R-value (h.ft2.F/Btu) ..... :          ¦
¦                      ; thickness (inches) ........ :          ¦
¦ 5. No Description    ; exterior insulation (Y/N) . :          ¦
¦                      ; air space or metal (Y/N) .. :          ¦
¦                      ; R-value (h.ft2.F/Btu) ..... :          ¦
¦                      ; thickness (inches) ........ :          ¦
¦ EXAMPLE5.INP              Up/Down/Enter              F7: Help ¦
+------------------------------------------------------------+
```

```
+--------------------- Results ---------------------+
¦                                                    ¦
¦    Results for  EXAMPLE5.INP :                     ¦
¦                                                    ¦
¦    Total R-value (h.ft2.F/Btu) =     11.72         ¦
¦                                                    ¦
¦    Total U-value (Btu/h.ft2.F) =     0.085         ¦
¦                                                    ¦
¦                                                    ¦
¦                                                    ¦
¦ Note: These results include inside & outside air films. ¦
+----------------------------------------------------+
```

Example #6 -- In roof assembly #3, equivalent to R-30 insulation is blown into a vented attic space. The ceiling has a square shape measuring 40 feet on each side.

Insulation (loose fill)	$L/K = R = 30.0$	h-ft²-°F/Btu
Broken down into:		
Exterior layer #1 -- 4" insulation	$L/K = R = 12$	h-ft²-°F/Btu
and Cavity insulation	$L/K = R = 18$	h-ft²-°F/Btu
Interior layer #1 -- 5/8" Gyp Board	$L/K = R = .56$	h-ft²-°F/Btu

(2×6) Steel Framing Member

Spacing	= 24	inches O.C.
Depth (actual)	= 6	inches
Flange Width (actual) = 1.625	inches	
Thickness	= .060 inches	

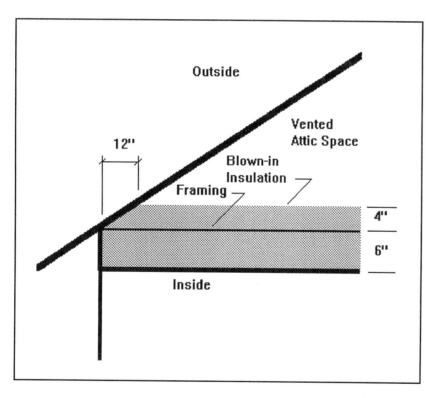

The insulation depth is 10 inches, therefore, the insulation covers the 2 by 6 joists by (10" - 6") = 4 inches.   The insulation is assumed to have a thermal resistance (R-value) of 3 h-ft²-°F/Btu per inch (30/10" deep). The roof assembly is considered to have a steel framed assembly with R-18 (6" × R-3 per inch) between framing members and an exterior insulation layer with R-12 (4" × R-3 per inch) insulation.

The attic space is vented and is considered to be ambient air, ie, the attic air space and the roof will not be included in the analysis. In the perimeter edge, the insulation thickness is reduced to 6 inches with no exterior insulation. The total ceiling area is 1600 ft² (40 ft by 40 ft area). The perimeter area where the exterior insulation averages 2" (R-6) is 160 ft² (40 ft × 4 × 1 ft). The remainder of the ceiling area (1600 ft²- 160 ft² = 1440 ft²) will have R-12 exterior insulation. The weighted average R-value of the exterior insulation will be

$$R_{exterior} = \frac{(\text{R-12} \times 1440 \text{ ft}^2) + (\text{R-6} \times 160 \text{ ft}^2)}{1600 \text{ ft}^2} = 11.4 \quad \text{h-ft}^2\text{-°F/Btu}$$

Average Exterior Insulation Thickness = (R-11.4) / (R-3 per inch) = 3.8"

This assembly will be modeled as follows:

```
+---------- Window #1: Frame Type & Layers Descriptions ----------+
|                                                                 |
| Project Name : ABC Commercial Building                          |
| Envelope Name : roof-3                                          |
|                                                                 |
| Frame Type (Metal or Wood) ........ : Metal                     |
|                                                                 |
| Interior layer #1 (adjacent to frame)..... : 5/8" gyp board     |
| Interior layer #2 (adjacent to layer #1).. :                    |
| Interior layer #3 (adjacent to layer #2).. :                    |
|                                                                 |
| Exterior layer #1 (adjacent to frame)..... : 3.8" insulation    |
| Exterior layer #2 (adjacent to layer #1).. :                    |
| Exterior layer #3 (adjacent to layer #2).. :                    |
| Exterior layer #4 (adjacent to layer #3).. :                    |
| Exterior layer #5 (adjacent to layer #4).. :                    |
|                                                                 |
| EXAMPLE6.INP            Up/Down/Enter              F7: Help      |
+-----------------------------------------------------------------+


     +--------- Window #2: Frame & Cavity Information ----------+
     |                                                         |
     | Envelope type (Wall, Floor, Roof or Ceiling) : Ceiling  |
     | Frame depth (inches) ..................... : 6          |
     | Flange width (inches) .................... : 1.625      |
     | Frame spacing (inches) ................... : 24         |
     | Web thickness (inches) ................... : .06        |
     | Percent knock-out in web ................. :            |
     | Cavity Insulation R-value ................ : 18         |
     |                                                         |
     | R-value   of insulation tape on interior flange :       |
     | Thickness of insulation tape on interior flange :       |
     | R-value   of insulation tape on exterior flange :       |
     | Thickness of insulation tape on exterior flange :       |
     |                                                         |
     |                                                         |
     | EXAMPLE6.INP           Up/Down/Enter         F7: Help   |
     +---------------------------------------------------------+
```

```
+---------- Window #3: Interior Layers Information -----------+
|                                                            |
| 1. 5/8" gyp board    ; interior insulation (Y/N) . : NO    |
|                      ; air space or metal (Y/N) .. : NO    |
|                      ; R-value (h.ft2.F/Btu) ..... : .56   |
|                      ; thickness (inches) ........ : .625  |
| 2. No Description     ; interior insulation (Y/N) . :      |
|                      ; air space or metal (Y/N) .. :       |
|                      ; R-value (h.ft2.F/Btu) ..... :       |
|                      ; thickness (inches) ........ :       |
| 3. No Description     ; interior insulation (Y/N) . :      |
|                      ; air space or metal (Y/N) .. :       |
|                      ; R-value (h.ft2.F/Btu) ..... :       |
|                      ; thickness (inches) ........ :       |
|                                                            |
| EXAMPLE6.INP            Up/Down/Enter           F7: Help |
+------------------------------------------------------------+


+---------- Window #4: Exterior Layers Information -----------+
| 1. 4" insulation      ; exterior insulation (Y/N) . : YES  |
|                      ; air space or metal (Y/N) .. : NO    |
|                      ; R-value (h.ft2.F/Btu) ..... : 11.4  |
|                      ; thickness (inches) ........ : 3.8   |
| 2. No Description     ; exterior insulation (Y/N) . :      |
|                      ; air space or metal (Y/N) .. :       |
|                      ; R-value (h.ft2.F/Btu) ..... :       |
|                      ; thickness (inches) ........ :       |
| 3. No Description     ; exterior insulation (Y/N) . :      |
|                      ; air space or metal (Y/N) .. :       |
|                      ; R-value (h.ft2.F/Btu) ..... :       |
|                      ; thickness (inches) ........ :       |
| 4. No Description     ; exterior insulation (Y/N) . :      |
|                      ; air space or metal (Y/N) .. :       |
|                      ; R-value (h.ft2.F/Btu) ..... :       |
|                      ; thickness (inches) ........ :       |
| 5. No Description     ; exterior insulation (Y/N) . :      |
|                      ; air space or metal (Y/N) .. :       |
|                      ; R-value (h.ft2.F/Btu) ..... :       |
|                      ; thickness (inches) ........ :       |
| EXAMPLE6.INP            Up/Down/Enter           F7: Help |
+------------------------------------------------------------+


+------------------- Results ---------------------+
|                                                 |
|   Results for  EXAMPLE6.INP :                   |
|                                                 |
|   Total R-value (h.ft2.F/Btu) =     21.43       |
|                                                 |
|   Total U-value (Btu/h.ft2.F) =     0.047       |
|                                                 |
|                                                 |
|                                                 |
| Note: These results include inside & outside air films. |
+-------------------------------------------------+
```

Example #7 -- Floor assembly #1 is a raised floor with a vented crawlspace. In both residential and nonresidential standards, it is assumed that the crawlspace has an effective R-value of 6.00 h-ft²-°F/Btu. Floor assembly #1 consists of the following layers:

Exterior layer #1 -- Vented crawlspace	R = 6.00	h-ft²-°F/Btu
Cavity Insulation	R = 19.0	h-ft²-°F/Btu
Interior layer #1 -- 5/8" Plywood	R = .78	h-ft²-°F/Btu
Interior layer #2 -- Carpet & pad	R = 2.08	h-ft²-°F/Btu

(2×8) Steel Framing Member
Spacing	= 16	inches O.C.
Depth (actual)	= 8	inches
Flange Width (actual) = 1.625		inches
Thickness	= .060	inches

When modeling this floor assembly, the vented crawlspace will be entered as an exterior air space. The carpet and pad will be entered as an interior insulation because of their insulative characteristics (high R-value). The input for the EZFRAME program is as follows:

```
+---------- Window #1: Frame Type & Layers Descriptions ----------+
|                                                                 |
| Project Name : My house                                         |
| Envelope Name : Floor #1                                        |
|                                                                 |
| Frame Type (Metal or Wood) ........ : Metal                     |
|                                                                 |
| Interior layer #1 (adjacent to frame)..... : 5/8" Plywood       |
| Interior layer #2 (adjacent to layer #1).. : carpet & pad       |
| Interior layer #3 (adjacent to layer #2).. :                    |
|                                                                 |
| Exterior layer #1 (adjacent to frame)..... : Crawlspace         |
| Exterior layer #2 (adjacent to layer #1).. :                    |
| Exterior layer #3 (adjacent to layer #2).. :                    |
| Exterior layer #4 (adjacent to layer #3).. :                    |
| Exterior layer #5 (adjacent to layer #4).. :                    |
|                                                                 |
| EXAMPLE7.INP          Up/Down/Enter               F7: Help      |
+-----------------------------------------------------------------+
```

```
+--------- Window #2: Frame & Cavity Information ----------+
¦                                                          ¦
¦ Envelope type (Wall, Floor, Roof or Ceiling) : Floor     ¦
¦ Frame depth (inches) ..................... : 8           ¦
¦ Flange width (inches) .................... : 1.625       ¦
¦ Frame spacing (inches) ................... : 16          ¦
¦ Web thickness (inches) ................... : .06         ¦
¦ Percent knock-out in web ................. : 0           ¦
¦ Cavity Insulation R-value ................ : 19          ¦
¦                                                          ¦
¦ R-value   of insulation tape on interior flange :        ¦
¦ Thickness of insulation tape on interior flange :        ¦
¦ R-value   of insulation tape on exterior flange :        ¦
¦ Thickness of insulation tape on exterior flange :        ¦
¦                                                          ¦
¦                                                          ¦
¦ EXAMPLE7.INP           Up/Down/Enter           F7: Help  ¦
+----------------------------------------------------------+
+----------- Window #3: Interior Layers Information ------------+
¦                                                              ¦
¦ 1. 5/8" Plywood        ; interior insulation (Y/N) . : NO    ¦
¦                        ; air space or metal (Y/N) .. : NO    ¦
¦                        ; R-value (h.ft2.F/Btu) ..... : .78   ¦
¦                        ; thickness (inches) ........ : .625  ¦
¦ 2. carpet & pad        ; interior insulation (Y/N) . : YES   ¦
¦                        ; air space or metal (Y/N) .. : NO    ¦
¦                        ; R-value (h.ft2.F/Btu) ..... : 2.08  ¦
¦                        ; thickness (inches) ........ : 1.0   ¦
¦ 3. No Description      ; interior insulation (Y/N) . :       ¦
¦                        ; air space or metal (Y/N) .. :       ¦
¦                        ; R-value (h.ft2.F/Btu) ..... :       ¦
¦                        ; thickness (inches) ........ :       ¦
¦                                                              ¦
¦ EXAMPLE7.INP           Up/Down/Enter           F7: Help      ¦
+--------------------------------------------------------------+


+----------- Window #4: Exterior Layers Information ------------+
¦ 1. Crawlspace         ; exterior insulation (Y/N) . : NO     ¦
¦                       ; air space or metal (Y/N) .. : YES    ¦
¦                       ; R-value (h.ft2.F/Btu) ..... : 6.0    ¦
¦                       ; thickness (inches) ........ : 18     ¦
¦ 2. No Description     ; exterior insulation (Y/N) . :        ¦
¦                       ; air space or metal (Y/N) .. :        ¦
¦                       ; R-value (h.ft2.F/Btu) ..... :        ¦
¦                       ; thickness (inches) ........ :        ¦
¦ 3. No Description     ; exterior insulation (Y/N) . :        ¦
¦                       ; air space or metal (Y/N) .. :        ¦
¦                       ; R-value (h.ft2.F/Btu) ..... :        ¦
¦                       ; thickness (inches) ........ :        ¦
¦ 4. No Description     ; exterior insulation (Y/N) . :        ¦
¦                       ; air space or metal (Y/N) .. :        ¦
¦                       ; R-value (h.ft2.F/Btu) ..... :        ¦
¦                       ; thickness (inches) ........ :        ¦
¦ 5. No Description     ; exterior insulation (Y/N) . :        ¦
¦                       ; air space or metal (Y/N) .. :        ¦
¦                       ; R-value (h.ft2.F/Btu) ..... :        ¦
¦                       ; thickness (inches) ........ :        ¦
¦ EXAMPLE7.INP          Up/Down/Enter            F7: Help      ¦
+--------------------------------------------------------------+
```

```
+---------------------- Results ----------------------+
|                                                     |
|   Results for  EXAMPLE7.INP :                       |
|                                                     |
|   Total R-value (h.ft2.F/Btu) =      18.20          |
|                                                     |
|   Total U-value (Btu/h.ft2.F) =      0.055          |
|                                                     |
|                                                     |
|                                                     |
| Note: These results include inside & outside air films. |
+-----------------------------------------------------+
```

Example #8 -- Floor assembly #2 is a raised floor without a vented crawlspace such as a floor over an unconditioned space or a floor exposed to the outside.   Floor assembly #2 consists of the following layers:

Exterior layer #1 -- None
Cavity Insulation   R = 19.0   h-ft²-°F/Btu
Interior layer #1 -- 5/8" Plywood   R = .78   h-ft²-°F/Btu
Interior layer #2 -- Carpet & pad   R = 2.08   h-ft²-°F/Btu

(2 × 8) Steel Framing Member
      Spacing   = 16   inches O.C.
      Depth (actual)   = 8   inches
      Flange Width (actual) = 1.625   inches
      Thickness   = .060 inches

When modeling this floor assembly, the carpet and pad will be entered as an interior insulation because of their insulative characteristics (high R-value).   The input for the EZFRAME program is as follows:

```
+---------- Window #1: Frame Type & Layers Descriptions ----------+
|                                                                 |
| Project Name : My House                                         |
| Envelope Name : Floor #2                                        |
|                                                                 |
| Frame Type (Metal or Wood) ........ : Metal                     |
|                                                                 |
| Interior layer #1 (adjacent to frame)..... : 5/8" Plywood       |
| Interior layer #2 (adjacent to layer #1).. : carpet & pad       |
| Interior layer #3 (adjacent to layer #2).. :                    |
|                                                                 |
| Exterior layer #1 (adjacent to frame)..... :                    |
| Exterior layer #2 (adjacent to layer #1).. :                    |
| Exterior layer #3 (adjacent to layer #2).. :                    |
| Exterior layer #4 (adjacent to layer #3).. :                    |
| Exterior layer #5 (adjacent to layer #4).. :                    |
|                                                                 |
| EXAMPLE8.INP          Up/Down/Enter              F7: Help |
+-----------------------------------------------------------------+
```

```
+--------- Window #2: Frame & Cavity Information ---------+
|                                                         |
| Envelope type (Wall, Floor, Roof or Ceiling) : Floor    |
| Frame depth (inches) ...................... : 8         |
| Flange width (inches) ..................... : 1.625     |
| Frame spacing (inches) .................... : 16        |
| Web thickness (inches) .................... : .06       |
| Percent knock-out in web .................. : 0         |
| Cavity Insulation R-value ................. : 19        |
|                                                         |
| R-value   of insulation tape on interior flange :       |
| Thickness of insulation tape on interior flange :       |
| R-value   of insulation tape on exterior flange :       |
| Thickness of insulation tape on exterior flange :       |
|                                                         |
|                                                         |
| EXAMPLE8.INP         Up/Down/Enter          F7: Help    |
+---------------------------------------------------------+
 +---------- Window #3: Interior Layers Information -----------+
 |                                                            |
 | 1. 5/8" Plywood    ; interior insulation (Y/N) . : NO      |
 |                    ; air space or metal (Y/N) .. : NO      |
 |                    ; R-value (h.ft2.F/Btu) ..... : .78     |
 |                    ; thickness (inches) ........ : .625    |
 | 2. carpet & pad    ; interior insulation (Y/N) . : YES     |
 |                    ; air space or metal (Y/N) .. : NO      |
 |                    ; R-value (h.ft2.F/Btu) ..... : 2.08    |
 |                    ; thickness (inches) ........ : 1.0     |
 | 3. No Description  ; interior insulation (Y/N) . :         |
 |                    ; air space or metal (Y/N) .. :         |
 |                    ; R-value (h.ft2.F/Btu) ..... :         |
 |                    ; thickness (inches) ........ :         |
 |                                                            |
 | EXAMPLE8.INP         Up/Down/Enter          F7: Help       |
 +------------------------------------------------------------+

 +---------- Window #4: Exterior Layers Information -----------+
 | 1. No Description  ; exterior insulation (Y/N) . :         |
 |                    ; air space or metal (Y/N) .. :         |
 |                    ; R-value (h.ft2.F/Btu) ..... :         |
 |                    ; thickness (inches) ........ :         |
 | 2. No Description  ; exterior insulation (Y/N) . :         |
 |                    ; air space or metal (Y/N) .. :         |
 |                    ; R-value (h.ft2.F/Btu) ..... :         |
 |                    ; thickness (inches) ........ :         |
 | 3. No Description  ; exterior insulation (Y/N) . :         |
 |                    ; air space or metal (Y/N) .. :         |
 |                    ; R-value (h.ft2.F/Btu) ..... :         |
 |                    ; thickness (inches) ........ :         |
 | 4. No Description  ; exterior insulation (Y/N) . :         |
 |                    ; air space or metal (Y/N) .. :         |
 |                    ; R-value (h.ft2.F/Btu) ..... :         |
 |                    ; thickness (inches) ........ :         |
 | 5. No Description  ; exterior insulation (Y/N) . :         |
 |                    ; air space or metal (Y/N) .. :         |
 |                    ; R-value (h.ft2.F/Btu) ..... :         |
 |                    ; thickness (inches) ........ :         |
 | EXAMPLE8.INP         Up/Down/Enter          F7: Help       |
 +------------------------------------------------------------+
```

```
+---------------------- Results ----------------------+
¦                                                     ¦
¦   Results for  EXAMPLE8.INP :                       ¦
¦                                                     ¦
¦   Total R-value (h.ft2.F/Btu) =     12.20           ¦
¦                                                     ¦
¦   Total U-value (Btu/h.ft2.F) =     0.082           ¦
¦                                                     ¦
¦                                                     ¦
¦                                                     ¦
¦ Note: These results include inside & outside air films. ¦
+-----------------------------------------------------+
```

Example #9 -- In this example, the wall assembly in example #1 is considered with wood framing.

Interior layer #1 -- 5/8" Gyp Board $L/K = R = .56$ h-ft²-°F/Btu
Exterior layer #1 -- 1" Foam Type Sheathing $L/K = R = 4.16$ h-ft²-°F/Btu
Exterior layer #2 -- 5/8" Plywood Exterior Facing $L/K = R = .78$ h-ft²-°F/Btu
Cavity insulation -- R-13 Batt Insulation $L/K = R = 13.00$ h-ft²-°F/Btu
Framing information -- (2 × 4) wood framing

      Spacing   = 16   inches O.C.
      Depth (actual)   = 3.5   inches
      Flange Width (actual)   = 1.5   inches

The representative area of this wall assembly is a section that is one foot high and 16 inches wide.

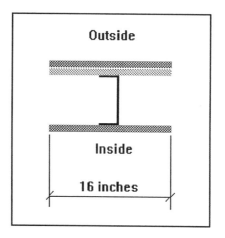

```
+---------- Window #1: Frame Type & Layers Descriptions ----------+
|                                                                 |
| Project Name : ABC Commercial Building                          |
| Envelope Name : wall-1                                          |
|                                                                 |
| Frame Type (Metal or Wood) ........ : Wood                      |
|                                                                 |
| Interior layer #1 (adjacent to frame)..... : 5/8" gyp board     |
| Interior layer #2 (adjacent to layer #1).. :                    |
| Interior layer #3 (adjacent to layer #2).. :                    |
|                                                                 |
| Exterior layer #1 (adjacent to frame)..... : 1" foam board      |
| Exterior layer #2 (adjacent to layer #1).. : 5/8" plywood       |
| Exterior layer #3 (adjacent to layer #2).. :                    |
| Exterior layer #4 (adjacent to layer #3).. :                    |
| Exterior layer #5 (adjacent to layer #4).. :                    |
|                                                                 |
| EXAMPLE9.INP          Up/Down/Enter              F7: Help |
+-----------------------------------------------------------------+
```

```
+--------- Window #2: Frame & Cavity Information ----------+
¦                                                          ¦
¦ Envelope type (Wall, Floor, Roof or Ceiling) : Wall      ¦
¦ Frame depth (inches) ...................... : 3.625      ¦
¦ Frame width (inches) ...................... : 1.625      ¦
¦ Frame spacing (inches) .................... : 16         ¦
¦ Cavity Insulation R-value ................. : 13         ¦
¦                                                          ¦
¦                                                          ¦
¦ EXAMPLE9.INP           Up/Down/Enter         F7: Help    ¦
+----------------------------------------------------------+
```

```
+----------- Window #3: Interior Layers Information ------------+
¦                                                              ¦
¦ 1. 5/8" gyp board      ; R-value (h.ft2.F/Btu)...... : .56   ¦
¦                        ; thickness (inches) ........ : .625  ¦
¦ 2. No Description       ; R-value (h.ft2.F/Btu) ..... :       ¦
¦                        ; thickness (inches) ....... :        ¦
¦ 3. No Description       ; R-value (h.ft2.F/Btu) ..... :       ¦
¦                        ; thickness (inches) ....... :        ¦
¦                                                              ¦
¦ EXAMPLE9.INP           Up/Down/Enter         F7: Help        ¦
+--------------------------------------------------------------+
```

```
+----------- Window #4: Exterior Layers Information ------------+
¦                                                              ¦
¦ 1. 1" foam board       ; R-value (h.ft2.F/Btu)...... : 4.16  ¦
¦                        ; thickness (inches) ....... : 1      ¦
¦ 2. 5/8" plywood        ; R-value (h.ft2.F/Btu) ..... : .78   ¦
¦                        ; thickness (inches) ....... : .625   ¦
¦ 3. No Description       ; R-value (h.ft2.F/Btu) ..... :       ¦
¦                        ; thickness (inches) ....... :        ¦
¦ 4. No Description       ; R-value (h.ft2.F/Btu) ..... :       ¦
¦                        ; thickness (inches) ....... :        ¦
¦ 5. No Description       ; R-value (h.ft2.F/Btu) ..... :       ¦
¦                        ; thickness (inches) ....... :        ¦
¦                                                              ¦
¦ EXAMPLE9.INP           Up/Down/Enter         F7: Help        ¦
+--------------------------------------------------------------+
```

```
+---------------------- Results ----------------------+
¦                                                     ¦
¦   Results for  EXAMPLE9.INP :                       ¦
¦                                                     ¦
¦   Total R-value (h.ft2.F/Btu) =      16.92          ¦
¦                                                     ¦
¦   Total U-value (Btu/h.ft2.F) =      0.059          ¦
¦                                                     ¦
¦                                                     ¦
¦                                                     ¦
¦ Note: These results include inside & outside air films. ¦
+-----------------------------------------------------+
```

```
+--------- Window #2: Frame & Cavity Information ----------+

_ Envelope type (Wall, Floor, Roof or Ceiling) : Floor    _
   _ Frame depth (inches) ........................ : 8        _
   _ Flange width (inches) ........................ : 1.625    _
   _ Frame spacing (inches) ..................... : 16       _
   _ Web thickness (inches) ..................... : .06      _
   _ Percent knock-out in web ................... : 0        _
   _ Cavity Insulation R-value .................. : 19       _

   _ R-value   of insulation tape on interior flange :       _
   _ Thickness of insulation tape on interior flange :       _
   _ R-value   of insulation tape on exterior flange :       _
   _ Thickness of insulation tape on exterior flange :       _


_ EXAMPLE7.INP      Up/Down/Enter          F7: Help _
   +----------------------------------------------------+
+----------- Window #3: Interior Layers Information ------------+

_ 1. 5/8" Plywood   ; interior insulation (Y/N) . : NO       _
   _              ; air space or metal (Y/N) .. : NO          _
   _              ; R-value (h.ft2.F/Btu) ..... : .78         _
   _              ; thickness (inches) ........ : .625        _
_ 2. carpet & pad   ; interior insulation (Y/N) . : YES       _
   _              ; air space or metal (Y/N) .. : NO          _
   _              ; R-value (h.ft2.F/Btu) ..... : 2.08        _
   _              ; thickness (inches) ........ : 1.0         _
_ 3. No Description  ; interior insulation (Y/N) . :          _
   _              ; air space or metal (Y/N) .. :             _
   _              ; R-value (h.ft2.F/Btu) ..... :             _
   _              ; thickness (inches) ........ :             _

_ EXAMPLE7.INP      Up/Down/Enter          F7: Help _
   +----------------------------------------------------+


+----------- Window #4: Exterior Layers Information ------------+
_ 1. Crawlspace     ; exterior insulation (Y/N) . : NO       _
   _              ; air space or metal (Y/N) .. : YES         _
   _              ; R-value (h.ft2.F/Btu) ..... : 6.0         _
   _              ; thickness (inches) ........ : 18          _
_ 2. No Description  ; exterior insulation (Y/N) . :          _
   _              ; air space or metal (Y/N) .. :             _
   _              ; R-value (h.ft2.F/Btu) ..... :             _
   _              ; thickness (inches) ........ :             _
_ 3. No Description  ; exterior insulation (Y/N) . :          _
   _              ; air space or metal (Y/N) .. :             _
   _              ; R-value (h.ft2.F/Btu) ..... :             _
   _              ; thickness (inches) ........ :             _
_ 4. No Description  ; exterior insulation (Y/N) . :          _
   _              ; air space or metal (Y/N) .. :             _
   _              ; R-value (h.ft2.F/Btu) ..... :             _
   _              ; thickness (inches) ........ :             _
_ 5. No Description  ; exterior insulation (Y/N) . :          _
   _              ; air space or metal (Y/N) .. :             _
   _              ; R-value (h.ft2.F/Btu) ..... :             _
   _              ; thickness (inches) ........ :             _
_ EXAMPLE7.INP      Up/Down/Enter          F7: Help _
   +----------------------------------------------------+
```

```
+------------------------ Results -----------------------+
_                                          _
_  Results for  EXAMPLE7.INP :                       _

_  Total R-value (h.ft2.F/Btu) =     18.20           _

_  Total U-value (Btu/h.ft2.F) =     0.055           _
         _                                _
         _                                _
_ Note: These results include inside & outside air films. _
       +--------------------------------------------------+
```

Example #8 -- Floor assembly #2 is a raised floor without a vented crawlspace such as a floor over an unconditioned space or a floor exposed to the outside.   Floor assembly #2 consists of the following layers:

Exterior layer #1 --   None
Cavity Insulation   R = 19.0   h-ft²-°F/Btu
Interior layer #1 -- 5/8" Plywood   R = .78   h-ft²-°F/Btu
Interior layer #2 -- Carpet & pad   R = 2.08   h-ft²-°F/Btu

(2 × 8) Steel Framing Member
        Spacing   = 16   inches O.C.
        Depth (actual)   = 8   inches
        Flange Width (actual) = 1.625   inches
        Thickness   = .060   inches

When modeling this floor assembly, the carpet and pad will be entered as an interior insulation because of their insulative characteristics (high R-value).   The input for the EZFRAME program is as follows:

```
+ ---------- Window #1: Frame Type & Layers Descriptions ---------- +
                                              _
    _ Project Name : My House
    _ Envelope Name : Floor #2                       _
                                                     _
    _ Frame Type (Metal or Wood) ........ : Metal    _
                                                     _
    _ Interior layer #1 (adjacent to frame)..... : 5/8" Plywood    _
    _ Interior layer #2 (adjacent to layer #1).. : carpet & pad    _
     _ Interior layer #3 (adjacent to layer #2).. :                _

    _ Exterior layer #1 (adjacent to frame)..... :                 _
    _ Exterior layer #2 (adjacent to layer #1).. :                 _
    _ Exterior layer #3 (adjacent to layer #2).. :                 _
    _ Exterior layer #4 (adjacent to layer #3).. :                 _
    _ Exterior layer #5 (adjacent to layer #4).. :                 _

  _ EXAMPLE8.INP        Up/Down/Enter              F7: Help _
      + ----------------------------------------------------------- +
```

```
+--------- Window #2: Frame & Cavity Information ----------+
                          _                          _
_ Envelope type (Wall, Floor, Roof or Ceiling) : Floor    _
    _ Frame depth (inches) ........................ : 8      _
    _ Flange width (inches) ...................... : 1.625   _
    _ Frame spacing (inches) .................... : 16       _
    _ Web thickness (inches) .................... : .06      _
    _ Percent knock-out in web ................... : 0       _
    _ Cavity Insulation R-value .................. : 19      _

    _ R-value   of insulation tape on interior flange :      _
    _ Thickness of insulation tape on interior flange :      _
    _ R-value   of insulation tape on exterior flange :      _
    _ Thickness of insulation tape on exterior flange :      _

         _                                   _
_ EXAMPLE8.INP      Up/Down/Enter        F7: Help _
   +-------------------------------------------------------+
+---------- Window #3: Interior Layers Information -----------+

_ 1. 5/8" Plywood    ; interior insulation (Y/N) . : NO      _
    _                 ; air space or metal (Y/N) .. : NO      _
    _                 ; R-value (h.ft2.F/Btu) ..... : .78     _
    _                 ; thickness (inches) ........ : .625    _
_ 2. carpet & pad    ; interior insulation (Y/N) . : YES     _
    _                 ; air space or metal (Y/N) .. : NO      _
    _                 ; R-value (h.ft2.F/Btu) ..... : 2.08    _
    _                 ; thickness (inches) ........ : 1.0     _
_ 3. No Description   ; interior insulation (Y/N) . :         _
    _                 ; air space or metal (Y/N) .. :         _
    _                 ; R-value (h.ft2.F/Btu) ..... :         _
    _                 ; thickness (inches) ........ :         _

_ EXAMPLE8.INP      Up/Down/Enter        F7: Help _
   +-------------------------------------------------------+

+---------- Window #4: Exterior Layers Information -----------+
_ 1. No Description   ; exterior insulation (Y/N) . :        _
    _                 ; air space or metal (Y/N) .. :         _
    _                 ; R-value (h.ft2.F/Btu) ..... :         _
    _                 ; thickness (inches) ........ :         _
_ 2. No Description   ; exterior insulation (Y/N) . :        _
    _                 ; air space or metal (Y/N) .. :         _
    _                 ; R-value (h.ft2.F/Btu) ..... :         _
    _                 ; thickness (inches) ........ :         _
_ 3. No Description   ; exterior insulation (Y/N) . :        _
    _                 ; air space or metal (Y/N) .. :         _
    _                 ; R-value (h.ft2.F/Btu) ..... :         _
    _                 ; thickness (inches) ........ :         _
_ 4. No Description   ; exterior insulation (Y/N) . :        _
    _                 ; air space or metal (Y/N) .. :         _
    _                 ; R-value (h.ft2.F/Btu) ..... :         _
    _                 ; thickness (inches) ........ :         _
_ 5. No Description   ; exterior insulation (Y/N) . :        _
    _                 ; air space or metal (Y/N) .. :         _
    _                 ; R-value (h.ft2.F/Btu) ..... :         _
    _                 ; thickness (inches) ........ :         _
_ EXAMPLE8.INP      Up/Down/Enter        F7: Help _
   +-------------------------------------------------------+
```

```
+----------------------- Results -----------------------+
  _ Results for  EXAMPLE8.INP :                 _       _

  _  Total R-value (h.ft2.F/Btu) =     12.20     _       _

  _  Total U-value (Btu/h.ft2.F) =     0.082     _       _

       _                                      _
       _                                      _

  _ Note: These results include inside & outside air films. _
    +-------------------------------------------------------+
```

Example #9 -- In this example, the wall assembly in example #1 is considered with wood framing.

Interior layer #1 -- 5/8" Gyp Board   L/K = R = .56   h-ft^2-°F/Btu
Exterior layer #1 -- 1" Foam Type Sheathing   L/K = R = 4.16   h-ft^2-°F/Btu
Exterior layer #2 -- 5/8" Plywood Exterior Facing L/K = R = .78   h-ft^2-°F/Btu
Cavity insulation -- R-13 Batt Insulation   L/K = R = 13.00   h-ft^2-°F/Btu
Framing information -- (2 × 4) wood framing

Spacing	= 16	inches O.C.
Depth (actual)	= 3.5	inches
Flange Width (actual)	= 1.5	inches

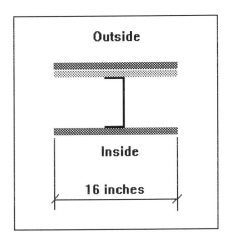

The representative area of this wall assembly is a section that is one foot high and 16 inches wide.

+ ---------- **Window #1: Frame Type & Layers Descriptions** ----------+

```
 _ Project Name : ABC Commercial Building              _
    _ Envelope Name : wall-1                       _

 _ Frame Type (Metal or Wood) ........ : Wood           _

 _ Interior layer #1 (adjacent to frame)..... : 5/8" gyp board   _
    _ Interior layer #2 (adjacent to layer #1).. :           _
    _ Interior layer #3 (adjacent to layer #2).. :           _

 _ Exterior layer #1 (adjacent to frame)..... : 1" foam board    _
 _ Exterior layer #2 (adjacent to layer #1).. : 5/8" plywood     _
    _ Exterior layer #3 (adjacent to layer #2).. :          _
    _ Exterior layer #4 (adjacent to layer #3).. :          _
    _ Exterior layer #5 (adjacent to layer #4).. :          _

 _ EXAMPLE9.INP       Up/Down/Enter            F7: Help _
     +--------------------------------------------------------+
```

```
 +--------- Window #2: Frame & Cavity Information ----------+

 _ Envelope type (Wall, Floor, Roof or Ceiling) : Wall   _
    _ Frame depth (inches) ........................ : 3.625   _
    _ Frame width (inches) ........................ : 1.625   _
    _ Frame spacing (inches) ..................... : 16    _
    _ Cavity Insulation R-value .................. : 13    _


 _ EXAMPLE9.INP       Up/Down/Enter            F7: Help _
     +--------------------------------------------------------+
```

```
 +----------- Window #3: Interior Layers Information ------------+

 _ 1. 5/8" gyp board    ; R-value (h.ft2.F/Btu)...... : .56    _
                  ; thickness (inches) ........ : .625   _
    _ 2. No Description    ; R-value (h.ft2.F/Btu) ..... :     _
                  ; thickness (inches) ........ :     _
    _ 3. No Description    ; R-value (h.ft2.F/Btu) ..... :     _
                  ; thickness (inches) ........ :     _

 _ EXAMPLE9.INP       Up/Down/Enter            F7: Help _
     +--------------------------------------------------------+
```

```
 +---------- Window #4: Exterior Layers Information ------------+

 _ 1. 1" foam board    ; R-value (h.ft2.F/Btu)...... : 4.16   _
                  ; thickness (inches) ........ : 1    _
    _ 2. 5/8" plywood    ; R-value (h.ft2.F/Btu) ..... : .78   _
                  ; thickness (inches) ........ : .625  _
    _ 3. No Description   ; R-value (h.ft2.F/Btu) ..... :     _
                  ; thickness (inches) ........ :     _
    _ 4. No Description   ; R-value (h.ft2.F/Btu) ..... :     _
                  ; thickness (inches) ........ :     _
    _ 5. No Description   ; R-value (h.ft2.F/Btu) ..... :     _
                  ; thickness (inches) ........ :     _
```

Glossary

active solar system a system, which uses solar heat collectors and mechanical devices to collect, store, and distribute solar energy.

ACCA abbreviation for The Air Conditioning Contractors of America.

AFUE abbreviation for the Annual Fuel Utilization Efficiency, a method of rating the efficiency of heating equipment.

air distribution system the system conveys air from the unit to the rooms, and back to the unit again. This system consists of supply ducts, supply air outlets in the rooms, and return intakes and ducts.

air conditioner a machine that controls the temperature, moisture, cleanliness and distribution of air.

AISI abbreviation for American Iron and Steel Institute.

ASHRAE abbreviation for the American Society of Heating, Refrigeration and Air-conditioning Engineers.

anchor bolt heavy, threaded bolt embedded in the foundation to secure sill to foundation wall or bottom plate of exterior wall to concrete floor slab.

ASTM formerly American Society for Testing and Materials, now ASTM, a nonprofit, national technical society that publishes definitions, standards, test methods, recommended installation practices and specifications for materials.

awning windows windows with hinges at the top.

axial load the longitudinal force acting on a member. Examples are the gravity loads carried by columns or studs.

beam load-bearing member spanning a distance between supports.

blocking a member in a floor, wall, or ceiling that serves primarily as a point for fastening sheathing, finishing materials or accessories.

bracing a method using horizontal, lateral braces or bridging to prevent the stud from bending under wind and axial loads and provides resistance to stud rotation.

bridging members attached between floor joists to distributed loads over more than one joist. Solid bridging consists of joist-depth lumber installed perpendicular to and between the joists. Cross bridging consists of pairs of braces set in an X form between joists.

Btu (British Thermal Unit) a heat measurement; the amount of heat needed to raise one pound of water one degree Fahrenheit.

buckling the bending, warping or crumpling of a member, such as a wall stud, subjected to axial, bending, bearing, or shear loads.

building codes community ordinances governing the manner in which a home can be constructed or modified.

CABO an abbreviation for the Council of American Building Officials which is made up of representatives from three model codes and issues National Research Board (NRB) reports.

camber curvature built in to a beam or truss to compensate for loads encountered when in place and full dead load is applied.

casement windows windows that open sideways with a hand crank

caulk a variety of different compounds used to seal seams and joints against the infiltration of water and air.

C-clamps tools used to clamp steel together, preventing the steel from separating during attachment.

ceiling joist a horizontal structural framing member which supports a ceiling and/or attic loads.

c.f.m. abbreviation for cubic feet per minute.

channels Z-furring furring channels are used to mechanically attach insulation blankets, rigid insulation or gypsum panels or base to the interior side of monolithic concrete or masonry walls. Sometimes also used for sound control and additional space for pipes, conduits or ducts. Hat channels, or drywall furring channels, are used to attach gypsum board or plywood to concrete, masonry and steel studs.

chlorofluorocarbons (CFCs) a blowing agent used in some exterior insulation that causes release of chlorine molecules into the atmosphere, contributing to ozone depletion.

clip angle an L shaped short piece of metal normally with a 90 degree bend typically used for connections.

cold-forming a process where light-gauge steel members are manufactured by (1) press-braking blanks sheared from sheets or cut length of coils or plates, or by (2) roll forming cold- or hot-rolled coils of sheet steel.

condensation the production of liquid which results when warm, moist air comes into contact with a colder surface and deposits moisture onto that surface. In colder climates, condensation may appear as frost.

conduction the transfer of heat through materials.

conduit, electrical a pipe, usually metal, in which wire is installed

convection the transfer of heat by movement of a fluid (liquid or gas).

cooling capacity the amount of heat (sensible and latent) removed from the structure by the cooling unit during one hour.

cooling load the rate at which heat must be removed from a space to maintain room air temperature at the constant value which was assumed when calculating heat gain.

corner brace structural framing member used to resist diagonal loads and strengthen the wall.

crawl space a shallow space below the living quarters of a house without a basement, sometimes enclosed.

C-section used for structural framing members, such as studs, joists headers, beams, girders, and rafters. The name comes from the member's "C" shaped cross-sectional configuration consisting of a web, flange, and lip.

density the mass of a substance in a unit volume.

depressurization a negative house pressure caused when the central air conditioning system's balance is disrupted, commonly by leaks in the supply or return ducts.

direct gain a passive solar heating system whereby solar radiation is admitted directly into the conditioned space.

DOE abbreviation for Department of Energy.

double framed walls a building method that creates an air gap surrounding the interior living space (or shell) with warm air. A primary benefit of double framing is the increased level of insulation that can be achieved by insulating both "shells."

drywall generic term for interior surfacing material, such as gypsum panels, applied to framing using dry construction methods (e.g., mechanical fasteners or adhesive).

duct a pipe or closed conduit made of sheet metal or other suitable material used for conducting air to and from an air-handling unit.

electric resistance heat electric baseboard or electric boiler used for space heating. Heat is generated by current flowing in a resistance wire similar to those used in a hair dryer.

endwall vertical wall constructed in line with the first or last frame of a structure.

energy efficiency ratio (EER) the cooling energy output from an air conditioner. The EER changes with the inside and outside conditions, falling as the temperature difference between inside and outside gets larger.

evaporation change of state from liquid to vapor.

exterior insulation and finish system (EIFS) a synthetic stucco system frequently used with exterior insulation.

EZFRAME software that allows computer users to experiment with innovative steel stud wall alternatives and quickly calculate the results.

fascia a board fastened to the ends of the rafters or joists forming part of a cornice.

fire-resistance a relative term used with a numerical rating or modifying adjective to indicate the extent to which a material or structure resists the effect of fire.

fire resistant refers to properties or designs that resist effects of any fire to which a material or structure might be expected to be subjected.

flange the part of a C-section or track that is perpendicular to the web.

flat straps straps used to reinforce joists and to brace steel studs laterally.

floor joist a horizontal structural framing member that supports floor loads.

foundation a component that transfers weight of building and occupants to the earth.

frame complete assembly of structural steel members.

frost-protected shallow foundations (FPSF) insulation around the foundation perimeter that conserves and redirects heat loss through the building slab toward the soils beneath the building foundation. At the same time, geothermal heat resources are directed toward the foundation, resulting in an elevated frost depth around the building.

furnace that part of an environmental system which converts gas, oil, electricity, or other fuel into heat for distribution within a structure.

gable the triangular vertical end of a building formed by the eaves and ridge of a sloped roof. A gable roof is characterized by two sections of roof of constant slope that meet at a ridge.

galvanized steel steel that has a zinc protective coating for resistance against corrosion.

gauge (gage) arbitrary scale of measurement for sheet metal thickness.

glazing a covering of transparent or translucent material (glass or plastic) used for admitting light.

greenhouse effect the heat build-up that occurs when a glazing material transmits solar radiation (short wave) into an enclosure and traps the long wave radiation.

header a horizontal built-up structural framing member used over wall or roof openings to carry loads across the opening

heating capacity the amount of heat added to the structure by the heating unit during one hour.

heat exchanger an extra circulation loop hooked into solar systems to prevent freezing of the water at night.

heat gain the amount of heat gained by a space from all its sources, including lights, equipment, and solar radiation. The total heat gain represents the amount of heat that must be removed from a space to maintain desired indoor conditions during cooling.

heat, latent heat energy which changes the form of a substance (e.g. ice to water) without changing its temperature.

heat, sensible heat which changes the temperature of a substance without changing its form.

heat loss a decrease in the amount of heat contained in a space, resulting from heat flow through floors, walls, doors, windows, roofs, and other building envelope components.

heat pump a refrigeration system designed so that the heat extracted at a low temperature and the heat rejected at a higher temperature may be used alternately for heating and cooling functions.

heat recovery reuse of heat which would otherwise be wasted.

heat transfer the methods by which heat may be conveyed from one place to another; e.g.: conduction, convection, radiation.

hopper windows windows with hinges at the bottom.

HSPF abbreviation for Heating System Performance Factor, a method of rating the efficiency of heating equipment.

HUD abbreviation for Housing and Urban Development.

HVAC system mechanical systems fueled by conventional sources of energy to provide controlled heating, ventilating, and air conditioning.

hydrochloro-fluorocarbons (HCFCs) a blowing agent used in some exterior insulation that causes release of chlorine molecules into the atmosphere, contributing to ozone depletion.

ICBO abbreviation for International Conference of Building Officials.

indirect gain system a solar heating system in which sunlight first strikes a thermal storage wall located between the sun and a living space. The sunlight absorbed by the mass is converted to heat and then transferred into the living space through convection and radiation.

infiltration the uncontrolled leakage of air into buildings through cracks and joints, and around windows and doors of a building.

Insu-Form™ a patented foundation system that includes styrene foam panels that recess into the ground, helping the slab retain heat.

insulation a material used to retard the flow of heat.

isolated gain system a system where solar collection and heat storage are isolated from the living spaces.

jack stud a vertical structural member that does not span the full height of the wall and supports vertical loads and/or transfers lateral loads. Jack studs are used to support headers.

jamb one of the finished upright sides of a door or window frame.

joint the space between the adjacent surfaces of two members or components joined and held together by screws, nails or other means.

joist small beam that supports part of the floor, ceiling, or roof of a building.

king stud a vertical structural member that spans the full height of the wall and supports vertical loads and lateral loads. Usually located at both ends of a header adjacent to the jack studs to resist lateral loads.

lip the part of a C-section which extends from the flange at the open end. The lip increases the strength characteristics of the member and acts as a stiffener to the flange.

load bearing or structural walls (i.e. transverse and/or axial load bearing). Steel framing systems that exceed the limits for a non-structural system (e.g. wall studs).

Manual J a widely accepted method of calculating the sensible cooling and heating loads under design conditions.

MECcheck™ a set of compliance materials that includes MECcheck™ software, the MECcheck Manual™ and MECcheck™ Prescriptive Packages.

mil a unit of measurement typically used in measuring the thickness of thin elements. One mil equals 1/1000 of an inch.

Model Energy Code (MEC) a regulatory document which provides requirements for the energy performance of residential buildings.

movable insulation insulation placed over windows when needed to prevent heat loss or gain, and removed for light, view, venting, or solar heat gain.

NAHB abbreviation for National Association of Home Builders.

non-structural or non-load bearing walls walls that do not support the structure above. Most interior walls are nonload-bearing; their main purpose is to support interior finished walls.

on center (o.c.) the measurement of spacing for studs, rafters, joists, etc., in a building from the center of one member to the center of the next.

orientation the position of a building relative to the influences of the natural environment, mainly the sun and wind.

ORNL abbreviation for Oak Ridge National Lab.

OSB abbreviation for oriented strand board.

OTW abbreviation for Omega Transworld, LTD, a subsidiary of Burrel Group that manufacturers concrete block and panels for residential construction.

passive solar system an assembly of natural and architectural components that are used to accomplish the transfer of thermal energy.

plenum the central collecting chamber for the conditioned air leaving the heating, cooling or year-round air conditioning unit. The conditioned air flows from the supply plenum to the supply ducts, which are connected to it.

Prest-on® clips a steel-framed fastening system that eliminates the use of back-up studs at gable ends, interior walls and building corners, thus reducing thermal bridging.

punch out a hole in the web of a steel framing member allowing for installation of plumbing, electrical and other trade installation.

radiation the direct transport of energy through a space by means of electromagnetic waves.

rafter one of a series of structural members of a roof designed to support roof loads.

rebar steel tensile reinforcements placed in concrete footings and walls.

resistance The tendency of a material to retard the flow of heat. The R-value is the unit of thermal resistance used for comparing insulating values of different materials.

R-value a measure of the resistance an insulating material offers to heat transfer.

SEER abbreviation for the Seasonal Energy Efficiency Rating.

shear wall a wall assembly capable of resisting lateral forces to prevent racking from wind or seismic loads.

sheathing plywood, gypsum, wood fiber, expanded plastic, or composition boards encasing walls, ceilings, floors, and roofs of framed buildings. Structural sheathing is used directly over structural members (eg. studs or joists) to distribute loads, brace walls, and strengthen assembly.

sheeting corrugated sheet metal or exterior finishing material.

siding the finish covering of the outside wall for a frame building.

slab a flat reinforced concrete element of a building that provides the base for the floor or roofing materials.

slab edge insulation insulation of concrete slab floors around the perimeter to reduce heat loss by conduction, and through exposed edges by convection and radiation.

slammer studs a steel framed fastening system that can reduce the number of studs, thus reducing thermal bridging.

soffit undersurface of a projection or opening. An economical method of filling above cabinets and housing overhead ducts, pipes, or conduits.

spacers used to separate multiple panes of glass within the window.

span the clear horizontal distance between bearing supports.

splice a joint at which two pieces are joined to each other, or the action of connecting the two pieces.

stud vertical structural element of a wall assembly which supports vertical loads and/or transfers lateral loads.

subfloor rough or structural floor placed directly on the floor joists or beams, to which the finished flooring is applied.

tension force that tends to pull the particles of a body apart.

thermal break a material of low thermal conductivity placed in such a way as to reduce the flow of heat between two materials of high thermal conductivity.

thermal bridging an effect which reduces the cavity insulation R-value that takes place when a highly conductive steel stud breaks the continuity of the cavity insulation. Heat has a tendency to choose the most conductive path, the path of least resistance, to bypass the insulation.

thermal mass a thermal absorptive building component used to store heat energy. Thermal mass can also refer to the amount of potential heat storage capacity available in a given system or assembly.

thermal resistance (R) resistance of a material or assembly to the flow of heat.

thermal storage wall a passive system in which the heat storage mass is a wall located between a window wall and living space(s) to be heated. The mass can be a variety of materials including masonry or water.

track used for applications such as top and bottom plate for walls and band joists for flooring systems. The track has a web and flanges, but no lips.

trim any material used to finish off corners between surfaces and around openings.

truss open, lightweight framework of members, usually designed to replace a large beam where spans are long.

UBC abbreviation for Uniform Building Code.

U-Value (Coefficient of Heat Transfer) the number of BTUs that flow through one square foot of roof, wall or floor, in one hour, when there is a one degree Fahrenheit difference in temperature between the inside and outside air under steady-state conditions.

vapor barrier a layer of material with low permeance to moisture used to prevent hidden condensation of water within building sections.

ventilation movement of fresh air through a building, either mechanically, induced, or naturally.

weatherstripping narrow or jamb-width sections of thin metal, fabric or other materials used to prevent infiltration of air and moisture around windows and doors.

web the part of a C-section or track that connects the two flanges.

web stiffener additional material that is attached to the web to strengthen the member against web crippling. Sometimes called bearing stiffener.

windbreaks a method of using trees and landscaping to divert winds and direct cool breezes toward the home.

Index

About the Authors

John H. Hacker, a retired civil engineer, has 40 years of experience in steel design. In his private practice he has designed steel framed building in the United States, Mexico, Canada, South America, and Thailand. He holds two U.S. patents for energy efficient steel framing systems and has won national awards for his design of energy efficient steel-framed homes. In addition, he has given seminars on steel framing in both the United States and Thailand.

Julie A. Gorges, a freelance writer for over seven years, has been published in dozens of magazines, including construction magazines such as *Metal Home Digest, Automated Builder,* and *Permanent Building and Foundations Magazine.* In addition, she has ten years experience as an energy consultant.